W0091360

Arpad Kelemen, Ajith Abraham and Yuehui Chen (Eds.)

Computational Intelligence in Bioinformatics

Studies in Computational Intelligence, Volume 94

Editor-in-chief
Prof. Janusz Kacprzyk
Systems Research Institute
Polish Academy of Sciences
ul. Newelska 6
01-447 Warsaw
Poland
E-mail: kacprzyk@ibspan.waw.pl

Further volumes of this series can be found on our homepage: springer.com

Vol. 73. Petra Perner (Ed.)
Case-Based Reasoning on Images and Signals, 2008
ISBN 978-3-540-73178-8

Vol. 74. Robert Schaefer
Foundation of Global Genetic Optimization, 2007
ISBN 978-3-540-73191-7

Vol. 75. Crina Grosan, Ajith Abraham and Hisao Ishibuchi (Eds.)
Hybrid Evolutionary Algorithms, 2007
ISBN 978-3-540-73296-9

Vol. 76. Subhas Chandra Mukhopadhyay and Gourab Sen Gupta (Eds.)
Autonomous Robots and Agents, 2007
ISBN 978-3-540-73423-9

Vol. 77. Barbara Hammer and Pascal Hitzler (Eds.)
Perspectives of Neural-Symbolic Integration, 2007
ISBN 978-3-540-73953-1

Vol. 78. Costin Badica and Marcin Paprzycki (Eds.)
Intelligent and Distributed Computing, 2008
ISBN 978-3-540-74929-5

Vol. 79. Xing Cai and T.-C. Jim Yeh (Eds.)
Quantitative Information Fusion for Hydrological Sciences, 2008
ISBN 978-3-540-75383-4

Vol. 80. Joachim Diederich
Rule Extraction from Support Vector Machines, 2008
ISBN 978-3-540-75389-6

Vol. 81. K. Sridharan
Robotic Exploration and Landmark Determination, 2008
ISBN 978-3-540-75393-3

Vol. 82. Ajith Abraham, Crina Grosan and Witold Pedrycz (Eds.)
Engineering Evolutionary Intelligent Systems, 2008
ISBN 978-3-540-75395-7

Vol. 83. Bhanu Prasad and S.R.M. Prasanna (Eds.)
Speech, Audio, Image and Biomedical Signal Processing using Neural Networks, 2008
ISBN 978-3-540-75397-1

Vol. 84. Marek R. Ogiela and Ryszard Tadeusiewicz
Modern Computational Intelligence Methods for the Interpretation of Medical Images, 2008
ISBN 978-3-540-75399-5

Vol. 85. Arpad Kelemen, Ajith Abraham and Yulan Liang (Eds.)
Computational Intelligence in Medical Informatics, 2008
ISBN 978-3-540-75766-5

Vol. 86. Zbigniew Les and Mogdalena Les
Shape Understanding Systems, 2008
ISBN 978-3-540-75768-9

Vol. 87. Yuri Avramenko and Andrzej Kraslawski
Case Based Design, 2008
ISBN 978-3-540-75705-4

Vol. 88. Tina Yu, David Davis, Cem Baydar and Rajkumar Roy (Eds.)
Evolutionary Computation in Practice, 2008
ISBN 978-3-540-75770-2

Vol. 89. Ito Takayuki, Hattori Hiromitsu, Zhang Minjie and Matsuo Tokuro (Eds.)
Rational, Robust, Secure, 2008
ISBN 978-3-540-76281-2

Vol. 90. Simone Marinai and Hiromichi Fujisawa (Eds.)
Machine Learning in Document Analysis and Recognition, 2008
ISBN 978-3-540-76279-9

Vol. 91. Horst Bunke, Kandel Abraham and Last Mark (Eds.)
Applied Pattern Recognition, 2008
ISBN 978-3-540-76830-2

Vol. 92. Ang Yang, Yin Shan and Lam Thu Bui (Eds.)
Success in Evolutionary Computation, 2008
ISBN 978-3-540-76285-0

Vol. 93. Manolis Wallace, Marios Angelides and Phivos Mylonas (Eds.)
Advances in Semantic Media Adaptation and Personalization, 2008
ISBN 978-3-540-76359-8

Vol. 94. Arpad Kelemen, Ajith Abraham and Yuehui Chen (Eds.)
Computational Intelligence in Bioinformatics, 2008
ISBN 978-3-540-76802-9

Arpad Kelemen
Ajith Abraham
Yuehui Chen
(Eds.)

Computational Intelligence in Bioinformatics

With 104 Figures and 34 Tables

Arpad Kelemen
Department of Neurology
Buffalo Neuroimaging Analysis Center
The Jacobs Neurological Institute
University at Buffalo, the State University of New York, 100 High Street, Buffalo
NY 14203, U.S.A.
akelemen@buffalo.edu

Yuehui Chen
School of Information Science and Engineering
Jinan University
Jiwei Road 106, Jinan 250022
P.R. China
yhchen@ujn.edu.cn

Ajith Abraham
Centre for Quantifiable Quality of Service in Communication Systems (Q2S)
Centre of Excellence
Norwegian University of Science and Technology
O.S. Bragstads plass 2E
N-7491 Trondheim
Norway
ajith.abraham@ieee.org

ISBN 978-3-540-76802-9 e-ISBN 978-3-540-76803-6

Studies in Computational Intelligence ISSN 1860-949X

Library of Congress Control Number: 2007940153

Cover Design: Deblik, Berlin, Germany

Printed on acid-free paper

9 8 7 6 5 4 3 2 1

springer.com

Preface

Bioinformatics involve the creation and advancement of algorithms using techniques including computational intelligence, applied mathematics and statistics, informatics, and biochemistry to solve biological problems usually on the molecular level. Major research efforts in the field include sequence analysis, gene finding, genome annotation, protein structure alignment analysis and prediction, prediction of gene expression, protein-protein docking/interactions, and the modeling of evolution.

Computational intelligence is a well-established paradigm, where new theories with a sound biological understanding have been evolving. Defining computational intelligence is not an easy task. In a nutshell, which becomes quite apparent in light of the current research pursuits, the area is heterogeneous with a combination of such technologies as neural networks, fuzzy systems, rough set, evolutionary computation, swarm intelligence, probabilistic reasoning, multi-agent systems etc. The recent trend is to integrate different components to take advantage of complementary features and to develop a synergistic system.

This book deals with the application of computational intelligence in bioinformatics. Addressing the various issues of bioinforatics using different computational intelligence approaches is the novelty of this edited volume. This volume comprises of 13 chapters including some introductory chapters giving the fundamental definitions and some important research challenges. Chapters were selected on the basis of fundamental ideas/concepts rather than the thoroughness of techniques deployed. The thirteen chapters are organized as follows.

In the introductory Chapter, *Tasoulis et al.* present neural networks, evolutionary algorithms and clustering algorithms and their application to DNA microarray experimental data analysis. Authors also discus different dimension reduction techniques.

Chapter 2 by *Kaderali and Radde* provide an overview of the reconstruction of gene regulatory networks from gene expression measurements. Authors present several different approaches to gene regulatory network inference, discuss their strengths and weaknesses, and provide guidelines on which models are appropriate under what circumstances.

Chapter 3 by *Donkers and Tuyls* introduce Bayesian belief networks and describe their current use within bioinformatics. The goal of the chapter is to help the reader to understand and apply belief networks in the domain of bioinformatics. Authors present the current state-of-the-art by discussing several real-world applications in bioinformatics, and also discuss some available software tools.

Das et al. in Chapter 4 explore the role of swarm intelligence algorithms in certain bioinformatics tasks like micro-array data clustering, multiple sequence alignment, protein structure prediction and molecular docking. This chapter begins with an overview of the basic concepts of bioinformatics along with their biological basis and then provides a detailed survey of the state of the art research centered around the applications of swarm intelligence algorithms in bioinformatics.

Liang and Kelemen in the fifth Chapter propose a time lagged recurrent neural network with trajectory learning for identifying and classifying the gene functional patterns from the heterogeneous nonlinear time series microarray experiments. Optimal network architectures with different memory structures were selected based on Akaike and Bayesian information criteria using two-way factorial design. The optimal model performance was compared to other popular gene classification algorithms, such as nearest neighbor, support vector machine, and self-organized map.

In Chapter 6, *Busa-Fekete et al.* suggest two algorithms for protein sequence classification that are based on a weighted binary tree representation of protein similarity data. *TreeInsert* assigns the class label to the query by determining a minimum cost necessary to insert the query in the (precomputed) trees representing the various classes. Then *TreNN* assigns the label to the query based on an analysis of the query's neighborhood within a binary tree containing members of the known classes. The algorithms were tested in combination with various sequence similarity scoring methods using a large number of classification tasks representing various degrees of difficulty.

In Chapter 7, *Smith* compares the traditional dynamic programming RNA gene finding methodolgy with an alternative evolutionary computation approach. Experiment results indicate that dynamic programming returns an exact score at the cost of very large computational resource usage, while the evolutionary computing approach allows for faster approximate search, but uses the RNA secondary structure information in the covariance model from the start.

Schaefer et al. in Chapter 8, illustrate how fuzzy rule-based classification can be applied successfully to analyze gene expression data. The generated classifier consists of an ensemble of fuzzy *if-then* rules, which together provide a reliable and accurate classification of the underlying data.

In Chapter 9, *Huang* and *Chow* overview the existing gene selection approaches and summarize the main challenges of gene selection. Using a typical gene selection model, authors further illustrate the implementation of these strategies and evaluate their contributions.

Cao et al. in Chapter 10, propose a fuzzy logic based novel gene regulatory network. The key motivation for this algorithm is that genes with regulatory relationships may be modeled via fuzzy logic, and the strength of regulations may be represented as the length of accumulated distance during a period of time intervals.

One unique feature of this algorithm is that it makes very limited a priori assumptions concerning the modeling.

In Chapter 11, *Haavisto and Hyötyniemi* apply linear multivariate regression tools for microarray gene expression data. Two examples comprising of yeast cell response to environmental changes and expression during the cell cycle, are used to demonstrate the presented subspace identification method for data-based modeling of genome dynamics.

Navas-Delgado et al. in Chapter 12 present an architecture for the development of Semantic Web applications, and the way it is applied to build an application for systems biology. The architecture is based on an ontology-based system with connected biomodules that could be globally analyzed as far as possible.

In the last Chapter *Han and Zhu* illustrate the various methods of encoding information in DNA strands and present the corresponding DNA algorithms, which will benefit the further research on DNA computing.

We are very much grateful to the authors of this volume and to the reviewers for their tremendous service by critically reviewing the chapters. The editors would like to thank Dr. Thomas Ditzinger (Springer Engineering Inhouse Editor) and Professor Janusz Kacprzyk (Editor-in-Chief, Springer Studies in Computational Intelligence Series) and Ms. Heather King (Springer Verlag, Heidelberg) for the editorial assistance and excellent cooperative collaboration to produce this important scientific work. We hope that the reader will share our excitement to present this volume on '*Computational Intelligence in Bioinformatics*' and will find it useful.

Arpad Kelemen, Ajith Abraham and Yuehui Chen (Editors)
September 2007

Contents

1

Computational Intelligence Algorithms and DNA Microarrays

D.K. Tasoulis[1], V.P. Plagianakos[2], and M.N. Vrahatis[2]

[1] Institute for Mathematical Sciences, Imperial College London, South Kensington, London SW7 2PG, United Kingdom
d.tasoulis@imperial.ac.uk

[2] Computational Intelligence Laboratory, Department of Mathematics, University of Patras Artificial Intelligence Research Center (UPAIRC), University of Patras, GR–26110 Patras, Greece
{vpp,vrahatis}@math.upatras.gr

Summary. In this chapter, we present Computational Intelligence algorithms, such as Neural Network algorithms, Evolutionary Algorithms, and clustering algorithms and their application to DNA microarray experimental data analysis. Additionally, dimension reduction techniques are evaluated. Our aim is to study and compare various Computational Intelligence approaches and demonstrate their applicability as well as their weaknesses and shortcomings to efficient DNA microarray data analysis.

1.1 Introduction

The development of microarray technologies gives scientists the ability to examine, discover and monitor the mRNA transcript levels of thousands of genes in a single experiment. The development of technologies capable to simultaneously study the expression of every gene in an organism has provided a wealth of biological insight. Nevertheless, the tremendous amount of data that can be obtained from microarray studies presents a challenge for data analysis.

This challenge is twofold. Primarily, discovering patterns hidden in the gene expression microarray data across a number of samples that are correlated with a specific condition is a tremendous opportunity and challenge for functional genomics and proteomics [1–3]. Unfortunately, employing any kind of pattern recognition algorithm to such data is hindered by the *curse of dimensionality* (limited number of samples and very high feature dimensionality). This is the second challenge. Usually to address this, one has to preprocess the expression matrix using a dimension reduction technique [4] and/or to find a subset of the genes that correctly characterizes the samples. Note that this is not similar to *"bi-clustering"*, which refers to the identification of genes that exhibit similar behavior across a subset of samples [5, 6]. In this chapter we examine the application of various Computational Intelligence

D.K. Tasoulis et al.: *Computational Intelligence Algorithms and DNA Microarrays*, Studies in Computational Intelligence (SCI) **94**, 1–31 (2008)
www.springerlink.com

methodologies to face problems arising from the twofold nature of the microarray data. We also examine various manners to combine and interact algorithms towards a completely automated system.

To this end the rest of this chapter is structured as follows. Sections 1.2 and 1.3 are devoted to a brief presentation of Neural Networks as classification tools, Evolutionary Algorithms that can be used for dimension reduction, and their synergy. In Section 1.4 various feature selection and dimension reduction techniques are presented, starting from the Principal Component Analysis, continuing with several clustering algorithms, and finally we analyze hybrid approaches. In Section 1.5 using well known and publicly available DNA microarray problems, we study and examine feasible solutions to many implementation issues and report comparative results of the presented algorithms and techniques. The chapter ends with a brief discussion and some concluding remarks.

1.2 Neural Networks

Feedforward Neural Networks (FNNs) are parallel computational models comprised of densely interconnected, simple, adaptive processing units, characterized by an inherent propensity for storing experiential knowledge and rendering it available for use. FNNs have been successfully applied in numerous application areas, including DNA microarray data analysis [7].

To train an FNN, supervised training is probably the most frequently employed technique. The training process is an incremental adaptation of connection weights that propagate information between neurons. A finite set of arbitrarily ordered examples is presented at the input of the network and associated to appropriate references through an error correction process. This can be viewed as the minimization of an error measure, which is usually defined as the sum-of-squared-differences error function E over the entire training set:

$$w^* = \min_{w \in \mathbb{R}^n} E(w), \tag{1.1}$$

where $w^* = (w_1^*, w_2^*, \ldots, w_n^*) \in \mathbb{R}^n$ is a minimizer of E. The rapid computation of such a minimizer is a rather difficult task since, in general, the number of network weights is high and the corresponding nonconvex error function possesses multitudes of local minima and has broad flat regions adjoined with narrow steep ones.

Let us consider the family of gradient–based supervised learning algorithms having the iterative form:

$$w^{k+1} = w^k + \eta^k d^k, \qquad k = 0, 1, 2, \ldots \tag{1.2}$$

where w^k is the current weight vector, d^k is a search direction, and η^k is a *global* learning rate, i.e. the same learning rate is used to update all the weights of the network. Various choices of the direction d^k give rise to distinct algorithms. A broad class of methods uses the search direction $d^k = -\nabla E(w^k)$, where the gradient $\nabla E(w)$

can be obtained by means of back–propagation of the error through the layers of the network [8]. The most popular training algorithm of this class, named batch Back–Propagation (BP), minimizes the error function using the steepest descent method [9] with constant, heuristically chosen, learning rate η. In practice, a small value for the learning rate is chosen ($0 < \eta < 1$) in order to secure the convergence of the BP training algorithm and to avoid oscillations in a direction where the error function is steep. It is well known that this approach tends to be inefficient. This happens, for example, when the search space contains long ravines that are characterized by sharp curvature across them and a gently slopping floor.

Next, we give an overview of two neural network training algorithms: the Rprop algorithm and the adaptive online algorithm. Both algorithms have been used on DNA microarray problems. Rprop is one of the fastest and most effective training algorithms. On the other hand, adaptive online seems more suitable for this kind of problems, due to its ability to train FNNs using extremely large training sets.

1.2.1 The Rprop Neural Network Training Algorithm

The Resilient backpropagation (Rprop) [10] algorithm is a local adaptive learning scheme performing supervised training of FNNs. To update each weight of the network, Rprop exploits information concerning the sign of the partial derivative of the error function. The size of the weight change, Δw_{ij}, is determined by a weight specific update value, $\Delta_{ij}^{(t)}$, given by the following formula:

$$\Delta w_{ij}^{(t)} = \begin{cases} -\Delta_{ij}^{(t)}, & \text{if } \frac{\partial E^{(t)}}{\partial w_{ij}} > 0, \\ +\Delta_{ij}^{(t)}, & \text{if } \frac{\partial E^{(t)}}{\partial w_{ij}} < 0, \\ 0, & \text{otherwise,} \end{cases}$$

where $\partial E^{(t)}/\partial w_{ij}$ denotes the summed gradient information over all patterns of the training set (batch training). The second step of the Rprop algorithm is to determine the new update values, using the following formula:

$$\Delta_{ij}^{(t)} = \begin{cases} \eta^{+}\Delta_{ij}^{(t-1)}, & \text{if } \frac{\partial E^{(t-1)}}{\partial w_{ij}} \frac{\partial E^{(t)}}{\partial w_{ij}} > 0, \\ \eta^{-}\Delta_{ij}^{(t-1)}, & \text{if } \frac{\partial E^{(t-1)}}{\partial w_{ij}} \frac{\partial E^{(t)}}{\partial w_{ij}} < 0, \\ \Delta_{ij}^{(t-1)}, & \text{otherwise,} \end{cases}$$

where $0 < \eta^{-} < 1 < \eta^{+}$, i.e. each time the partial derivative with respect to w_{ij} changes its sign, which is an indication that the last update was too big and the algorithm has possibly overshot a local minimizer, the update value $\Delta_{ij}^{(t)}$ is decreased by η^{-}. If the derivative retains its sign, the update value is slightly increased to further accelerate convergence in shallow regions of the weight space.

In our experiments, the five parameters of the Rprop method were initialized using values commonly encountered in the literature. More specifically, the increase factor was set to $\eta^{+} = 1.2$; the decrease factor was set to $\eta^{-} = 0.5$; the initial update value is set to $\Delta_0 = 0.07$; the maximum step, which prevents the weights from becoming too large, was $\Delta_{\max} = 50$; and the minimum step, which is used to avoid too small weight changes, was $\Delta_{\min} = 10^{-6}$.

1.2.2 The Adaptive Online Neural Network Training Algorithm

Despite the abundance of methods for learning from examples, there are only a few that can be used effectively for on–line learning. For example, the classic batch training algorithms cannot straightforwardly handle non–stationary data. Even when some of them are used in on–line training there exists the problem of "catastrophic interference", in which training on new examples interferes excessively with previously learned examples, leading to saturation and slow convergence [11].

Methods suited to on–line learning are those that can efficiently handle non–stationary and time–varying data, while at the same time, require relatively little additional memory and computation to process one additional example. The Adaptive Online Backpropagation (AOBP) algorithm [12, 13] belongs to this class and can be used in on–line neural networks training. A high level description of the algorithm is given in Algorithm 1.

In the algorithm model η is the learning rate, K is the meta–learning rate and $\langle \cdot, \cdot \rangle$ stands for the usual inner product in $\mathbb{R}^n$. As the termination condition the classification error, or an upper limit to the error function evaluations can be used. The key features of this method are the low storage requirements and the inexpensive computations. Moreover, in order to calculate the learning rate for the next iteration, it uses information from the current, as well as, the previous iteration. This seems to provide some kind of stabilization in the calculated values of the learning rate, and previous experiments show that it helps the method to exhibit fast convergence and high success rate.

THE TRAINING ALGORITHM

0: Initialize the weights w^0, η^0, and K.
1: **Repeat**
2: Set $k = k + 1$
3: Randomly choose a pattern from the training set.
4: Using this pattern, calculate the error, $E(w^k)$ and then the gradient, $\nabla E(w^k)$.
5: Calculate the new weights using:
$w^{k+1} = w^k - \eta^k \nabla E(w^k)$
6: Calculate the new learning rate using:
$\eta^{k+1} = \eta^k + K \langle \nabla E(w^{k-1}), \nabla E(w^k) \rangle$
7: **Until** the *termination condition* is met.
8: **Return** the final weights w^{k+1}.

Algorithm 1: The Online Training Algorithm in Pseudocode

1.3 Evolutionary Algorithms

Evolutionary Algorithms (EAs) are stochastic search methods that mimic the metaphor of natural biological evolution. They operate on a population of potential solutions applying the principle of survival of the fittest to produce better and better approximations to a solution. At each generation, a new set of approximations is created by the process of selecting individuals according to their level of fitness in the problem domain and breeding them together using operators borrowed from natural genetics [14]. Many attempts have been made within the Artificial Intelligence community to integrate EAs and ANNs. We examine the application of EAs to microarray classification to determine the optimal, or near optimal, subset of predictive genes on the complex and large spaces of possible gene sets. Next we outline the Differential Evolution algorithm and its search operators.

The Differential Evolution Algorithm

Differential Evolution [15] is an optimization method, capable of handling non differentiable, nonlinear and multimodal objective functions. To fulfill this requirement, DE has been designed as a stochastic parallel direct search method, which utilizes concepts borrowed from the broad class of evolutionary algorithms. The method typically requires few, easily chosen, control parameters. Experimental results have shown that DE has good convergence properties and outperforms other well known evolutionary algorithms [15]. DE has been applied on numerous optimization tasks. It has successfully solved many artificial benchmark problems [16], as well as, hard real–world problems (see for example [17]). In [18] it was employed to train neural networks and in [19, 20] we have proposed a method to efficiently train neural networks having arbitrary, as well as, constrained integer weights. The DE algorithm has also been implemented on parallel and distributed computers [21, 22].

DE is a population–based stochastic algorithm that exploits a population of potential solutions, *individuals*, to effectively probe the search space. The population of the individuals is randomly initialized in the optimization domain with NP, n–dimensional vectors, following a uniform probability distribution and is evolved over time to explore the search space. NP is fixed throughout the training process. At each iteration, called *generation*, new vectors are generated by the combination of randomly chosen vectors from the current population. This operation in our context is referred to as *mutation*. The resulting vectors are then mixed with another predetermined vector – the *target* vector – and this operation is called *recombination*. This operation yields the so–called *trial* vector. The trial vector is accepted for the next generation depending on the value of the fitness function. Otherwise, the target vector is retained in the next generation. This last operator is referred to as *selection*.

The search operators efficiently shuffle information among the individuals, enabling the search for an optimum to focus on the most promising regions of the solution space. The first operator considered is mutation. For each individual x_g^i,

$i = 1, \ldots, NP$, where g denotes the current generation, a new individual v^i_{g+1} (mutant vector) is generated according to one of the following equations:

$$v^i_{g+1} = x^{\text{best}}_g + \mu(x^{r1}_g - x^{r2}_g), \tag{1.3}$$

$$v^i_{g+1} = x^{r1}_g + \mu(x^{r2}_g - x^{r3}_g), \tag{1.4}$$

$$v^i_{g+1} = x^i_g + \mu(x^{\text{best}}_g - x^i_g) + \mu(x^{r1}_g - x^{r2}_g), \tag{1.5}$$

$$v^i_{g+1} = x^{\text{best}}_g + \mu(x^{r1}_g - x^{r2}_g) + \mu(x^{r3}_g - x^{r4}_g), \tag{1.6}$$

$$v^i_{g+1} = x^{r1}_g + \mu(x^{r2}_g - x^{r3}_g) + \mu(x^{r4}_g - x^{r5}_g), \tag{1.7}$$

where x^{best}_g is the best member of the previous generation; $\mu > 0$ is a real parameter, called *mutation constant*, which controls the amplification of the difference between two individuals so as to avoid the stagnation of the search process; and $r_1, r_2, r_3, r_4, r_5 \in \{1, 2, \ldots, i-1, i+1, \ldots, NP\}$, are random integers mutually different.

Trying to rationalize the above equations, we observe that Equation (1.4) is similar to the crossover operator used by some Genetic Algorithms and Equation (1.3) derives from it, when the best member of the previous generation is employed. Equations (1.5), (1.6) and (1.7) are modifications obtained by the combination of Equations (1.3) and (1.4). It is clear that more such relations can be generated using the above ones as building blocks.

The recombination operator is subsequently applied to further increase the diversity of the mutant individuals. To this end, the resulting individuals are combined with other predetermined individuals, called the target individuals. Specifically, for each component l ($l = 1, 2, \ldots, n$) of the mutant individual v^i_{g+1}, we choose randomly a real number r in the interval $[0, 1]$. We then compare this number with the *recombination constant*, ρ. If $r \leqslant \rho$, we select, as the l–th component of the trial individual u^i_{g+1}, the l–th component of the mutant individual v^i_{g+1}. Otherwise, the l–th component of the target vector x^i_{g+1} becomes the l–th component of the trial vector. This operation yields the trial individual. Finally, the trial individual is accepted for the next generation only if it reduces the value of the objective function.

One problem when applying EAs, in general, is to find a set of control parameters which optimally balances the exploration and exploitation capabilities of the algorithm. There is always a trade off between the efficient exploration of the search space and its effective exploitation. For example, if the recombination and mutation rates are too high, much of the search space will be explored, but there is a high probability of losing good solutions. In extreme cases the algorithm has difficulty to converge to the global minimum due to the insufficient exploitation. Fortunately, the convergence properties of DE typically do not depend heavily on its control parameters.

Although, DE performs stably across the space of possible parameter settings, different operators may exhibit different convergence properties. More specifically, DE operators that use the best individual as a starting point for the computation of the mutant vector, constantly push the population closer to the location of the best computed point. On the other hand, operators that utilize many randomly chosen individuals for the computation of the mutant individual, greatly enhance the exploration capability of the algorithm. In [23] we present a detailed study and experimental results on exploration vs. exploitation issues.

1.4 Feature Selection and Dimension Reduction Techniques

An important issue in any classification task is to define those features that significantly contribute to the classification of interest, while at the same time discarding the least significant and/or erroneous ones. This procedure is also referred to as dimension reduction. The problem of high dimensionality is often tackled by user specified subspaces of interest. However, user–identification of the subspaces is error–prone, especially when no prior domain knowledge is available. Another way to address high dimensionality is to apply a dimensionality reduction method to the dataset. Methods such as the Principal Component Analysis [24], optimally transform the original data space into a lower dimensional space by forming dimensions that are linear combinations of given attributes. The new space has the property that distances between points remain approximately the same as before. Alternatively, one can apply a clustering algorithm to the data set in order to reduce the dimensionality of the problem. To this end, the Principal Component Analysis, as well as several clustering algorithms used for dimension reduction are presented below.

1.4.1 Principal Component Analysis

In general, the Principal Component Analysis (PCA) is a powerful multivariate data analysis method [4]. Its main purpose is to reduce and summarize large and high dimensional datasets by removing redundancies and identifying correlation among a set of measurements or variables. It is a useful statistical technique that has found many applications in different scientific fields such as face recognition, image processing and compression, molecular dynamics, information retrieval, and gene expression analysis. PCA is mainly used in gene expression analysis in order to find an alternative representation of the data using a much smaller number of variables, as well as, to detect characteristic patterns in noisy data of high dimensionality. More specifically, PCA is a way of identifying patterns in data and expressing the data in such a way as to highlight their similarities and differences. Since patterns in high dimensional data can be hard to find, PCA is a powerful tool of analysis, especially when the visualization of the data is impossible.

Although PCA may succeed in reducing the dimensionality, the new dimensions can be difficult to interpret. Moreover, to compute the new set of dimensions information from all the original dimensions is required. The selection of a subset of attributes in the context of clustering is studied in [25, 26]. In the context of classification, subset selection has also been studied [24].

1.4.2 Reducing the Dimensions Using Clustering

Clustering can be defined as the process of "grouping a collection of objects into subsets or clusters, such that those within one cluster are more closely related to each other than objects assigned to different clusters" [27]. Clustering is applied in various fields including data mining [28], statistical data analysis and social sciences [29], compression and vector quantization [30], global optimization [31, 32], image analysis, and others. Clustering techniques have been successfully applied to gene expression data [33–36] and have proved useful for identifying biologically relevant groupings of genes and samples [37].

Cluster analysis is one key step in understanding how the activity of genes varies during biological processes and is affected by disease states and cellular environments. In particular clustering can be used either to identify sets of genes according to their expression in a set of samples [34, 38], or to cluster samples into homogeneous groups that may correspond to particular macroscopic phenotypes [39]. The latter is in general more difficult, but is very valuable in clinical practice.

Identifying sets of genes that have a similar expression in a set of samples can lead to a successful dimension reduction technique. Aiming to this, clustering methodology can be applied to identify meaningful clusters of features (genes), and subsequently feature selection can be accomplished by selecting one or more representatives from each cluster. Such a selection can be based on the distance among the feature values and the identified cluster center. The feature with the minimum such distance from the cluster center can be a valid selection.

Although numerous clustering algorithms exist [40], mostly hierarchical clustering methods have been applied to microarray data. Hierarchical clustering algorithms construct hierarchies of clusters in a top–down (agglomerative) or bottom–up (divisive) fashion. This kind of algorithms have proved to give high quality results. One of the most representative hierarchical approaches is the one developed by Eisen et al. [34]. In that work, the authors employed an agglomerative algorithm and adopted a method for the graphical representation of the clustered dataset. This method has been widely used by many biologists and has become the most widely used tool in gene expression data analysis [33, 41, 42]. Nonetheless, the high sensitivity of agglomerative methods to small variations of the inputs and the high computational requirements, their usage is hindered in real applications, where the number of samples and their dimensionality is expected to be high (the cost is quadratic to the number of samples).

Partitioning clustering algorithms, start from an initial clustering (that may be randomly formed) and create flat partitionings by iteratively adjusting the clusters based on the distance of the data points from a representative member of the cluster. The most commonly used partitioning clustering algorithm is k–means. This algorithm initializes k centers and iteratively assigns each data point to the cluster whose centroid has the minimum Euclidean distance from the data point. Although, k–means type algorithms can yield satisfactory clustering results at a low cost, as their running time is proportional to kn, where n is the number of samples, they heavily depend on the initialization. Additionally, there is no automatic technique able to select the number of clusters k, but most of the times this is achieved by examining the results of successive re-executions of the algorithm.

Graph theoretical clustering approaches construct a proximity graph, in which each data point corresponds to a vertex, and the edges among vertices model their proximity. Xing and Karp [43], developed a sample–based clustering algorithm named CLIFF (CLustering via Iterative Feature Filtering), which iteratively employs sample partitions as a reference to filter genes. The selection of genes through this approach relies on the outcome of an NCut algorithm, which is not robust to noise and outliers.

Another graph theoretical algorithm, CLICK (CLuster Identification via Connectivity Kernels) [35], tries to recognize highly connected components in the proximity graph as clusters. The authors demonstrated the superior performance of CLICK to the approaches of Eisen et al. [34], and the Self Organizing Map [44] based clustering approach. However, as claimed in [1], CLICK has little guarantee of not generating highly unbalanced partitions. Furthermore, in gene expression data, two clusters of co–expressed genes, C1 and C2, may be highly intersected with each other. In such situations, C1 and C2 are not likely to be split by CLICK, but would be reported as one highly connected component.

Finally, Alter et al. [45], by examining the projection of the data to a small number of principal components obtained through a Principal Component Analysis, attempt to capture the majority of gene variations. However, the large number of irrelevant genes does not guarantee that the discriminatory information will be highlighted to the projected data. For an overview of the related literature see [1–3, 46].

Below, we briefly describe five well–known clustering algorithms, namely, a) the unsupervised k–windows clustering algorithm [47, 48]. (UkW) b) the Density–Based Spatial Clustering of Applications with Noise (DBSCAN) clustering algorithm [49], c) the Principal Direction Divisive Partitioning (PDDP) clustering algorithm [50], d) the fuzzy c–means (FCM) clustering algorithm [51], and e) the Growing Neural Gas (GNG) [52]. Note that the UkW, DBSCAN and GNG, apart from identifying the clusters, are also able to approximate the number of clusters present in the data set; thus no special knowledge about the data is required. However, PDDP and FCM need explicit determination of the cluster number. The PDDP, algorithm has also the ability to endogenously handle the large dimensionality since it is based on the PCA technique.

Unsupervised k–Windows Clustering Algorithm

One of the most important class of clustering algorithms are the density based methods [53–55], especially for data of low attribute dimensionality [56–58]. These methods operate by identifying regions of high density in dataset objects, surrounded by regions of low density. One recently proposed technique in this class is the "Unsupervised k–Windows" (UkW) [48], that utilizes hyperrectangles to discover clusters. The algorithm makes use of techniques from computational geometry and encapsulates clusters using linear containers in the shape of d–dimensional hyperrectangles that are iteratively adjusted with movements and enlargements until a certain termination criterion is satisfied [48, 59]. Furthermore, with proper tuning, the algorithm is able to detect clusters of arbitrary shapes [59].

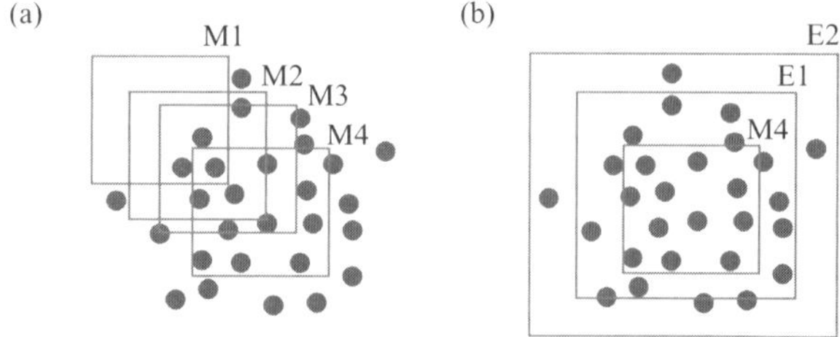

Fig. 1.1. (a) Sequential movements M2, M3, M4 of initial window M1. (b) Sequential enlargements E1, E2 of window M4

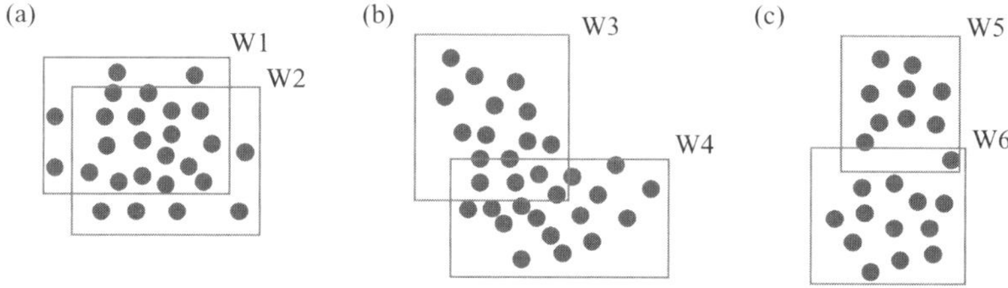

Fig. 1.2. (a) W1 and W2 satisfy the similarity condition and W1 is deleted. (b) W3 and W4 satisfy the merge operation and are considered to belong to the same cluster. (c) W5 and W6 have a small overlap and capture two different clusters

UkW aims at capturing all objects that belong to one cluster within a d–dimensional window. Windows are defined to be hyperrectangles (orthogonal ranges) in d dimensions [48]. UkW employs two fundamental procedures: *movement* and *enlargement*. The movement procedure aims at positioning each window as close as possible to the center of a cluster. The enlargement process attempts to enlarge the window so that it includes as many objects from the current cluster as possible. The two steps are illustrated in Figure 1.1.

A fundamental issue in cluster analysis, independent of the particular clustering technique applied, is the determination of the number of clusters present in a dataset. For instance well–known and widely used iterative techniques, such as the k–means algorithm [60] as well as the fuzzy c–means algorithm [51], require from the user to specify the number of clusters present in the data prior to the execution of the algorithm. UkW provides an estimate for the number of clusters that describe a dataset. The key idea is to initialize a large number of windows. When the movement and enlargement of all windows terminate, all overlapping windows are considered for merging by considering their intersection. An example of this operation is exhibited in Figure 1.2. For a detailed description of the algorithm see [59].

The DBSCAN Clustering Algorithm

The DBSCAN [49] clustering algorithm relies on a density–based notion of clusters and is designed to discover clusters of arbitrary shape as well as to distinguish

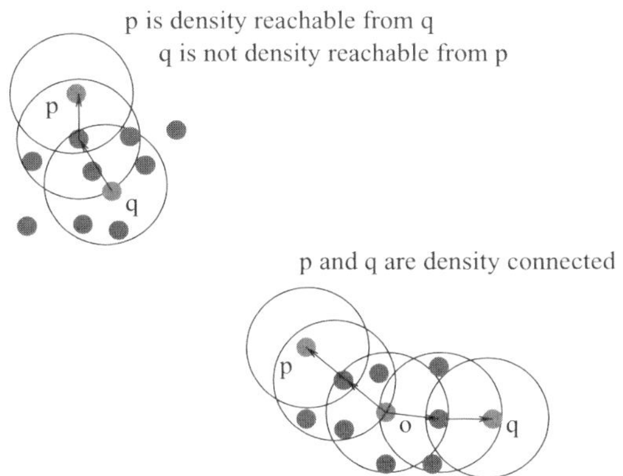

Fig. 1.3. An example of "Density–Reachable" and "Density Connected" points

noise. More specifically, the algorithm is based on the idea that in a neighborhood of a given radius (*Eps*) for each point in a cluster at least a minimum number of objects (*MinPts*) should be contained. Such points are called core points and each point in their neighborhood is considered as "Directly Density–Reachable" from that. Consequently the algorithm uses the notion of density reachable chains of objects; i.e. a point q is "Density–Reachable" from a point p, if there is a chain of objects $p_1,\ldots,p_k$ such that $p_1 = q$, $p_k = p$ and p_{i+1} is "Directly Density–Reachable" from p_i for $i = 1,\ldots,k$. Finally, a point p is defined as "Density Connected" to a point q, if there is a point o that both p, q are "Density–Reachable" from that. Fig 1.3, illustrates an example of these definitions.

Using the above described definitions, the algorithms considers as a cluster the subset of points from the dataset that are "Density–Reachable" from each other and additionally each pair of points inside the cluster is "Density Connected". Any point of the dataset not in a cluster is considered as noise.

To discover the clusters the algorithm retrieves density–reachable points from the data by iteratively collecting directly density–reachable objects. The algorithm scans the *eps*, neighborhood of each point in the database. If that neighborhood has more than $MinPts$ points a new cluster C containing them is created. Then, the neighborhood of all points q in C which have not yet been processed is checked. If the points in neighborhood of q are more than $MinPts$, then those which are not already contained in C are added to the cluster and their neighborhood will be checked in a subsequent step. This procedure is iterated until no new point can be added to the current cluster C. In Fig 1.4, an example of the result of the DBSCAN algorithm is demonstrated, for three clusters of different sizes, with convex and non–convex shape. Additionally, some of the neighborhoods are depicted, to better illustrate the operation of the algorithm. For a detailed description of the algorithm see [54].

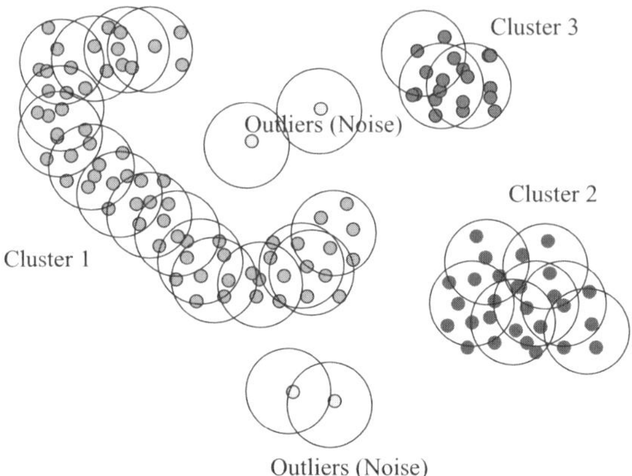

Fig. 1.4. An example of the result of the DBSCAN algorithm

The Fuzzy c–Means Clustering Algorithm

The Fuzzy c–Means (FCM) algorithm [51], considers each cluster as a fuzzy set. It firstly initializes a number of c prototype vectors (centroids) p^j over the dataset. The centroids represent the center of the clusters. Next it computes a degree of membership for every data vector x^i at each cluster using the membership function:

$$\mu_j(x^i) = \left(\sum_{l=1}^{c} \left(\frac{\|x^i - p^j\|}{\|x^i - p^l\|} \right)^{1/r-1} \right)^{-1},$$

which takes values in the interval $[0,1]$, where $r \in (1,\infty)$ determines the fuzziness of the partition. If r tends to 1_+, then the resulting partition asymptotically approaches a crisp partition. On the other hand, if r tends to infinity, the partition becomes a maximally fuzzy partition. Finally, the c prototypes are updated using the following equation:

$$p^j = \frac{\sum_{i=1}^{n} \left[m_j(x^i)\right]^r x^i}{\sum_{i=1}^{n} \left[m_j(x^i)\right]^r}.$$

This procedure is iteratively repeated until the measure of the distortion:

$$d = \sum_{j=1}^{c} \sum_{i=1}^{n} \left[m_j(x^i)\right]^r \|x^i - p^l\|^2,$$

changes less than a user defined threshold.

1.4.3 The PDDP Clustering Algorithm

The PDDP algorithm [50], is a divisive clustering algorithm. The key component in this algorithm is the computation of the principal directions of the data. Starting with an initial cluster of all the data points, the algorithm iteratively splits the clusters. The use of a distance or similarity measure is limited to deciding which cluster should be split next, but the similarity measure is not used to perform the actual splitting. In detail, all the data points are projected onto the leading eigenvector of the covariance matrix of the data. Based on the sign of that projection the algorithm splits an initial cluster into two. This fact enables the algorithm to operate on extremely high dimensional spaces. PDDP, as well as PDDP(l) [61], which is a recent generalization of PDDP, does not provide a direct estimation for the number of clusters. Proposed methods that provide such estimations through these algorithms are based on scattering of the data around their centroids. Nonetheless, they tend to overestimate the true number of clusters resulting in rigid clustering [50, 61].

1.4.4 Growing Neural Gas

GNG [52] is an incremental neural network. It can be described as a graph consisting of k nodes, each of which has an associated weight vector, w_j, defining the node's position in the data space and a set of edges between the node and its neighbors. During the clustering procedure, new nodes are introduced into the network until a maximal number of nodes is reached. GNG starts with two nodes, randomly positioned in the data space, connected by an edge. Adaptation of weights, i.e. the nodes position, is performed iteratively. For each data object the closest node (winner), s_1, and the closest neighbor of a winner, node s_2, are determined. These two nodes are connected by an edge.

An age variable is associated with each edge. At each learning step the ages of all edges emanating from the winner are increased by 1. When the edge connecting s_1 and s_2 is created its age is set to 0. By tracing the changes of the age variable inactive nodes are detected. Any nodes having no emanating edges and edges exceeding a maximal age are removed.

The neighborhood of the winner is limited to its topological neighbors. The winner and its topological neighbors are moved in the data space toward the presented object by a constant fraction of the distance, defined separately for the winner and its topological neighbors. There is no neighborhood function or ranking concept. Thus, all topological neighbors are updated in the same manner.

1.4.5 A Hybrid Approach

The PCA technique optimally transforms the data set, with limited loss of information, to a space of significantly lower dimension. However, it is a global technique in the sense that does not deal with special characteristics that might exist in different parts of the data space.

To deal with this it is possible to hybridize a clustering algorithm and PCA. Firstly, the entire data set is partitioned into clusters of features, and next, each feature cluster can be independently transformed to a lower dimension space through the PCA technique. This application of the PCA is local and has the potential of better adapting to the special characteristics that might exist in the data set.

The techniques reported in this section should not be confused with any kind of *bi-clustering* approach [5,6]. In the latter case the aim is to find subsets of genes that exhibit a similar behavior for a subset of samples. However the techniques reported below aim to either organize the genes in groups and infer compact representation for each group. Either they aim to recognize clusters of samples that have a physical common character. Sometimes to achieve this the compact representations inferred from the initial procedure are employed but this is quite different to the bi-clustering point of view.

1.5 Experimental Analysis

In this Section we initially describe the microarray problems used in the remaining of this chapter. Then, we perform an extensive evaluation of various clustering algorithms for supervised as well as unsupervised classification of the data sets. Subsequently, we implement and test FNN classifiers combined with clustering methods and the PCA dimension reduction technique. Finally, we report results from a hybrid approach that utilizes EAs for gene selections and FNNs for classification.

1.5.1 DNA Microarray Problems

The evaluation of all the Computational Intelligence algorithms presented in this chapter is performed through the following well–known and publicly available data sets:

(a) The ALL–AML data set [39]. This study examines mRNA expression profiles from 72 leukemia patients to develop an expression–based classification method for acute leukemia. In the data set each sample is measured over 7129 genes. The first 38 samples were used for the clustering process (train set), while the remaining 34 were used to evaluate the clustering result (test set). The initial 38 samples contained 27 acute myeloid leukemia (ALL) samples and 11 acute lymphoblastic leukemia (AML) samples. The test set contained 20 ALL samples and 14 AML samples. The data set is available at:
http://www.broad.mit.edu/cancer/pub/all_aml
(b) The COLON data set [33] consists of 40 tumor and 22 normal colon tissues. For each sample there exist 2000 gene expression level measurements. The data set is available at:
http://microarray.princeton.edu/oncology
(c) The PROSTATE data set [62] contains 52 prostate tumor samples and 50 non-tumor prostate samples. For each sample there exist 6033 gene expression level measurements. It is available at:
http://www.broad.mit.edu/cgi-bin/cancer/datasets.cgi
(d) The LYMPHOMA dataset [41] that contains 62 samples of the 3 lymphoid malignancies samples types. The samples are measured over 4026 gene expression levels. This dataset is available at:
http://genome-www.stanford.edu/

All the data sets contain a relatively large number of patients and have been well characterized and studied. Notice that no additional preprocessing or alteration was performed to the data, except for the application of the methods described in this chapter.

1.5.2 Evaluation of the Clustering Algorithms

In the literature, both supervised and unsupervised classifiers have been used to build classification models from microarray data. Supervised classifiers employ predefined information about the class of the data to build the classification model. On the other hand, no class information is necessary to the unsupervised methods.

To investigate the performance of the clustering algorithms on gene expression microarray data we primarily employ the data set from the ALL–AML microarray problem. We performed two independent sets of experiments. In the first set, the clustering methodology was applied on two previously published gene subsets as well as their union. The comparative results assess the comparative performance of the clustering algorithms.

In the second set of experiments, we do not use class information for the gene selection. To this end, to reduce the dimensionality of the problem we use the PCA technique, as well as, dimension reduction through clustering. This second set of experiments is closer to real life applications where no class information is a priori known. Moreover, the hybridization of clustering and the PCA is evaluated. The hybrid scheme seems also able to provide results equivalent to those obtained with the supervised gene selection. Thus, this scheme is also applied on the remaining three data sets for further evaluation.

Clustering Based on Supervised Gene Selection

Generally, in a typical biological system, it is often not known how many genes are sufficient to characterize a macroscopic phenotype. In practice, a working mechanistic hypothesis that is testable and largely captures the biological truth, seldom involves more than a few dozens of genes. Therefore, identifying the relevant genes is critical [43]. Initially, we intended to study the performance of the UkW clustering algorithm, so we applied it over the complete ALL–AML train set. The algorithm was applied to the measurements of the 7129 genes, as well as various randomly selected gene subsets having from 10 to 2000 genes each. The algorithm produced clusters that often contained both AML and ALL samples. Typically, at least 80% of all the samples that were assigned to a cluster were characterized by the same leukemia type.

To improve the quality of the clustering, it proved essential to identify sets of genes that significantly contribute to the partition of interest. Clearly, there exist many such sets and it is difficult to determine the best one. To this end, we tested the UkW clustering algorithm on two previously discovered subsets of significant genes. The first set has been published in the original paper of Golub et al. [39] (we call it $GeneSet_1$), while the second set was proposed by Thomas et al. [63] ($GeneSet_2$).

Each dataset contains 50 genes. Furthermore, we tested the clustering algorithms on the union of the above gene sets ($GeneSet_3$), consisting of 72 genes.

In [39] $GeneSet_1$ was constructed by electing 50 highly correlated genes with the ALL–AML class distinction. Next, the authors used a Self Organizing Map [64] based clustering approach, to discover clusters on the training set. SOM automatically grouped the 38 samples into two classes, one containing 24 out of the 25 ALL samples, and the other containing 10 out of the 13 AML samples.

Regarding the second set of genes ($GeneSet_2$), the 50 most highly correlated genes with the ALL–AML class distinction (top 25 differentially expressed probe sets in either sample group) have been selected. More specifically, the selection approach is based on well–defined assumptions, uses rigorous and well–characterized statistical measures, and tries to account for the heterogeneity and genomic complexity of the data. The modelling approach uses known sample group membership to focus on expression profiles of individual genes in a sensitive and robust manner, and can be used to test statistical hypotheses about gene expression. For more information see [63].

Applying the UkW algorithm on those 3 gene train sets, each produced 6 clusters containing ALL or AML samples. Table 1.1 exhibits the results. More specifically, the algorithm using $GeneSet_1$ discovered 4 ALL clusters and 2 AML clusters (3 misclassifications), while using $GeneSet_2$ discovered 4 clusters containing only ALL samples and 2 clusters containing only AML samples (0 misclassifications). The algorithm discovered 4 ALL clusters and 2 AML clusters (1 misclassification) when applied to $GeneSet_3$. $GeneSet_2$ yielded the best results in the training set (followed by $GeneSet_3$).

Table 1.1. The performance of the UkW algorithm for the different train sets

Clustering result for the train set $GeneSet_1$ ALL accuracy: 87.5% — AML accuracy: 100%						
Leukemia type	ALL Clusters				AML Clusters	
	Cluster 1	Cluster 2	Cluster 3	Cluster 4	Cluster 1	Cluster 2
ALL	4	4	12	4	3	0
AML	0	0	0	0	4	7

Clustering result for the train set $GeneSet_2$ ALL accuracy: 100.0% — AML accuracy: 100%						
Leukemia type	ALL Clusters				AML Clusters	
	Cluster 1	Cluster 2	Cluster 3	Cluster 4	Cluster 1	Cluster 2
ALL	10	3	10	4	0	0
AML	0	0	0	0	8	3

Clustering result for the train set $GeneSet_3$ ALL accuracy: 95.83% — AML accuracy: 100%						
Leukemia type	ALL Clusters				AML Clusters	
	Cluster 1	Cluster 2	Cluster 3	Cluster 4	Cluster 1	Cluster 2
ALL	8	9	5	4	0	1
AML	0	0	0	0	7	4

Table 1.2. The performance of the UkW algorithm for the different test sets

Clustering result for the test set $GeneSet_1$ ALL accuracy: 60.00% — AML accuracy: 92.85%						
Leukemia type	ALL Clusters				AML Clusters	
	Cluster 1	Cluster 2	Cluster 3	Cluster 4	Cluster 1	Cluster 2
ALL	2	0	7	3	8	0
AML	1	0	0	0	8	5
Clustering result for the test set $GeneSet_2$ ALL accuracy: 100% — AML accuracy: 78.57%						
Leukemia type	ALL Clusters				AML Clusters	
	Cluster 1	Cluster 2	Cluster 3	Cluster 4	Cluster 1	Cluster 2
ALL	8	0	9	3	0	0
AML	0	0	3	0	8	3
Clustering result for the test set $GeneSet_3$ ALL accuracy: 90% — AML accuracy: 100%						
Leukemia type	ALL Clusters				AML Clusters	
	Cluster 1	Cluster 2	Cluster 3	Cluster 4	Cluster 1	Cluster 2
ALL	10	4	3	1	0	2
AML	0	0	0	0	5	9

Table 1.3. Comparative results for the test set $GeneSet_3$

	Misclassified samples		Number of clusters		Accuracy (%)	
	train set	test set	train set	test set	AML	ALL
DBSCAN	1	3	4	4	78.5	100
FCM	1	2	4	4	85.7	100
GNG	1	3	3	3	78.5	100
PDDP	2	4	6	6	71.4	100
UkW	1	2	6	6	100.0	90.0

To further evaluate the clustering results each sample from each test set was assigned to one of the clusters discovered in the train set according to its distance from the cluster center. Specifically, if an ALL (AML) sample from the test set was assigned to an ALL (AML, respectively) cluster then that sample was considered correctly classified. From the results exhibited in Table 1.2 it is evident that using the clustering from $GeneSet_1$ one AML and eight ALL samples from the test set were misclassified, resulting in a 73.5% correct classification. The clusters discovered using $GeneSet_2$ resulted in three misclassified AML samples (91.2% correct classification), while $GeneSet_3$ clusters yielded the best performance with only two misclassified ALL samples (94.1% correct classification).

In Table 1.3 we present comparative results from the test set $GeneSet_3$ only, as all the clustering algorithms exhibited improved classifications performance on this dataset. The best performance was achieved by the UkW algorithm and the FCM, followed by the DBSCAN and GNG algorithms. Notice that the FCM requires from the user to supply the number of clusters (supervised clustering) and that the

DBSCAN algorithm did not classify seven samples of the train set and five samples of the test set (all of them belonging in the AML class), since it characterized them as outliers.

Although the PDDP algorithm exhibited the worst classification performance, it must be noted that it was the only algorithm capable of using all the 7129 genes to cluster the samples. Using the complete set of genes, the PDDP algorithm misclassified two training set samples and eight test set samples.

Clustering Based on Unsupervised Gene Selection

Not using the class information to perform gene selection, we have to resort to unsupervised methods. We employ the UkW algorithm, for this task since it proved quite successful in the previous set of experiments. More specifically, the UkW algorithm was applied over the entire data set to select clusters of genes. Feature selection was accomplished by extracting from each cluster one representative feature (gene), based on the Euclidean distance among the feature values and the identified cluster center. The feature with the minimum distance from the cluster center was selected. This approach produced a new subset containing 293 genes ($GeneSet_4$).

The UkW algorithm was then applied on $GeneSet_4$ to group the samples. The results are illustrated in Table 1.4. From this table it is evident that high classification accuracy is possible even when class information is not known. Specifically, UkW exhibited accuracy of 93.6% and 76% for the ALL and the AML samples, respectively.

A second set of experiments is performed using the PCA technique for dimension reduction. A common problem when using PCA is that there is no clear answer to the question of how many factors should be retained for the new data set. A rule of thumb is to inspect the *scree plot*, i.e. plot all the eigenvalues in decreasing order. The plot looks like the side of a hill and "scree" refers to the debris fallen from the top and lying at its base. The scree test suggests to stop analysis at the point the mountain (signal) ends and the debris (error) begins. However, for the considered problem the scree plot was indicative, but not decisive. The scree plot, exhibited in Figure 1.5 (left), suggests that the contributions are relatively low after approximately ten components. In our experiments, we tried all the subsets using factors from 2 to 70. The classification accuracy is shown in Figure 1.5 (right). The best performance was attained when 25 factors were used (84.72%).

Although, the PCA technique tries to limit the loss of information, the classification accuracy is significantly lower, compared to the results obtained by supervised

Table 1.4. The performance of the UkW algorithm for the $GeneSet_4$ data set

Clustering result for the set $GeneSet_4$ ALL accuracy: 93.61% — AML accuracy: 76%						
Leukemia type	ALL Clusters					AML Cluster
	Cluster 1	Cluster 2	Cluster 3	Cluster 4	Cluster 5	Cluster 1
ALL	12	5	8	16	3	3
AML	2	0	3	0	1	19

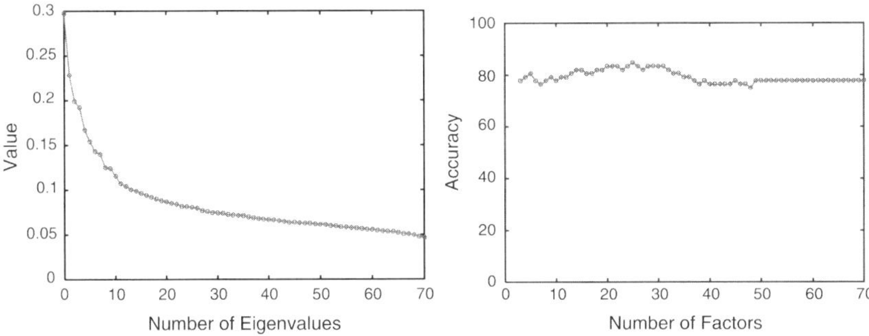

Fig. 1.5. Plot of the 70 first eigenvalues in decreasing order (*left*) and the corresponding classification accuracies (*right*)

Table 1.5. The performance of the UkW algorithm for the $GeneSet_5$ data set

Clustering result for the set $GeneSet_5$ ALL accuracy: 97.87% — AML accuracy: 88%						
Leukemia type	ALL Clusters				AML Clusters	
	Cluster 1	Cluster 2	Cluster 3	Cluster 4	Cluster 1	Cluster 2
ALL	7	14	14	11	1	0
AML	0	0	3	0	13	9

gene selection, Next, we study the hybridization of the clustering the PCA technique, with the aim to obtain more informative representations of the data.

To this end, the entire data set is firstly partitioned into clusters of features using the UkW algorithm. Next, each feature cluster is independently transformed to a lower dimension space through the PCA technique. Regarding the number of factors selected from each cluster many approaches could be followed. In our experiments only two factors from each cluster were selected, resulting in $GeneSet_5$. Experiments conducted using scree plots exhibited identical results. Our experience is that the number of selected factors from each cluster is not critical, since the entire data set has already been partitioned to a specific cluster number determined by the algorithm itself. Finally, the algorithm is again applied to group the samples into clusters and the results are exhibited in Table 1.5. The UkW exhibited accuracy 97.87% and 88% for the ALL and the AML samples, respectively.

Overall, the obtained experimental results regarding the various gene subsets indicate that using $GeneSet_1$ and $GeneSet_2$, yields very satisfactory results. The best results were obtained using the union of the genes in $GeneSet_1$ and $GeneSet_2$. The drawback of this feature selection scheme is that it relies on human expertise ($GeneSet_1$) and requires class information ($GeneSet_2$) to construct the final dataset ($GeneSet_3$). On the other hand, performing unsupervised gene selection using either PCA or UkW may result in a lower classification accuracy.

The hybridization of the two approaches yielded results comparable to those obtained through the first three gene sets. The main drawback of this approach is that it requires information from all the genes.

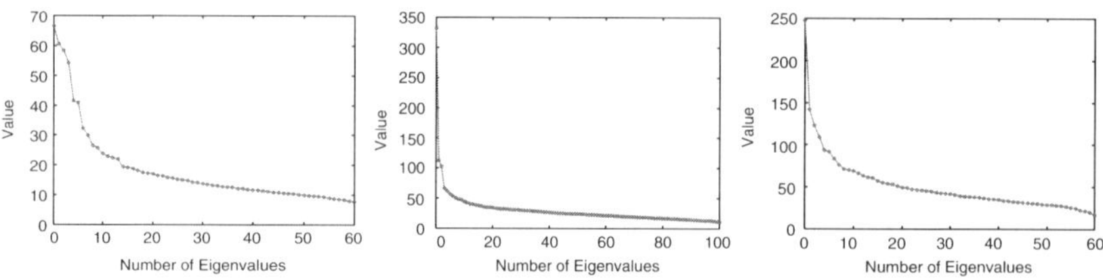

Fig. 1.6. Plot of the first eigenvalues in decreasing order, for the COLON (*left*), PROSTATE (*middle*) and the LYMPHOMA (*right*) datasets

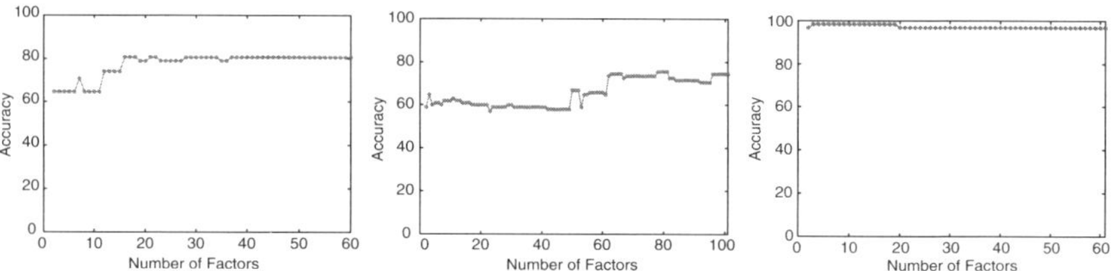

Fig. 1.7. Classification accuracy for all the factors, for the COLON (*left*), PROSTATE (*middle*) and the LYMPHOMA (*right*) datasets

To further investigate the efficiency of the hybridization scheme we compare it against the PCA dimension reduction technique on the COLON, the PROSTATE and the LYMPHOMA microarray data sets. While the hybrid approach automatically determines the number of reduced dimensions only the screen plot can provide such an information for the PCA technique. Although the scree plots, reported in Figure 1.6, provide an indication they are not conclusive. Generally, in all three cases the contributions are relatively low after approximately twenty components.

In our experiments, we tried all available factors for each dataset and utilized the UkW algorithm for the classification. The classification accuracy of the UkW clustering algorithm for each of the three datasets and all the available factors is reported in Figure 1.7. For the COLON dataset the best classification accuracy obtained was 80.64% employing 16 factors. For the PROSTATE dataset the best result was 82.35% classification accuracy, using 71 factors. Finally, for the LYMPHOMA dataset the best result was 98.38% classification accuracy using only 3 factors.

The results of the hybrid approach, for the three datasets, are presented in Table 1.6. As it is evident, the classification accuracy of the resulting partitions increases in all three cases. The high number of factors that the hybrid scheme decides to use, does not impose a problem to the algorithm since they originate in different clusters, and they are not correlated to each other. Furthermore, the additional advantage of the automatic determination of the required factors, exhibits a robust result that is not possible through the PCA technique. The classification accuracies obtained are considered very high, in comparison to other previously published approaches [65].

Table 1.6. The performance of the hybrid approach for the COLON, PROSTATE and LYMPHOMA datasets

Dataset	Number of Factors Used	Classification Accuracy (%)
COLON	229	82.25
PROSTATE	84	83.3
LYMPHOMA	103	99.01

1.5.3 Using Clustering and Feedforward Neural Networks Classifiers

Although the Feedforward Neural Networks (FNNs) trained using the PCA projection of the dataset can provide high classification accuracy, there is no straightforward interpretation of the new dimensions. Consequently, to compute features for a new patient, information from all the genes is required. On the other hand, the clustering algorithms identify a subset of genes that significantly contribute to the partition of interest. Thus, only the expression levels of the selected genes are needed for the future operation of the system. Unfortunately, there exist many such subsets and it is difficult for any clustering algorithm to determine the best one.

The first step towards the implementation of such a system is to apply a clustering algorithm over the entire training sets. Dimension reduction is performed by selecting a representative feature from each identified feature cluster, as usual. The representative features will be used as input to the FNN classifier. To this end, the (supervised) FCM and the (unsupervised) UkW clustering algorithms were applied on the three data sets mentioned above. Since the number of clusters, c, present in each data set is unknown, all possible values from 3 to 30 were tried for the FCM algorithm. On the other hand, the UkW clustering algorithm was executed only once and it provided 14 features for the COLON data set, 18 features for the PROSTATE data set, and 22 features for the ALL–AML data set.

Consequently, an FNN having two hidden layers consisting of 5 neurons each, was trained using the Rprop and the AOBP training algorithms to classify the features of the data sets. In the experiments, we performed random splitting of the data into learning and test sets. Specifically, the data was partitioned randomly into a learning set consisting of two-thirds of the whole set and a test set consisting of the remaining one-third. To reduce the variability, the splitting was repeated 50 times as in [65]. For each splitting 50 independently initialized FNNs were trained, resulting in a total of 2500 experiments. The comparative results for the three problems considered here are illustrated using boxplots in Figures 1.8, 1.9, and 1.10, respectively. Each boxplot depicts the obtained values for the classification accuracy, in the 2500 experiments. The box has lines at the lower quartile, median, and upper quartile values. The lines extending from each end of the box (whiskers) indicate the range covered by the remaining data. The outliers, i.e. the values that lie beyond the ends of the whiskers, are represented with crosses. Notches represent a robust estimate of the uncertainty about the median. From these figures it is evident that the UkW algorithm exhibited the best performance. The mean classification success for each problem was 65.9%, 73.5%, and 69.2%, clearly above the mean classification success of FCM regardless

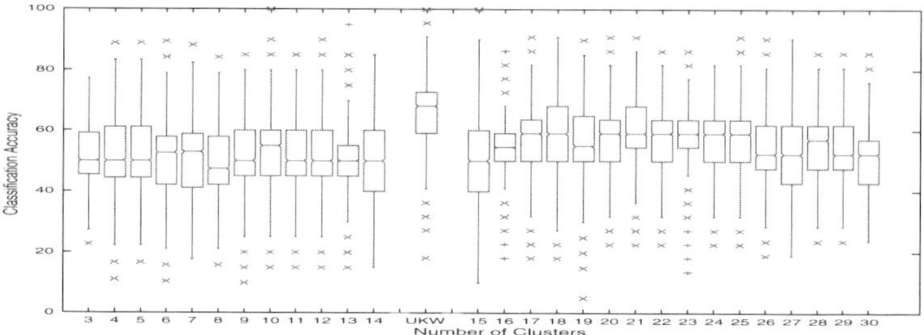

Fig. 1.8. COLON: Classification accuracy of FNNs incorporating the FCM and the UkW clustering algorithms

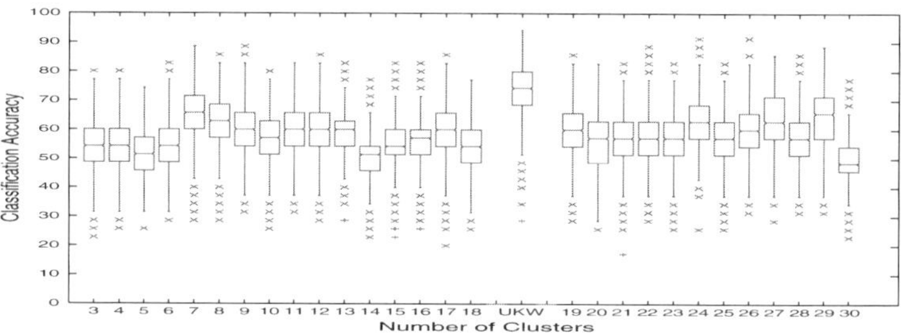

Fig. 1.9. PROSTATE: Classification accuracy of FNNs incorporating the FCM and the UkW clustering algorithms

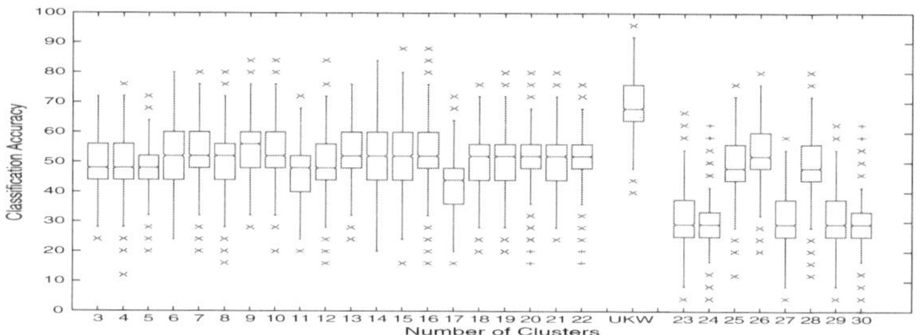

Fig. 1.10. ALL–AML: Classification accuracy of FNNs incorporating the FCM and the UkW clustering algorithms

the value of c. Moreover, FCM's results were heavily dependent on the number of features selected.

In spite of UkW algorithm's good results, this first set of experiments revealed the limitation of the direct application of any clustering algorithm, since even better classification accuracy is possible (see for example [65]). As a next step, we examine the classification accuracy on PCA derived features. Since for the for the considered problems the scree plots were indicative, but not decisive (see Figure 1.6) for the number of factors to use, we tried all of them from 3 to 30. As above, 50

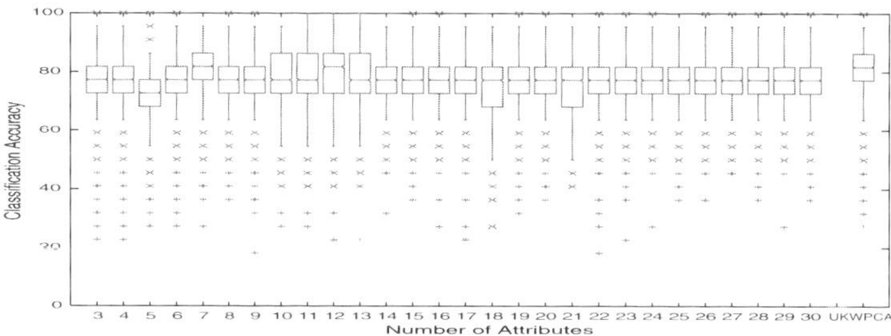

Fig. 1.11. COLON: Classification accuracy of FNNs incorporating the PCA technique and the proposed UkWPCA scheme

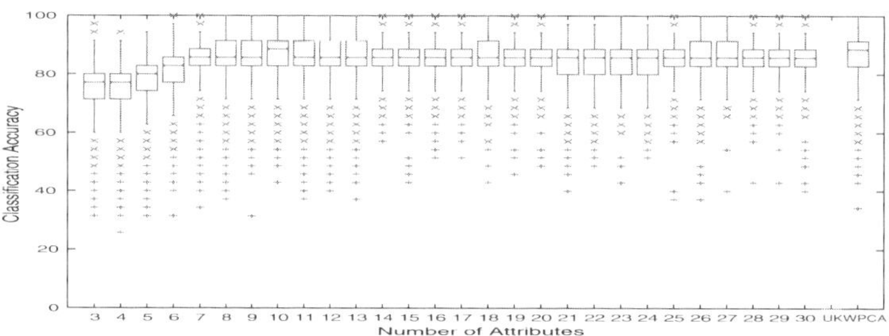

Fig. 1.12. PROSTATE: Classification accuracy of FNNs incorporating the PCA technique and the proposed UkWPCA scheme

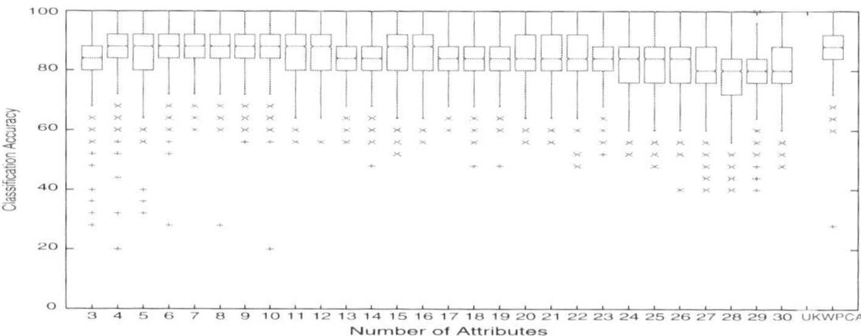

Fig. 1.13. ALL–AML: Classification accuracy of FNNs incorporating the PCA technique and the proposed UkWPCA scheme

random splittings of the data were performed and 50 independently initialized FNNs were trained. The results from the 2500 experiments for each problem are illustrated in Figures 1.11, 1.12, and 1.13, respectively. The results show that the classification accuracy depends on the number of factors used and that the best results do not exactly match the scree plot indication. Although, the FNNs trained using the PCA projection of the data set, in general, provide high classification accuracy, there is no straightforward way to select the right number of factors for each problem. FNNs using features computed by the PCA technique exhibited mean classification accuracy 79.1%, 86.5%, and 88.5%, for the optimal selection of the number of factors.

The above discussion suggests that the UkW algorithm is capable of automatically identifying meaningful groups of features, while the PCA technique optimally transforms the data set, with limited loss of information, to a space of significantly lower dimension. Since both properties are desirable for an automatic classification system, we next examine the classification accuracy using the hybrid approach to reduce the dimension.

As in the previous sets of experiments, we performed 50 random splittings of the data set and consequently 50 independently initialized FNNs were trained using the Rprop algorithm. The classification accuracy of the proposed system (UkWPCA) is illustrated in the last column of Figures 1.11, 1.12, and 1.13, respectively. To summarize, the UkW algorithm automatically provided a good approximation of the number of clusters present in the data sets, while the PCA technique transformed the discovered clusters resulting in the most informative features. FNNs trained using these features had the highest classification accuracy and the most robust performance. Specifically, the mean classification accuracies for the three problems considered here were 80.3%, 87.1%, and 87.4%, respectively.

1.5.4 Using Evolutionary Algorithms and Feedforward Neural Networks Classifiers

Here, we propose the application of EAs to microarray classification to determine the optimal, or near optimal, subset of predictive genes on the complex and large space of possible gene sets. Although a vast number of gene subsets are evaluated by the EA, selecting the most informative genes is a non trivial task. Common problems include the existence of: a) relevant genes that are not included in the final subset, because of the insufficient exploration of the gene pool, b) significantly different subsets of genes being the most informative as the evolution progresses, and c) many subsets that perform equally well, as they all predict the test data satisfactorily. From a practical point of view, the lack of a unique solution does not seem to present a problem.

The EA approach we propose utilizes the DE algorithm and maintains a population of trial gene subsets, imposes random changes on the genes that compose those subsets, and incorporates selection (driven by a neural network classifier) to determine which are the most informative ones. Only those genes are maintained in successive generations; the rest are removed from the trial pool. At each iteration, every subset is given as input to an FNN classifier and the effectiveness of the trained FNN determines the fitness of the subset of genes. The size of the population and the number of features in each subset are parameters that we explore experimentally. For the experiments reported in this chapter we employed Equation (1.3) as the main DE search operator.

For the approach discussed above, each population member represents a subset of genes, so a special representation and a custom fitness function must be designed. When seeking subsets containing n genes, each individual consists of n integers. The first integer is the index of the first gene to be included in the subset, the second integer denotes the number of genes to skip until the second gene to be included is reached, the third integer component denotes the number of genes to skip until the third included gene, and so on. This representation was necessary in order to avoid multiple inclusion of the same gene. Moreover, a version of DE that uses integer vectors has been proposed and thoroughly studied in previous studies [19, 20, 22].

Let us now focus on the custom fitness function. Initially, the k–nearest neighbors (KNN) classifier was used as a fitness function to evaluate the fitness of each gene subset. KNN classification is based on a distance function such as the Euclidean distance or Pearson's correlation that is computed for pairs of samples in n–dimensional space. Each sample is classified according to the class memberships of its k–nearest neighbors, as these are determined by the distance function. KNN has the advantage of simplicity and it usually performs well on data sets that are not linearly separable. However, our preliminary experimental results indicated that although the evolutionary algorithm produces gene subsets that help the KNN classifier to achieve high classification accuracy on the training samples, KNN fails to correctly classify the test data.

Thus we decided to use FNNs instead of the KNN classifier. The utilization of FNNs as fitness function greatly improved the classification accuracy of this approach. An FNN was trained using each subset of genes and the fitness of the subset is scored by analyzing how well the FNN separates the training data into separate classes. One third of the data set is used as a training set for the FNN and one third is used to measure the classification accuracy of the FNN classifier. The remaining patterns of the data set are kept to estimate the classification capability of the final gene subset.

Below, we report the experimental results. We have tested and compared the performance of the this approach on many publicly available microarray data sets. Here we report results from the COLON and the PROSTATE data sets. Since the appropriate size of the most predictive gene set is unknown, DE was employed for various gene set sizes ranging from 10 to 100 with a step of 10. The FNN used at the fitness function consisted of two hidden layers with eight and seven neurons, respectively. The input layer contained as many neurons as the size of the gene set. One output neuron was used at the output layer whose value for each sample determined the network classification decision. Since both problems had two different classes for the patterns, a value lower than 0.5 regarded the pattern to belong to the first class; otherwise regarded it to belong to the second class.

For each different gene set size the data were partitioned randomly into a learning set consisting of two–thirds of the whole set and a test set consisting of the remaining one third, as already mentioned. The one third of the training set was used by the Rprop and the AOBP algorithms to train the FNNs. The performance of the respective gene set was measured according to the generalization of the trained FNN on the rest of the training set. Both the Rprop and the AOBP training algorithms exhibited stable performance and are suitable for this kind of tasks. Note that, the test set was only used to evaluate the classification accuracy that can be obtained using the final gene set discovered by the DE algorithm. To reduce the variability,

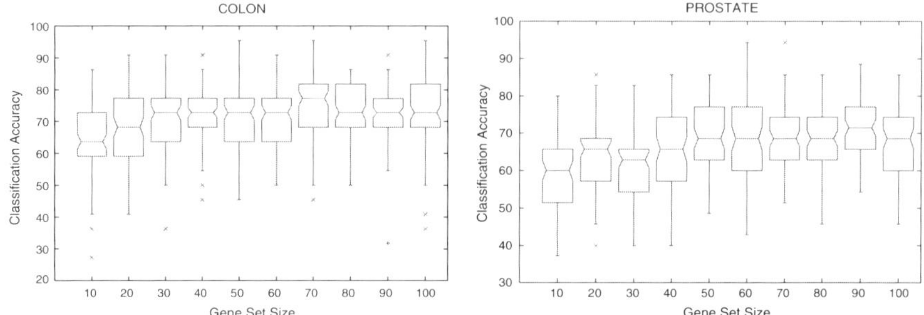

Fig. 1.14. Classification accuracy obtained by FNNs trained using the DE selected gene set for the COLON (*left*) and PROSTATE (*right*) datasets

the splitting was repeated 10 times and 10 independent runs were performed each time, resulting in a total of 100 experiments, for gene set size.

The classification accuracy of the proposed system is illustrated using boxplots in Figure 1.14. Each boxplot depicts the obtained values for the classification accuracy, in the 100 experiments. As demonstrated, using a gene set size of 50–80 for the COLON dataset the algorithm managed to achieve the best results. The same is achieved for the PROSTATE dataset for a gene set size ranging from 40 to 60. The experimental results are comparable to those obtained by other approaches [65, 66].

1.6 Concluding Remarks

Although the classification of the data obtained from microarray studies is very important in medical diagnosis of many diseases, it still presents a challenge for data analysis. This is due to the tremendous amount of available data (typically several Gigabytes of data), the redundant, erroneous or incomplete data sets, and the high dimensionality. Thus, the application of techniques for dimension reduction and/or selection of subsets of informative genes are essential to counter this very difficult problem. The selection of gene subsets that retain high predictive accuracy for certain cell–type classification, poses a central problem in microarray data analysis. The application and combination of various Computational Intelligence methods holds a great promise for automated feature selection and classification.

To summarize, in this chapter we have presented, implemented and tested supervised clustering algorithms, unsupervised clustering algorithms, the Principal Component Analysis dimension reduction technique, Feedforward Artificial Neural Networks, Evolutionary Algorithms, and hybrid approaches. Our goal was to evaluate and compare various approaches in an attempt to investigate their weaknesses and their shortcomings with respect to DNA microarray data analysis and classification.

Neural Networks have traditionally been used by the research community due their easiness of implementation and their high quality results. Their application to the microarray data, needs the existence of a preprocessing phase that would reduce the dimension of the learning space. Using either Evolutionary techniques in a supervised manner or unsupervised cluster analysis significant results can be obtained. However, the unsupervised characteristics of the latter approach provide an intuitive advantage. On the other hand, using cluster analysis directly to infer knowledge without resorting to an external trainer as in the Neural Network case, also seems quite promising.

Among the different clustering algorithms studied, the density based approaches provided the best results. The experimental results of the DBSCAN, the UkW and the GNG algorithms are indicative of how effective is their feature of automatic discovery of the cluster number. However, in the case of GNG the numerous parameters seem to deteriorate its clustering accuracy. All the above comments are true in the case that user-defined information about the most important subset of genes is used.

In the case, that no such information is available our first resort is the PDDP clustering algorithm, that can be directly applied to the original high dimensional space. However, it results in low performance clustering, which can be attributed to its crude splitting technique. Nevertheless, by borrowing the idea of PDDP, that is to apply PCA to different parts of the data space, we can design a hybrid method that is completely automated and does not require any kind of external steering. To this end, we examined how the hybrid techniques can further improve the classification accuracy of traditional classifiers such as Neural Networks. These results are not restrictive to FNNs, but can be straightforwardly extended to other types of classifiers, such as Support Vector Machines, Probabilistic Neural Networks, etc.

Briefly, we can claim that the reported experimental results indicate that there exists no unique and clear solution to this hard real–life problem. One must try different approaches in order to gain insight and better analyze the DNA microarray data. However, Computational Intelligence techniques are clearly capable of:

(a) exhibiting high classification success rates,
(b) having completely automatic operation,
(c) discovering the subsets of features that contribute significantly,
(d) constructing non–linear relationships between the input and the output.

Thus, even when compared against the best known alternative methods, Computational Intelligence techniques seem to prevail. Extensive experiments on publicly available microarray datasets indicate that the approaches proposed and studied here are fast, robust, effective and reliable. However, further testing on bigger data sets from new microarray studies is necessary before we can establish a general, flexible all-purpose methodology.

References

1. Jiang, D., Tang, C., Zhangi, A.: Cluster analysis for gene expression data: A survey. IEEE Transactions on Knowledge and Data Engineering **16**(11) (2004) 1370–1386
2. Larranaga, P., Calvo, B., Santana, R., Bielza, C., Galdiano, J., Inza, I., Lozano, J.A., Armananzas, R., Santafe, G., Perez, A., Robles, V.: Machine learning in bioinformatics. Briefings in Bioinformatics **7**(1) (2006) 86–112

3. Statnikov, A., Aliferis, C.F., Tsamardinos, I., Hardin, D., Levy, S.: A comprehensive evaluation of multicategory classification methods for microarray gene expression cancer diagnosis Bioinformatics **21**(5) (2005) 631–643
4. Wall, M., Rechtsteiner, A., Rocha, L.: Singular value decomposition and principal component analysis. In: A Practical Approach to Microarray Data Analysis. Kluwer (2003) 91–109
5. Van Mechelen, I., Bock, H.H., De Boeck, P.: Two-mode clustering methods:a structured overview. Statistical Methods in Medical Research **13**(5) (2004) 363–394
6. Kung, S.Y., Mak, M.W.: A Machine Learning Approach to DNA Microarray Biclustering Analysis. In: Proceedings of the IEEE International Workshop on Machine Learning for Signal Processing, (2005) 314–321
7. Wang, Z., Wang, Y., Xuan, J., Dong, Y., Bakay, M., Feng, Y., Clarke, R., Hoffman, E.P.: Optimized multilayer perceptrons for molecular classification and diagnosis using genomic data. Bioinformatics **22**(6) (2006) 755–761
8. Rumelhart, D., Hinton, G., Williams, R.: Learning internal representations by error propagation. MIT Press Cambridge, MA, USA (1986)
9. Gill, P., Murray, W., Wright, M.: Practical optimization. London: Academic Press, (1981)
10. Riedmiller, M., Braun, H.: A direct adaptive method for faster backpropagation learning: The RPROP algorithm. In: Proceedings of the IEEE International Conference on Neural Networks, San Francisco, CA. (1993) 586–591
11. Sutton, R., Whitehead, S.: Online learning with random representations. Proceedings of the Tenth International Conference on Machine Learning (1993) 314–321
12. Magoulas, G., Plagianakos, V.P., Vrahatis, M.N.: Development and convergence analysis of training algorithms with local learning rate adaptation. In: IEEE International Joint Conference on Neural Networks (IJCNN'2000), **1** (2000) 21–26.
13. Plagianakos, V.P., Magoulas, G., Vrahatis, M.N.: Global learning rate adaptation in on-line neural network training. In: Second International ICSC Symposium on Neural Computation (NC'2000). (2000)
14. Bäck, T., Schwefel, H.: An overview of evolutionary algorithms for parameter optimization. Evolutionary Computation **1**(1) (1993) 1–23
15. Storn, R., Price, K.: Differential evolution – a simple and efficient adaptive scheme for global optimization over continuous spaces. Journal of Global Optimization **11** (1997) 341–359
16. Storn, R., Price, K.: Minimizing the real functions of the icec'96 contest by differential evolution. In: IEEE Conference on Evolutionary Computation. (1996) 842–844
17. DiSilvestro, M., Suh, J.K.: A cross-validation of the biphasic poroviscoelastic model of articular cartilage in unconfined compression, indentation, and confined compression. Journal of Biomechanics **34** (2001) 519–525
18. Ilonen, J., Kamarainen, J.K., Lampinen, J.: Differential evolution training algorithm for feed forward neural networks. Neural Processing Letters **17**(1) (2003) 93–105
19. Plagianakos, V.P., Vrahatis, M.N.: Neural network training with constrained integer weights. In Angeline, P., Michalewicz, Z., Schoenauer, M., Yao, X., Zalzala, A., eds.: Proceedings of the Congress of Evolutionary Computation (CEC'99). IEEE Press (1999) 2007–2013
20. Plagianakos, V.P., Vrahatis, M.N.: Training neural networks with 3–bit integer weights. In Banzhaf, W., Daida, J., Eiben, A., Garzon, M., Honavar, V., Jakiela, M., Smith, R., eds.: Proceedings of the Genetic and Evolutionary Computation Conference (GECCO'99). Morgan Kaufmann (1999) 910–915

21. Tasoulis, D.K., Pavlidis, N.G., Plagianakos, V.P., Vrahatis, M.N.: Parallel differential evolution. In: IEEE Congress on Evolutionary Computation (CEC 2004), **2** (2004) 2023–2029
22. Plagianakos, V.P., Vrahatis, M.N.: Parallel evolutionary training algorithms for 'hardware-friendly' neural networks. Natural Computing **1** (2002) 307–322
23. Tasoulis, D.K., Plagianakos, V.P., Vrahatis, M.N.: Clustering in evolutionary algorithms to efficiently compute simultaneously local and global minima. In: IEEE Congress on Evolutionary Computation. Volume 2., Edinburgh, UK (2005) 1847–1854
24. John, G., Kohavi, R., Pfleger, K.: Irrelevant features and the subset selection problem. In: International Conference on Machine Learning. (1994) 121–129
25. Aggarwal, C., Wolf, J., Yu, P., Procopiuc, C., Park, J.: Fast algorithms for projected clustering. In: 1999 ACM SIGMOD international conference on Management of data, ACM Press (1999) 61–72
26. Agrawal, R., Gehrke, J., Gunopulos, D., Raghavan, P.: Automatic subspace clustering of high dimensional data for data mining applications. In: 1998 ACM SIGMOD international conference on Management of data, ACM Press (1998) 94–105
27. Hastie, T., Tibshirani, R., Friedman, J.: The Elements of Statistical Learning. Springer-Verlag (2001)
28. Fayyad, U., Piatetsky-Shapiro, G., Smyth, P.: Advances in Knowledge Discovery and Data Mining. MIT Press (1996)
29. Aldenderfer, M., Blashfield, R.: Cluster Analysis. Volume 44 of Quantitative Applications in the Social Sciences. SAGE Publications, London (1984)
30. Ramasubramanian, V., Paliwal, K.: Fast *k*-dimensional tree algorithms for nearest neighbor search with application to vector quantization encoding. IEEE Transactions on Signal Processing **40**(3) (1992) 518–531
31. Becker, R., Lago, G.: A global optimization algorithm. In: Proceedings of the 8th Allerton Conference on Circuits and Systems Theory. (1970) 3–12
32. Torn, A., Zilinskas, A.: Global Optimization. Springer-Verlag, Berlin (1989)
33. Alon, U., Barkai, N., Notterman, D., K.Gish, Ybarra, S., Mack, D., Levine, A.: Broad patterns of gene expression revealed by clustering analysis of tumor and normal colon tissues probed by oligonucleotide array. Proc. Natl. Acad. Sci. USA **96**(12) (1999) 6745–6750
34. Eisen, M., Spellman, P., Brown, P., Botstein, D.: Cluster analysis and display of genome-wide expression patterns. Proc. Natl. Acad. Sci. USA **95** (1998) 14863–14868
35. Shamir, R., Sharan, R.: Click: A clustering algorithm for gene expression analysis. In: 8th International Conference on Intelligent Systems for Molecular Biology (ISMB 00), AAAI Press (2000)
36. Tavazoie, S., Hughes, J., Campbell, M., Cho, R., Church, G.: Systematic determination of genetic network architecture. Nature Genetics volume **22** (1999) 281–285
37. Tasoulis, D.K., Plagianakos, V.P., Vrahatis, M.N.: Unsupervised clustering in mRNA expression profiles. Computers in Biology and Medicine **36**(10) (2006)
38. Wen, X., Fuhrman, S., Michaels, G., Carr, D., Smith, S., Barker, J., Somogyi, R.: Large-scale temporal gene expression mapping of cns development. Proceedings of the National Academy of Science USA **95** (1998) 334–339
39. Golub, T., Slomin, D., Tamayo, P., Huard, C., Gaasenbeek, M., Mesirov, J., Coller, H., Loh, M., Downing, J., Caligiuri, M., Bloomfield, C., Lander, E.: Molecular classification of cancer: Class discovery and class prediction by gene expression monitoring. Science **286** (1999) 531–537
40. Jain, A., Murty, M., Flynn, P.: Data clustering: a review. ACM Computing Surveys **31**(3) (1999) 264–323

41. Alizadeh, A., et al.: Distinct types of diffuse large b-cell lymphoma identified by gene expression profiling. Nature **403**(6769) (2000) 503–511
42. Perou, C., Jeffrey, S., de Rijn, M.V., Rees, C., Eisen, M., Ross, D., Pergamenschikov, A., Williams, C., Zhu, S., J.C. Lee, D.L., Shalon, D., Brown, P., Botstein, D.: Distinctive gene expression patterns in human mammary epithelial cells and breast cancers. Proc. Natl. Acad. Sci. USA **96** (1999) 9212–9217
43. Xing, E., Karp, R.: Cliff: Clustering of high–dimensional microarray data via iterative feature filtering using normalized cuts. Bioinformatics Discovery Note **1** (2001) 1–9
44. Tamayo, P., Slonim, D., Mesirov, Q., Zhu, J., Kitareewan, S., Dmitrovsky, E., Lander, E., Golub, T.: Interpreting patterns of gene expression with self-organizing maps: Methods and application to hematopoietic differentiation. Proc. Natl. Acad. Sci. USA **96** (1999) 2907–2912
45. Alter, O., Brown, P., Bostein, D.: Singular value decomposition for genome-wide expression data processing and modeling. Proc. Natl. Acad. Sci. USA **97**(18) (2000) 10101–10106
46. Szallasi, Z., Somogyi, R.: Genetic network analysis – the millennium opening version. In: Pacific Symposium of BioComputing Tutorial. (2001)
47. Tasoulis, D.K., Vrahatis, M.N.: Unsupervised distributed clustering. In: Proceedings of the IASTED International Conference on Parallel and Distributed Computing and Networks, Innsbruck, Austria (2004) 347–351
48. Vrahatis, M.N., Boutsinas, B., Alevizos, P., Pavlides, G.: The new *k*-windows algorithm for improving the *k*-means clustering algorithm. Journal of Complexity **18** (2002) 375–391
49. Sander, J., Ester, M., Kriegel, H.P., Xu, X.: Density-based clustering in spatial databases: The algorithm gdbscan and its applications. Data Mining and Knowledge Discovery **2**(2) (1998) 169–194
50. Boley, D.: Principal direction divisive partitioning. Data Mining and Knowledge Discovery **2**(4) (1998) 325–344
51. Bezdek, J.: Pattern Recognition with Fuzzy Objective Function Algorithms. Kluwer Academic Publishers (1981)
52. Fritzke, B.: Growing cell structures a self-organizing network for unsupervised and supervised learning. Neural Netw. **7**(9) (1994) 1441–1460
53. Ankerst, M., Breunig, M.M., Kriegel, H.P., Sander, J.: Optics: Ordering points to identify the clustering structure. In: Proceedings of ACM-SIGMOD International Conference on Management of Data. (1999)
54. Ester, M., Kriegel, H., Sander, J., Xu, X.: A density-based algorithm for discovering clusters in large spatial databases with noise. In: Proceedings of the 2nd Int. Conf. on Knowledge Discovery and Data Mining. (1996) 226–231
55. Procopiuc, C., Jones, M., Agarwal, P., Murali, T.: A Monte Carlo algorithm for fast projective clustering. In: Proc. 2002 ACM SIGMOD, New York, NY, USA, ACM Press (2002) 418–427
56. Berkhin, P.: A survey of clustering data mining techniques. In Kogan, J., Nicholas, C., Teboulle, M., eds.: Grouping Multidimensional Data: Recent Advances in Clustering. Springer, Berlin (2006) 25–72
57. Jain, A.K., Murty, M.N., Flynn, P.J.: Data clustering: a review. ACM Computing Surveys **31**(3) (1999) 264–323
58. Tan, P.N., Steinbach, M., Kumar, V.: Introduction to Data Mining. Pearson Addison-Wesley, Boston (2005)

59. Tasoulis, D.K., Vrahatis, M.N.: Novel approaches to unsupervised clustering through the k-windows algorithm. In Sirmakessis, S., ed.: Knowledge Mining. Volume 185 of Studies in Fuzziness and Soft Computing. Springer-Verlag (2005) 51–78
60. Hartigan, J., Wong, M.: A k-means clustering algorithm. Applied Statistics **28** (1979) 100–108
61. Zeimpekis, D., Gallopoulos, E.: PDDP(l): Towards a Flexing Principal Direction Divisive Partitioning Clustering Algorithms. In Boley, D., Dhillon, I., Ghosh, J., Kogan, J., eds.: Proc. IEEE ICDM '03 Workshop on Clustering Large Data Sets, Melbourne, Florida (2003) 26–35
62. Singh, D., et al.: Gene expression correlates of clinical prostate cancer behavior. Cancer Cell **1** (2002) 203–209
63. Thomas, J., Olson, J., Tapscott, S., Zhao, L.: An efficient and robust statistical modeling approach to discover differentially expressed genes using genomic expression profiles. Genome Research **11** (2001) 1227–1236
64. Kohonen, T.: Self–Organized Maps. Springer Verlag, New York, Berlin (1997)
65. Ye, J., Li, T., Xiong, T., Janardan, R.: Using uncorrelated discriminant analysis for tissue classification with gene expression data. IEEE/ACM Transactions on Computational Biology and Bioinformatics **1**(4) (2004) 181–190
66. Plagianakos, V.P., Tasoulis, D.K., Vrahatis, M.N.: Hybrid dimension reduction approach for gene expression data classification. In: International Joint Conference on Neural Networks 2005, Post-Conference Workshop on Computational Intelligence Approaches for the Analysis of Bioinformatics Data. (2005)

2

Inferring Gene Regulatory Networks from Expression Data

Lars Kaderali[1] and Nicole Radde[2]

[1] University of Heidelberg, Viroquant Research Group Modeling, Bioquant BQ26, Im Neuenheimer Feld 267, 69120 Heidelberg, Germany, lars.kaderali@bioquant.uni-heidelberg.de

[2] University of Leipzig, Institute for Medical Informatics, Statistics and Epidemiology, Härtelstr. 16-18, 04107 Leipzig, Germany, nicole.radde@imise.uni-leipzig.de

Summary. Gene regulatory networks describe how cells control the expression of genes, which, together with some additional regulation further downstream, determines the production of proteins essential for cellular function. The level of expression of each gene in the genome is modified by controlling whether and how vigorously it is transcribed to RNA, and subsequently translated to protein. RNA and protein expression will influence expression rates of other genes, thus giving rise to a complicated network structure.

An analysis of regulatory processes within the cell will significantly further our understanding of cellular dynamics. It will shed light on normal and abnormal, diseased cellular events, and may provide information on pathways in dire diseases such as cancer. These pathways can provide information on how the disease develops, and what processes are involved in progression. Ultimately, we can hope that this will provide us with new therapeutic approaches and targets for drug design.

It is thus no surprise that many efforts have been undertaken to reconstruct gene regulatory networks from gene expression measurements. In this chapter, we will provide an introductory overview over the field. In particular, we will present several different approaches to gene regulatory network inference, discuss their strengths and weaknesses, and provide guidelines on which models are appropriate under what circumstances. In addition, we sketch future developments and open problems.

2.1 Introduction

Biology has undergone a seminal shift in the last decade, with a transition from focusing on simple, small components of cells, such as DNA, RNA and proteins, to the analysis of relationships and interactions between various parts of a biological system. The traditional approach to much of molecular biology breaks up a system into its various parts, analyzes each part in turn, and hopes to reassemble the parts back into a whole system. In contrast, the systems biology approach aims at understanding and modeling the entire system quantitatively, proposing that the system is more than the sum of its parts and can only be understood as a whole.

L. Kaderali and N. Radde: *Inferring Gene Regulatory Networks from Expression Data*, Studies in Computational Intelligence (SCI) **94**, 33–74 (2008)
www.springerlink.com

Gene regulatory networks control a cell at the genomic level, they orchestrate which genes and how vigorously these genes are transcribed to RNA, which in turn functions as a template for protein synthesis. Genes and proteins do not act independently. Instead, they interact with each other and form complicated regulatory networks. Proteins which function as transcription factors can positively or negatively influence the expression of another gene, and thus the production of other proteins. Some proteins act independently, others only become active in a complex. Gene regulatory networks describe these regulatory processes, and thus the molecular reaction of a cell to various stimuli. High throughput experimental techniques to measure RNA and protein concentrations enable new approaches to the analysis of such networks. The analysis of these data requires sophisticated techniques particularly tailored to the task. New statistical, qualitative and quantitative methods are being developed for this purpose.

At the modeling side, several levels of detail have traditionally been used to describe gene regulation. Starting with very simple models which allow for qualitative statements only, in recent years there is a tendency to describe the dynamic response of a system in more detail. Also, besides the analysis of given network models, the inference of parameters of a gene regulatory network from experimental data has become one of the big challenges in computational biology. As the number of parameters usually far exceeds the number of measurements available for this purpose, leading to under-determined problems, modelers have begun to use heterogeneous data sources for network inference and to include biological knowledge into the parameter estimation.

In the following, we will give an overview over different models and describe the challenges and current developments, with a focus on mathematical and computational techniques. In addition, we will present a novel method particularly suitable for the typical setting where one has only a low number of data points to estimate model parameters, but when still quantitative modeling is desired. We will show how inference from data can be carried out using the models discussed, and we will present algorithms for the computations involved.

Modeling of gene regulatory networks is a quickly evolving field, with new developments and algorithms being published almost daily. We can thus only provide an introduction to the subject matter with a rough overview, and in no way cover the field exhaustively. We will provide links to further literature where appropriate throughout the chapter, providing the reader with references for additional and more detailed information.

Before going into detail with the mathematical modeling of regulatory networks, we will briefly review the biological background in the following section. For more details see for example Alberts *et al.* [5], Cooper [28], Berg and Singer [13], or Collado-Vides and Hofestädt [27].

2.1.1 Biological Background

To understand the role regulatory networks play, we will start with the main players in a cell, the *proteins*. They consist of long folded chains of amino acids and attend various tasks essential for survival of the cell. For example, they function as transporters, induce chemical reactions as enzymes, take part in metabolic pathways, recognize and transmit external signals, or act as ion channels in the cell membrane [5]. Proteins are permanently produced, this process is called *gene expression*. It consists of two stages, *transcription* and *translation*, and is highly regulated at different levels.

The information which proteins a cell can generally produce is encoded in its genome, the entirety of *genes* located on the DNA. During transcription, information from a gene is transcribed into an intermediate product called *messenger RNA*, or shortly *mRNA*. It serves as a template to produce a protein in the second step, the translation. The velocity and rate of this process is highly regulated and can vary in a wide range, making the organism flexible to adapt to external influences such as nutrition supply and to changes in environmental conditions such as temperature or salinity. It also enables the cell to respond to various stimuli and to maintain basic metabolic processes necessary for survival [27].

Regulation happens at different levels in the cell. We start with probably the most important mechanism, the regulation of transcription initiation. This is the main regulatory mechanism in prokaryotes. In eukaryotic cells, regulation is complicated by other effects such as alternative splicing or transport processes, we will neglect this here for simplicity. In transcription, an enzyme called *RNA-polymerase* (RNAP) is needed to catalyze the production of mRNA from an existing DNA template. This is initiated by binding of RNAP to the *promoter*, a regulatory region in front of the gene's coding region. Promoters contain specific binding sites for *transcription factors*, that is, for proteins regulating gene expression. Binding of RNAP and thus transcription initiation are facilitated by these transcription factors. *Operators* are DNA regions with binding sites for *repressors*, transcription factors which inhibit binding of the polymerase. A repressor-operator complex can influence the expression rates of multiple genes simultaneously. Some genes which encode for proteins involved in the same regulatory process are organized in *operons*, they are located side by side and are regulated by one single promoter. Their expression patterns are thus highly correlated. Transcription factors can also affect the process of RNA production by inducing conformational changes of the DNA, which can either activate or inhibit the polymerase [5].

Transcription factors do not always act independently, they can influence each other. When this influence is positive, one says that the transcription factors cooperatively enhance each other, their collective influence exceeds the sum of single influences. For example, some transcription factors are inactive until they form an active complex with other proteins. A transcription factor bound to DNA can facilitate the binding of another transcription factor by electrostatic attraction. Transcription factors can also inhibit each other. This is the case, for example, when several transcription factors compete for the same binding site, or when an occupied binding site prevents binding at another binding site, because the sequences of both sites overlap or because two transcription factors repel each other.

Regulation also happens after the gene is transcribed to mRNA. This is called *post-transcriptional regulation*. An example is the binding of a protein to mRNA, thus changing the secondary structure of the molecule, and hence stabilizing it or marking it for degradation. Analogously, regulation of protein concentration after translation is called *post-translational modification*. Mostly, a chemical group is appended to the protein, which induces a conformational change and activates or inactivates the protein. Many transcription factors taking part in signal transduction pathways have to be chemically modified to become active. These chemical modifications happen at a much faster time scale than the time scale for gene expression, which has consequences for quantitative models.

In addition to the production of RNA and protein, chemical degradation also affects concentrations of these molecules. RNA is quite unstable, and proteins are also degraded after some time. This is usually described as a first order decay process, thus degradation is assumed to be proportional to the component's concentration. Degradation rates are sometimes measured, and may then be included in models of gene regulation.

Figure 2.1 shows an example for regulation of gene expression at different levels. The four genes X, Y, Z_1 and Z_2 encode proteins which function as transcription factors. Protein X and the chemically modified protein Z_2 compete for the same binding site within an operator O. The repressor-operator complex inhibits transcription of the genes X and Y. Proteins Y and Z_1 form a complex that acts as a transcription factor for the operon Z containing the genes Z_1 and Z_2.

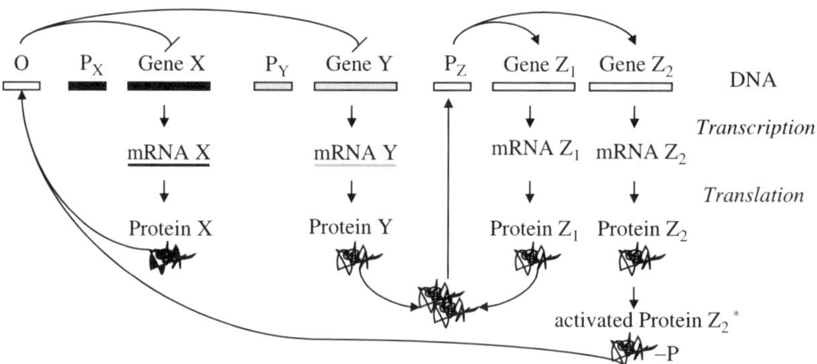

Fig. 2.1. Sample regulatory network consisting of four genes X, Y, Z_1 and Z_2. Regulation of gene expression happens at different levels: Protein X binds to an operator O and has thus a negative influence on the transcription rates of genes X and Y. Protein X and the chemically modified protein Z_2 compete for the same binding site. The proteins Y and Z_1 form an active complex, this complex acts as a transcription factor promoting expression of the operon Z, which in turn contains the genes Z_1 and Z_2

2.1.2 Modeling Gene Regulatory Networks

How can gene regulatory processes be represented in a computer? Our aim is twofold — inference of regulatory networks from data on the one hand, but also the simulation of a network's behavior on the other hand.

Recent advances in high-throughput biological techniques provide the basis for large scale analysis, which gives new insight into activities of cellular components under various biochemical and physiological conditions. DNA chips make the simultaneous measurement of concentrations of thousands of different RNA molecules possible, fermentation experiments yield data series of hundreds of metabolites, and large-scale measurements of protein concentrations are gradually becoming feasible. Moreover, the amount of protein-protein interaction and transcription factor binding site data is rapidly growing.

In computer models, gene regulatory networks are usually represented as directed graphs, with nodes corresponding to genes, and edges indicating interactions between the genes. In this chapter, we will discuss four different classes of models. In each section, we introduce a specific model or model class, and treat the inference problem. Subsequently, advantages and limitations of the models as well as possible extensions are discussed.

Boolean networks, described in Section 2.2, are probably the simplest models conceivable for regulatory networks. They assume that each gene is in one of two states, either *active* or *inactive*. Interactions between genes are modeled through Boolean logic functions, and updates are carried out simultaneously for all genes in discrete time steps. The updates are deterministic, and Boolean networks provide only a qualitative description of a system.

Relevance networks are described in Section 2.3. These approaches are based on pairwise distances (or similarities) between gene expression measurements, and try to reconstruct the networks using a threshold on the distance between genes.

Bayesian networks, discussed in Section 2.4, are probabilistic models. They model the conditional independence structure between genes in the network. Edges in a network correspond to probabilistic dependence relations between nodes, described by conditional probability distributions. Distributions used can be discrete or continuous, and Bayesian networks can be used to compute likely successor states for a given system in a known state.

Finally, *differential equation models*, described in Sections 2.5 to 2.7, provide a quantitative description of gene regulatory networks. Models used here range from simple linear differential equation models to complicated systems of nonlinear partial differential equations and stochastic kinetic approaches. In Section 2.5, we describe ordinary differential equation models. In Section 2.6, we present a novel method combining Bayesian networks and differential equations, and show first results on data from the yeast cell cycle network. Differential equation models going beyond ordinary differential equations are described in Section 2.7.

Finally, the last Section 2.8 gives a summary and an outlook, and provides a comparison between the model classes introduced.

2.2 Boolean Networks

Boolean networks offer a binary, discrete-time description of a system. They can be seen as a generalization of Boolean cellular automata [102], and have been introduced as models of genetic regulatory networks by Kauffman [52] in 1969. Let us start by stating a formal definition of a Boolean network:

Definition 1 (Boolean Network). *A* Boolean network *is defined as a tuple* $G = (X, B)$*, where* $X = (x_1, x_2, ..., x_n) \in \{0,1\}^n$ *is a vector of Boolean variables, and B is a set of Boolean functions* $B = \{f_1, f_2, ..., f_n\}$*,* $f_i : \{0,1\}^n \mapsto \{0,1\}$*.*

In gene expression networks, the x_i correspond to the genes and the f_i describe the interactions between them. In Boolean network models, one assumes that each gene can be modeled as being in one of two states, *on* (expressed, 1) or *off* (not expressed, 0). The functions B are used to update the nodes at discrete time-steps, all nodes X are updated synchronously using the Boolean functions B, that is, $x_i(t+1) = f_i(x_1(t), ..., x_n(t))$. We call a snapshot of the values of the nodes $x(t) = (x_1(t), x_2(t), ..., x_n(t))$ at time t the *expression pattern* or *state* of the network at the respective time point.

A Boolean network can be graphically represented in several ways, emphasizing different aspects of the network. An example is shown in Figure 2.2 for a small sample network consisting of three nodes A, B and C. The graph representation in Figure 2.2A shows how the nodes influence each other. Pointed arrows indicate an activation, see for example the positive regulation of node A by node B with the corresponding Boolean logic rule $A' = B$. In this example, the next value of node A, denoted A', will be equal to the current value of node B. Flat arrows indicate an inhibition, see for example the rule $B' = \neg A$. Here, the next value of node B will be the negation of node A, that is, $B' = 1$ if $A = 0$ and $B' = 0$ if $A = 1$. The value C' is computed from the current values of A and B together using the logical "OR" operation, hence $C' = 1$ if $A = 1$ or $B = 1$, and $C' = 0$ otherwise. The corresponding logical Boolean rules are given in Figure 2.2B. Figure 2.2C shows the *state transition table* of the network, it is a tabular representation of all possible "input" states of the network and, for each input, the resulting "output" or subsequent state. Figure 2.2D shows this table in a graph representation, visualizing the networks *state space* and its dynamics by connecting each input state with its corresponding output state. In this latter graph, it can be seen that the particular network in this example converges to a cycle of size four from any initial state.

2.2.1 Inferring a Boolean Network from Data

We will now discuss the problem of inferring a Boolean network from time series data. To formalize this, we define the *Consistency Problem:*

Definition 2 (Consistency Problem). *Let (I,O) be a pair of observed expression patterns of an unknown network* $G = (X, B)$*, such that* $O = B(I)$*, that is, O is the expression pattern of the network G after one time step when starting at state I.*

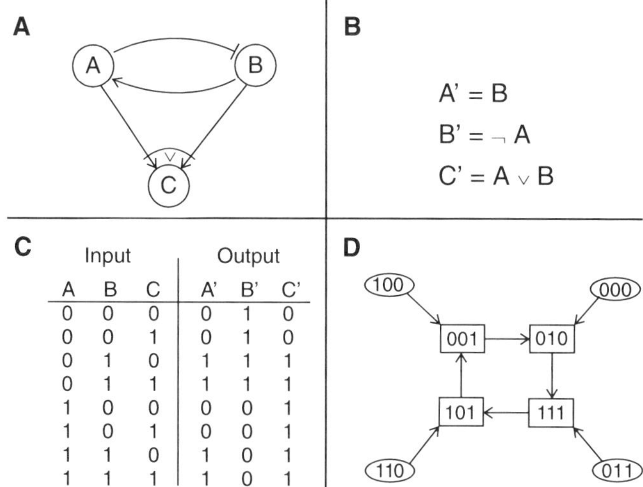

Fig. 2.2. Different representations of a sample Boolean network consisting of three nodes. (A) Graph representation, (B) logical Boolean rules, (C) state transition table and (D) state transition graph. In (A), pointed arrows indicate an activation, for example, gene A will be activated if gene B is active. Flat arrows indicate an inhibition, for example, gene B will be deactivated if gene A is active. Gene C is activated if either gene A or gene B is active, as denoted by the "or" symbol "∨" in the figure. In Figure 2.2B, the same relationship is expressed in boolean logical rules. Figure 2.2C shows a tabular representation of all possible input states and the resulting next states of the network. Figure 2.2D visualizes the state space in a graphical form, showing how the eight possible states of the network are interconnected. For example, if the network is in state $(A = 1, B = 0, C = 0)$, then the next state of the network will be $(A = 0, B = 0, C = 1)$

Then, a network $G' = (X', B')$ *is* consistent *with* (I, O)*, if* $O = B'(I)$*.* G' *is consistent with a set of expression pairs* $D = \{(I_j, O_j)\}_{j=1,\ldots,m}$*, if it is consistent with each pair* (I_j, O_j) *in D. The* Consistency Problem *is the problem to decide, whether a Boolean network consistent with given data D exists, and output one if it exists [2].*

The *Identification Problem* for Boolean networks in addition asks whether the network is unique:

Definition 3 (Identification Problem). *Given the number of genes n and a set of m input-output pairs D, the* Identification Problem *is the problem to decide whether a* unique *Boolean network consistent with D exists, and output it if it does [2].*

The number of possible networks with a given number of nodes n is huge, hence exhaustive search is usually prohibitive. For a network of n nodes, for each node, there will be 2^{2^n} possible functions of n inputs. Even if we restrict the Boolean functions to functions with at most $k < n$ inputs, there will be 2^{2^k} possible functions, and each node has $n!/(n-k)!$ possible ordered combinations of k different inputs.

The number of possible networks for given n and k will thus be

$$\left(2^{2^k}\frac{n!}{(n-k)!}\right)^n, \tag{2.1}$$

which grows exponentially with the number n of nodes in the network. However, if the indegree k of the network is fixed, the following can be shown:

Theorem 1 (Akutsu, 1999). *The consistency problem and the identification problem can be solved in polynomial time for Boolean networks with maximum indegrees bounded by a constant k.*

If k is close to n, the consistency and identification problems are NP-hard [3].

Also the number of data points required to estimate the Boolean functions from data grows exponentially with the network size. Surprisingly, for networks of fixed indegree k, $O(\log n)$ input/output patterns are sufficient on average for the network identification, with constant around $k2^k$ in front of the $\log n$ [2]. This is why much effort has been spent on devising learning algorithms for Boolean networks with fixed maximum indegree. Several algorithms have been proposed for network inference, for example [2, 3, 54]. In the following, we will sketch the REVEAL algorithm by Liang, Fuhrmann and Somogyi [58], which is based on information theoretic principles.

The REVerse Engineering ALgorithm REVEAL

The strategy employed in the REVEAL algorithm is to infer regulatory interactions between nodes from measures of *mutual information* in state transition tables. The observed data D is considered a random variable, and information theoretic properties are then used to derive the network topology.

Given a random variable X with k possible, discrete outcomes $x_1, \ldots, x_k$, the *Shannon entropy* H of X is defined in terms of the probabilities $p(x_i)$ of the possible outcomes as

$$H(X) = -\sum_{i=1}^{k} p(x_i)\log p(x_i), \tag{2.2}$$

where the sum is over the different outcomes x_i with associated probabilities $p(x_i)$ [80]. The entropy is a measure of the *uncertainty* associated with a random variable. In a system with two binary random variables X and Y, the *individual* and *combined entropies* are defined as

$$H(X) = -\sum_{x\in\{0,1\}} p(x)\log p(x) \tag{2.3}$$

$$H(Y) = -\sum_{y\in\{0,1\}} p(y)\log p(y) \tag{2.4}$$

$$H(X,Y) = -\sum_{\substack{(x,y)\in\\ \{0,1\}\times\{0,1\}}} p(x,y)\log p(x,y), \tag{2.5}$$

where $p(x)$, $p(y)$ and $p(x,y)$ are the individual and combined probability distributions of the random variables X and Y, respectively. Note that, for sets $a = \{X_1, X_2, \ldots, X_n\}$ of random variables X_i, we will use the notation $H(a)$ to denote the joint entropy $H(X_1, X_2, \ldots, X_n)$, derived by naturally extending equation (2.5) for more than two variables. Similarly, for two sets a and b of random variables, $H(a,b) = H(a \cup b)$.

The *conditional entropy* $H(X|Y)$ is a measure of the remaining uncertainty associated with a random variable X, given that the value of a second random variable Y is known. The conditional entropies $H(X|Y)$ and $H(Y|X)$ are related to the individual and combined entropies through

$$H(X,Y) = H(Y|X) + H(X) = H(X|Y) + H(Y), \tag{2.6}$$

or, in words, the combined entropy of X and Y is the sum of the individual entropy of a single variable and the information contained in the second variable that is not shared with the first. The *mutual information* is then defined as

$$M(X,Y) = H(X) - H(X|Y) = H(Y) - H(Y|X), \tag{2.7}$$

it is a measure of the information about one variable, that is shared by the second variable. Mutual information measures, how much knowing one of the variables X and Y reduces our uncertainty about the other.

REVEAL extracts relationships between genes from mutual information in gene expression measurements. The idea is, that when $M(X,Y) = H(X)$, then Y completely determines X. Rewriting $M(X,Y)$ according to equation (2.7), it follows that

$$Y \text{ completely determines } X \Longleftrightarrow H(Y) = H(X,Y), \tag{2.8}$$

hence the computation of $M(X,Y)$ is not even necessary.

Now let a set of m input-output patterns $D = \{(I_1, O_1), (I_2, O_2), \ldots, (I_m, O_m)\}$ be given. REVEAL then estimates the entropies from the data, and compares the single and combined entropies $H(b)$ and $H(a,b)$ for each node a and each subset of the genes b. If b exactly determines a, that is, if $H(b) = H(a,b)$, then a corresponding rule is added to the network. The pseudocode for REVEAL is given in Algorithm 1.

The worst-case running time of REVEAL is $O(mn^{k+1})$: Time $O(m)$ to estimate the entropies from the input data, and this must be done for each node and all subsets of the nodes of size up to k (lines 1–3).

2.2.2 Advantages and Disadvantages of the Boolean Network Model

As we have seen, the Boolean network model provides a straightforward model of regulatory networks, and under the condition of bounded indegree, efficient algorithms for network inference exist. Boolean networks are attractive due to their simplicity, they are easily applied and quickly implemented. The underlying assumptions

Algorithm 1 REVEAL

```
for each node a do
   for i = 1 to k do
      for each subset b of size i of the nodes X do
         compute the entropy H(b) from the inputs
         compute the joint entropy H(a,b) from the inputs b and outputs a
         if H(b) = H(a,b) then
            b exactly determines a, add a corresponding rule to the inferred network
            proceed with the next node a
         end if
      end for
   end for
end for
```

however seem very strict, in particular, modeling genes as being in one of only two states, either *on*, or *off*, certainly is an oversimplification of true biological networks. Similarly, true networks are time-continuous and asynchronous, whereas Boolean networks assume time-discrete, synchronous updates.

Still, recent research results indicate that many biologically relevant phenomena can be explained by this model, and that relevant questions can be answered using the Boolean formalism [82]. Focusing on fundamental, generic principles rather than quantitative biochemical detail, Boolean networks can capture many biological phenomena, such as switch-like behavior, oscillations, multi-stationarity, stability and hysteresis [48, 94, 96], and they can provide a qualitative description of a system [92]. Recent modeling results combined with the first experimental techniques to validate genetic models with data from living cells show that models as simple as Boolean networks can indeed predict the overall dynamic trajectory of a biological genetic circuit [16]. It seems that for understanding the general dynamics of a regulatory network, it is the wiring that is most important, and often detailed dynamic parameters are not needed [103]. For example, Albert and Othmer [4] have predicted the trajectory of the segment polarity network in *Drosophila melanogaster* solely on the basis of discrete binary models. Similarly, Li *et al.* [57] have modeled the genetic network controlling the yeast cell cycle using a binary model.

A serious limitation of the Boolean network approach is that, although a steady state of a Boolean network will qualitatively correspond to a steady state of an equivalent continuous model based on differential equations, not all steady states of the continuous model will necessarily be steady states of the Boolean model [40]. Conversely, periodic solutions in the Boolean model may not occur in the continuous model. This problem limits the utility of Boolean modeling of gene networks [85].

Clearly, Boolean networks are not suitable when detailed kinetic parameters are desired, and the focus is on the quantitative behavior of a system. Their key advantage and limitation at the same time is their simplicity, enabling them to capture the overall behavior of a system, but limiting the analysis to qualitative aspects. On the other hand, this simplicity allows the model to be applied to relatively large regulatory networks, when more detailed methods would be infeasible simply due to the lack of sufficient experimental data. At the same time, the simple two-state structure of each node in the Boolean network poses the problem that experimental data, which are usually measured on a continuous scale, need to be binarized, requiring delicate decisions about how this is best done.

Another shortcoming of Boolean networks is that they are deterministic in nature. However, true biological networks are known to have stochastic components, for example, proteins are produced from an activated promoter in short bursts that seem to occur at random time intervals, and probabilistic outcomes in switching mechanisms can be observed [65]. Furthermore, in realistic situations, we are usually dealing with noisy inputs and experimental measurement errors, which may lead to inconsistent data.

Finally, the dynamics of gene networks strongly depends on whether and how intra-cellular transport and diffusion of RNA and protein are modeled [60,61], which seems to play a particularly important role in eukaryotic cells [85]. The incorporation of such processes in Boolean network models is difficult, if not impossible [85].

2.2.3 Extensions of the Boolean Model

Several extensions of the Boolean network model have been proposed to overcome some of its limitations. To overcome the problems stemming from noisy and inconsistent data, from a learning-theoretic perspective, one relaxes the consistency problem to find a network that makes as few errors as possible. The resulting problem is known as the best-fit problem [17, 82] and is underlying many algorithms in machine learning.

To deal with the probabilistic nature of gene expression data, a popular extension of Boolean networks are the so-called *Probabilistic Boolean Networks* (PBN) [81]. The basic idea of PBNs is to aggregate several Boolean functions together, so that each can make a prediction of the target genes. One then randomly selects one of the functions, with probability being proportional to some weights assigned to the functions. PBNs can be interpreted as several Boolean networks operating in parallel, and one gets selected at random for a given time step. Thereafter, all networks are synchronized to the new state, so that each can make the next transition should it be selected [81].

Silvescu and Honavar [83] describe a generalization of Boolean networks to address dependences of genes that span over more than one time unit. Their model allows each gene to be controlled by a Boolean function of expression levels of at most k genes at T different time points, and they describe an algorithm for the inference of such networks from gene expression data. Other generalizations allow multi-valued networks, where each gene can be in one of several discrete states, and not just on or off [86].

2.3 Relevance Networks and Information Theoretic Approaches

While Boolean network models are based on the assumption that genes can only be in one of two states, *expressed* or *not expressed*, *relevance network* approaches [20] look at similarity or dissimilarity between pairs of genes on a continuous scale. Two steps are involved in network reconstruction using a relevance network approach:

1. All pairs of genes are compared using some measure of similarity or dissimilarity. For example, all genes can be compared against each other using pairwise correlation coefficients, or information theoretic measures such as mutual information can be used.
2. The complete set of pairwise comparisons is filtered to determine the relevant connections, corresponding to either positive or negative associations between genes.

The resulting network can then be represented in a graphical form. We will only briefly present one representative algorithm based on the relevance network approach, the ARACNe algorithm by Basso *et al.* [11,63].

2.3.1 The ARACNe Algorithm

Similar to REVEAL, ARACNe (Algorithm for the Reconstruction of Accurate Cellular NEtworks) [11, 63] is based on *mutual information* to identify regulations between genes. In a first step, it also identifies statistically significant gene-gene coregulation by mutual information. ARACNe can do this for discrete and continuous random variables, mutual information is estimated using Gaussian kernel estimators [12]. The algorithm is hence not limited to Boolean networks such as REVEAL. A statistical test is then used to determine relevant edges in the network, Monte Carlo randomization of the data is used for the computation of p-values, and edges are filtered based on a p-value threshold.

In a further step, ARACNe then prunes the network to eliminate indirect relationships, in which two genes are coregulated by one or more intermediary genes. This is done using the *data processing inequality (DPI)*, which essentially states that if three random variables X, Y and Z depend from one another in a linear fashion $X \to Y \to Z$, then the mutual information $M(X,Z) \leq \min[M(X,Y), M(Y,Z)]$. This is used to find and remove indirect edges $X \to Z$ from the network.

The authors of ARACNe claim that relationships in the final reconstructed network have a high probability of representing direct regulatory interactions or interactions mediated by post-transcriptional modifiers. They show results on microarray gene expression data from human B cells, reconstructing a network with approximately $129,000$ interactions from 336 expression profiles [11].

2.3.2 Advantages and Disadvantages of Relevance Network Approaches

Similar to Boolean networks, relevance networks are relatively simple models of gene regulatory networks. They use straightforward and easy to compute measures of pairwise similarity or dissimilarity between genes to reconstruct the network, such as correlation coefficients or information theoretic measures. In contrast to Boolean networks however, they are continuous models, that is, genes can have expression values on a quantitative scale.

One of the disadvantages of these approaches is, that they do not consider time, and thus disregard dynamic aspects of gene expression. Hence, these models can not infer causality, and it is not clear how to carry out simulations with an inferred network. Although algorithms such as ARACNe operate on a continuous scale for the gene expression levels, the method does not return any kinetic parameters, and is not based on chemical reaction kinetics. Furthermore, the relevance network approach is based on pairwise similarity only, and it may thus miss interactions between multiple genes. Finally, the choice of threshold for the inclusion of edges is somewhat arbitrary, and varying threshold parameters slightly may change the network considerably.

On the other hand, depending on the similarity measure used, relevance network approaches are less sensitive to noise than differential equations models. Although the data processing inequality used in ARACNe is not sufficient to identify indirect regulations, and hence the algorithm may sometimes remove direct relations as well, the pruning step helps the algorithm to derive sparse networks.

The simplicity of the relevance network approach makes it applicable to large networks. ARACNe, for example, is an algorithm with polynomial time complexity, and the authors report its use on networks with several hundred genes [11, 64]. It remains to be seen, how reliable the inferred interactions are for such large-scale applications.

2.4 Bayesian Networks

While Boolean networks assume a fixed functional dependence between different nodes, *conditional models* look at statistical correlation between genes. Conditional models try to explain the correlation between two genes by other genes in the network. These models are particularly simple in the Gaussian setting, since in this case networks can be learned from data using classical statistical tests [33, 84]. The most popular conditional model is the Bayesian network model, which is widely used to model and infer gene regulatory networks [69].

Definition 4 (Bayesian Network). *A* Bayesian Network *is a directed, acyclic graph* $G = (X, A)$*, together with a set of local probability distributions P. The vertices* $X = \{X_1, ..., X_n\}$ *correspond to variables, and the directed edges A represent probabilistic dependence relations between the variables. If there is an arc from variable* X_i *to* X_j*, then* X_j *depends probabilistically on* X_i*. In this case,* X_i *is called a* parent *of* X_j*. A node with no parents is* unconditional*. P contains the local probability distributions of each node* X_i *conditioned on its parents,* $p(X_i|parents(X_i))$*.*

Figure 2.3 shows a simple Bayesian example network with three nodes A, B and C, each assumed to be in one of two states, either *on* or *off*. The conditional

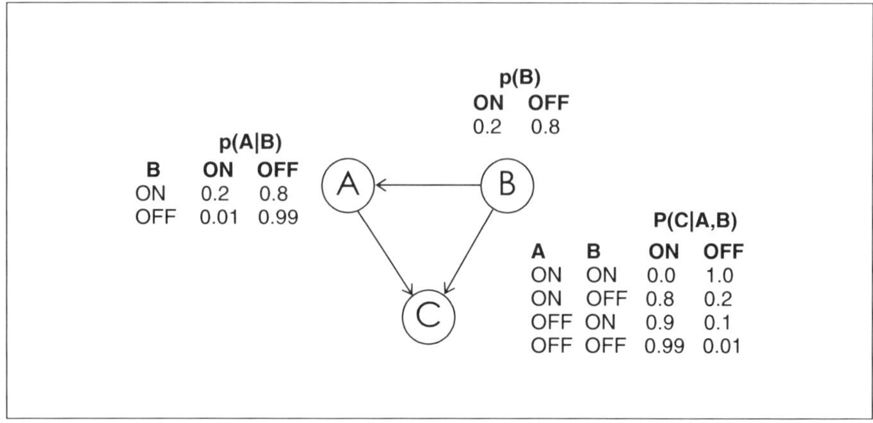

Fig. 2.3. Sample Bayesian Network with three nodes with two possible states each (ON and OFF). Given next to each node are the conditional distributions for the node, conditioned on its parents, as indicated by the arcs. For example, the probability that A is *off* given that B is *on*, $p(A = \text{off}|B = \text{on})$ is 0.8

probabilities $p(A|B)$, $p(C|A,B)$ and the unconditional probability $p(B)$ in this binary case are easily tabulated, as shown in the figure.

Note that the probability distributions of the nodes in Bayesian networks can be of any type, and need not necessarily be restricted to discrete or even binary values as in our example.

Given a Bayesian network, it is easy to compute the joint probability distribution of all variables in the network:

Definition 5 (Joint Distribution). *The* joint distribution *of a set of variables* $X_1, X_2, \ldots, X_n$ *is the product of the local distributions,*

$$p(X_1, X_2, \ldots, X_n) = \prod_{i=1}^{n} p(X_i|parents(X_i)). \tag{2.9}$$

In our example, the joint probability distribution is given by

$$p(A,B,C) = p(B)p(A|B)p(C|A,B), \tag{2.10}$$

and, for example, the joint probability that all nodes are *on* is $p(A = \text{on}, B = \text{on}, C = \text{on}) = p(B = \text{on})p(A = \text{on}|B = \text{on})p(C = \text{on}|A = \text{on}, B = \text{on}) = 0.2 \times 0.2 \times 0.0 = 0.0$. It is important to note at this point that the joint probability distribution can only be resolved this way if the network does not contain any directed cycles.

Bayesian networks provide a graphical representation of statistical dependences between variables, but more importantly, they also visualize independence relations among variables. Conditional independence of variables is represented in the graph by the property of *d-separation*, for *directional separation* [70]:

Definition 6 (d-separation). *Let a Bayesian network $G = (X,A)$ with local probability distributions P be given. Two nodes X_i and X_j, $i \neq j$, are* d-separated *in the graph G by a given set $S \subseteq X \setminus \{X_i, X_j\}$ of the nodes X, if and only if the variables X_i and X_j are independent given the values of the nodes in S.*

Informally, d-separation means that no information can flow between nodes X_i and X_j, when the nodes in S are observed. X_i and X_j are independent conditional on S if knowledge about X_i yields no extra information about X_j, once the values of the variables in S are known.

Two Bayesian network structures may actually represent the same constraints of conditional independence – the two networks are equivalent. For example, the structures $X_1 \to X_2 \to X_3$ and $X_1 \leftarrow X_2 \leftarrow X_3$ both represent the assertion that X_1 and X_3 are conditionally independent given X_2 [46]. When inferring networks from data, we cannot distinguish between equivalent networks, that is, causal relationships cannot be derived. This should be kept in mind when working with Bayesian networks – the best we can hope for is to recover a structure that is in the same equivalence class as the true network. Formally, Bayesian networks in the same equivalence class can be characterized as having the same underlying undirected graph, but may disagree on the direction of some edges. See for example [72] for details.

With these definitions and precautions at hand, we now come to the problem of learning a Bayesian network from given data.

2.4.1 Learning a Bayesian Network from Data

Learning a Bayesian network from given data requires estimating the conditional probability distributions and independence relations from the data. In order to do this, we would have to test independence of a given gene pair from every subset of the other genes. Examples for such *constraint based learning* approaches are given, for example, in [70] for networks involving only a few genes. For bigger networks, this approach quickly becomes infeasible, simply because of the number of tests that would be required.

The difficult issue is the decomposition of the joint probability distribution into conditional distributions among the relevant variables. This decomposition yields the network topology, estimating the distinct conditional probability distributions given the dependence structure is then relatively easy. In fact, given that the identification problem for Boolean networks is NP-hard, it is probably no surprise that inferring the dependence structure of a Bayesian network from given data is NP-hard as well. For this reason, the inference problem is usually tackled using heuristic approaches. Methods used include Bayesian and quasi-Bayesian approaches [19, 46, 55, 88, 91] as well as non-Bayesian methods [71, 89]. In the following, we will focus on the Bayesian approach. We will discuss the problems of structure learning and parameter learning for Bayesian networks in turn, starting with the easier parameter learning problem.

Learning Probabilities for a given Network Topology

Assume we are given the graph $G = (X, A)$, and all we ask for is details of the conditional distributions P. Let us furthermore assume that the conditional distributions are parameterized by some parameter vector ω, and that the general form of the distribution is known. Hence, we are asking for the values of ω given example data $\mathcal{D}$ assumed to have been generated by an underlying Bayesian network with topology G. The Bayesian approach then is to ask for the *posterior* distribution of the parameters, given the network topology and the data. Using Bayes' theorem,

$$p(\omega|\mathcal{D},G) = \frac{p(\mathcal{D}|\omega,G)p(\omega|G)}{p(\mathcal{D}|G)}. \tag{2.11}$$

The *evidence* $p(\mathcal{D}|G) = \int p(\mathcal{D}|\omega,G)p(\omega|G)\,d\omega$ averages over all possible parameters ω and normalizes equation (2.11). It can be neglected when scoring parameter values relative to one another, since it is independent of ω. The *likelihood* $p(\mathcal{D}|\omega,G)$ describes the probability that a network with given structure G and parameters ω has generated the data $\mathcal{D}$, and will depend on the functional form of the local distributions P used in the Bayesian network, for example normal distributions or discrete distributions. Finally, $p(\omega|G)$ is a *prior* distribution on the network parameters ω, and is often chosen to be conjugate to the likelihood for computational reasons. If prior knowledge is available here, this can easily be included in the Bayesian network framework through $p(\omega|G)$.

Heckerman [45] gives the example of multinomial distributions $p(\mathcal{D}|\omega,G)$, hence each node is assumed to be discrete, having r_i possible values $x_i^1,\ldots,x_i^{r_i}$. Under the assumption that there are no missing data in $\mathcal{D}$ and furthermore assuming that the parameters of the multinomial distributions for the different nodes are independent from one another, the computation of the posterior distribution is easy when a Dirichlet prior is used. In this case, the posterior distribution can be shown to be a Dirichlet distribution as well. One can then maximize the posterior to find the most likely parameters ω of the network, or average over possible configurations of ω to obtain predictions for the next state of the network.

Learning the Network Topology

Let us now consider the problem of learning the structure of a Bayesian network from given data. To evaluate different structures, we consider the posterior probability of a network topology G given the data $\mathcal{D}$:

$$p(G|\mathcal{D}) = \frac{p(\mathcal{D}|G)p(G)}{p(\mathcal{D})}. \tag{2.12}$$

The term $p(\mathcal{D})$ is the evidence, and can be written as an average $p(\mathcal{D}) = \int p(\mathcal{D}|G)p(G)\,dG$ over all possible model structures. Again, when scoring network structures relative to one another, we need not compute it and can neglect this term.

The likelihood $p(\mathscr{D}|G)$ can be computed by marginalizing $p(\mathscr{D}|\omega,G)$ over all possible parameters ω of the local distributions,

$$p(\mathscr{D}|G) = \int p(\mathscr{D}|\omega,G)p(\omega|G)\,d\omega, \tag{2.13}$$

hence the local parameters ω are treated as nuisance parameters and are integrated out.

Finally, $p(G)$ is a prior distribution over network structures. In principle, this prior can be used to encode any biological knowledge that is available on the system under consideration. The simplest conceivable structure prior is to assume that every structure is equally likely. Alternatively, a structure prior can be defined by assigning confidences $0 < w(x,y) \leq 1$ to the edges (x,y) of the fully connected graph, and scoring structures using the prior

$$p(G) = \frac{1}{N} \prod_{(x,y)\in A} w(x,y), \tag{2.14}$$

where N is a normalizing constant to make the right hand side a proper distribution, and A is the set of directed edges (arcs) of the network. Many alternative structure prior distributions have been proposed in the literature. For example, Heckerman *et al.* [47] suggest using a prior network and penalizing the prior probability of any structure according to some measure of deviation between the prior network and the topology of interest. Madigan *et al.* [59] describe an approach to elicit prior knowledge from experts and encode it into the prior. Bernard *et al.* [14] use transcription factor binding site information to define a prior distribution, thus including knowledge from other data sources into the network inference.

Different strategies can then be employed to search the model space for the network topology with highest posterior probability given the data. Exhaustive search is usually prohibitive, since the number of possible network topologies with n variables is equal to the number of acyclic directed graphs with n nodes, which is growing exponentially with n [76]. This is why researchers have used heuristic search algorithms, such as greedy search, simulated annealing, gradient descent procedures, genetic algorithms and Monte Carlo methods [25].

2.4.2 Advantages and Disadvantages of the Bayesian Network Model

Bayesian networks are attractive models for gene regulatory networks since they are stochastic in nature. They can thus deal with noisy measurements and stochastic aspects of gene expression in a natural way [29, 65], and they are easily extended to deal with missing data [45]. Furthermore, they provide an intuitive and easy to grasp visualization of the conditional dependence structure in given data, and are much easier for humans to understand than full conditional distributions. At the same time, depending on the probability distributions used (continuous or discrete), they can model quantitative aspects of gene regulatory networks.

Still, the level of detail they provide on the system modeled is rather coarse [29]. Furthermore, learning Bayesian networks from data is NP-hard, hence heuristic search methods have to be used, which do not guarantee that the globally optimal solution is found [29]. Probably their main disadvantage is that they disregard dynamical aspects completely, and that they require the network structure to be acyclic, since otherwise the joint distribution cannot be decomposed as in equation (2.9). However, feedback loops are known to play key roles in causing certain kinds of dynamic behavior such as oscillations or multi-stationarity [44,48,85,94,96], which cannot be captured by the Bayesian network model. In spite of these limitations, Bayesian networks have been used for example to infer regulatory interactions in the yeast cell cycle [36,87].

2.4.3 Extensions of Bayesian Networks

Efforts have been made to overcome the mentioned limitations. Bayesian networks can be extended to capture the dynamic aspects of regulatory networks by assuming that the system evolves over time. Thus, gene expression is modeled as a time series, and one considers different vectors $X(1),...,X(T)$ at T consecutive time points. One then assumes that a variable $X_i(t)$ of a particular gene i at time t can have parents only at time $t-1$. The cycles in the Bayesian network then unroll, and the resulting network is acyclic and the joint probability in equation (2.9) becomes tractable again. The resulting networks are called *Dynamic Bayesian Networks* [37,67,106].

Dynamic Bayesian Networks have been combined with *hidden variables* to capture non-transcriptional effects [73]. Similarly aiming at the inclusion of information from additional data sources into the Bayesian network learning process, Bernard and Hartemink [14] include transcription factor binding location data through the prior distribution, while evidence from gene expression data is considered through the likelihood.

Other extensions of Bayesian networks try to deal with the typical setting encountered with microarray data – where many genes are measured, but only few time points are available. *Regularization* approaches are used to avoid overfitting in this situation, different methods have been proposed for Bayesian networks. For example, Steck and Jaakkola [90] discuss parameter choices for a Dirichlet prior for the marginal likelihood (2.13), and show that sparse networks are learned for specific choices of parameters. Bulashevska and Eils [18] achieve regularization by constraining the form of the local probability distributions, they restrict interactions to Boolean logic semantics, and utilize Gibbs sampling to learn the model from the data.

2.5 Quantitative Models using Ordinary Differential Equations

We have seen that Bayesian networks highlight the stochastic nature of gene regulation, but are static models since they comprise no explicit time dependence in their definition. In contrast, we will now turn to *ordinary differential equations* (ODEs), which provide a deterministic, quantitative description of the time evolution of a system. ODEs are used in many scientific fields to describe a system's dynamic behavior. They provide a detailed time- and state continuous description of the system under consideration. In recent years, ODEs have also been established as models for gene regulatory networks, ranging from simple linear models to complex nonlinear systems.

We start with a formal definition of a continuous dynamical system:

Definition 7 (Continuous dynamical system). A continuous dynamical system *is a triple* (U, Φ^t, T). *The* state space U *is an open subset of* $\mathbb{R}^n$ *and the set* $T \in \mathbb{R}$ *is the set of time points. The function* Φ^t *is called* evolution function *and maps for every time point* $t \in T$ *a state* $x \in U$ *onto a state* $x \in U$, *hence* $\Phi^t : T \times U \to U$. Φ^t *is assumed to be a smooth function.*

In our models, $T = \mathbb{R}$, and then Φ^t is called a *flow*. It is assumed to be the solution of an autonomous first order differential equation of the form

$$\dot{x}(t) = f(x(t)), \qquad \text{where } x(t) \in U,\ f \in \mathscr{C}^1. \tag{2.15}$$

We assume the vector field $f(x(t))$ to be continuously differentiable, that is, $f(x(t)) \in \mathscr{C}^1$, since this guarantees uniqueness of a solution of equation (2.15), given an initial state $x(t_0)$. In gene regulatory network models, the state vector $x(t)$ contains concentrations of all n network components at time t. Hence, the state space U is often restricted to the positive quadrant $U = \mathbb{R}^n_+$.

Several suggestions have been made how to choose the function $f(x(t))$, we will highlight the main models in the following.

Linear Models

Chen *et al.* [24] in 1999 were among the first to use ordinary differential equations to model gene expression networks. They used a simple linear function $f(x(t)) = Ax(t)$ with an $n \times n$-matrix A with constant entries. Here, every regulation in the network is described by one single parameter a_{ij}, one thus has to estimate n^2 parameters to infer the corresponding network structure. Linear ODEs have the advantage of being analytically tractable, thus time-consuming numerical integration can be avoided. On the other hand, systems of the form $\dot{x}(t) = Ax(t) + b$ do not show a rich variety of dynamic behavior. They only have one isolated *stationary state* $x_s = -A^{-1}b$ in which the temporal change of x vanishes. Once reaching this state, the concentrations of the network components remain constant. (This is the usual case when A is invertible. If A^{-1} does not exist, the situation is more complicated, since the equation $\dot{x}(t) = 0$ then either has no solution or many non-isolated stationary states). If x_s is stable in the sense that small perturbations of the system at rest in x_s disappear and the system returns to x_s after some time, it is globally stable, that is, the system eventually approaches this state from any initial concentration vector $x(t_0)$. If x_s is not stable, then the solution x is not bounded, leading to infinitely increasing or decreasing concentrations. For these reasons, linear models are not well suited for regulatory networks, in which the concentrations are expected to be bounded and should not become negative. Furthermore, *oscillations* or *multi-stationarity*, which are both important properties of true biological networks, are nonlinear phenomena and cannot be captured with linear models.

Nevertheless, linear models are still used to reverse engineer gene regulatory networks from experimental data, in particular for large systems including a lot of genes (see, for example, [10, 43, 56, 77, 99]). Gustafsson *et al.* [43] try to infer the regulatory network of the yeast cell cycle using microarray measurements of the whole yeast genome, which contains about 6000 genes. They argue, that even if the nature of interactions between genes is nonlinear, it can be approximated around a specific working point with its linearization, which then provides a good starting point for further considerations.

Additive Models based on Chemical Reaction Kinetics

For smaller networks, more detailed and complex models with more free parameters are feasible. Thus instead of linear models, for networks containing only a few components, ODEs of the form

$$\dot{x}_i(t) = \sum_{j=1}^{n} f_{ij}(x_j(t)) - \gamma_i x_i(t) \qquad i = 1, \ldots, n \tag{2.16}$$

with nonlinear *regulation functions* $f_{ij} : \mathbb{R} \to \mathbb{R}$ and a first order degradation term $\gamma_i x_i(t)$ are frequently used (see, e.g. [31]).

Like linear models, these are *additive models*, where the influences of different regulators are added and are thus assumed to act independently. This is often a necessary simplification to keep the number of variables tractable, but in fact numerous effects within a cell are non-additive. For example, some proteins form multimers and only become functional in these complexes, several different transcription factors can compete for a single binding site, or they act in cooperation and amplify each other. Efforts have been made to overcome these limitations of additive models, and cooperative effects are described as logical AND and OR gates, respectively [6, 79]. However, including interactions between different regulators makes the model far more complicated since the regulation functions $f_{ij}(x_j)$ then become multi-dimensional.

The regulation functions $f_{ij}(x_j)$ describe the effect of a regulator j on the temporal change of the concentration of component i. According to equation (2.16), a gene regulatory network is characterized when all individual dependences between regulated components and regulators, that is, between $\dot{x}_i$ and x_j, are known. Many efforts have therefore been made to derive an appropriate parameterization of a regulation function. These approaches are often based on chemical reaction kinetics, in which the binding process of a transcription factor TF to a specific binding site BS of the DNA is considered a reversible chemical reaction with reaction constant K:

$$TF + BS \overset{K}{\rightleftharpoons} \underbrace{TF\text{-}BS\text{-complex}}_{C} \tag{2.17}$$

The temporal changes of concentrations over time are expressed with differential equations:

$$\frac{d}{dt}[TF] = -k_1[TF][BS] - k_2[C] \tag{2.18}$$

$$\frac{d}{dt}[BS] = \frac{d}{dt}[TF] \tag{2.19}$$

$$\frac{d}{dt}[C] = -\frac{d}{dt}[TF] \tag{2.20}$$

Here, $[\cdot]$ denote concentrations, and k_1 and k_2 are rates for complex formation and dissociation, respectively. Solving for the reaction constant $K = k_1/k_2$ in equilibrium leads to the following relation between K and the steady state concentrations of all components involved in the reaction, known as the *law of mass action*:

$$K = \frac{[C_s]}{[TF_s][BS_s]} \tag{2.21}$$

Rewriting and substituting the difference between the total concentration of binding sites and that of the free binding sites for the complex concentration $[C_s]$, that is, inserting $[C_s] = [BS_t] - [BS_s]$, leads to

$$1 - \frac{[BS_s]}{[BS_t]} = \frac{[BS_b]}{[BS_t]} = \frac{[TF_s]}{K^{-1} + [TF_s]}. \tag{2.22}$$

The fraction of occupied binding sites $[BS_b]$ thus increases hyperbolically with the transcription factor concentration.

For one single binding site, the left hand side of equation (2.22) can be interpreted as the probability of this site to be occupied by a transcription factor. Therefore, when the number of free transcription factors far exceeds the number of bound ones, and thus the number of free transcription factors can be approximated with the total amount of transcription factors, $[TF_s] \approx [TF_t]$, the probability P_C of the binding site to be occupied can be written in terms of $[TF_t]$ as

$$P_C([TF_t]) = \frac{[TF_t]}{[TF_t] + K^{-1}}. \tag{2.23}$$

This probability is proportional to the effect on the transcription rate of x_i and also to the amount of protein, provided that mRNA lifetime and translation rates are constant, leading to the following parameterization:

$$\dot{x}_i(x_j) = k_{ij} \frac{x_j}{x_j + K^{-1}}. \tag{2.24}$$

Here, we have changed the notation according to equation (2.15). Relation (2.24) is known as the *Michaelis Menten kinetics* [6]. Taking also cooperative effects between several transcription factors x_j into account, we can write the regulation function as a *Hill function*

$$f_{ij}(x_j) = k_{ij} \frac{x_j^{m_{ij}}}{x_j^{m_{ij}} + \theta_{ij}^{m_{ij}}} \tag{2.25}$$

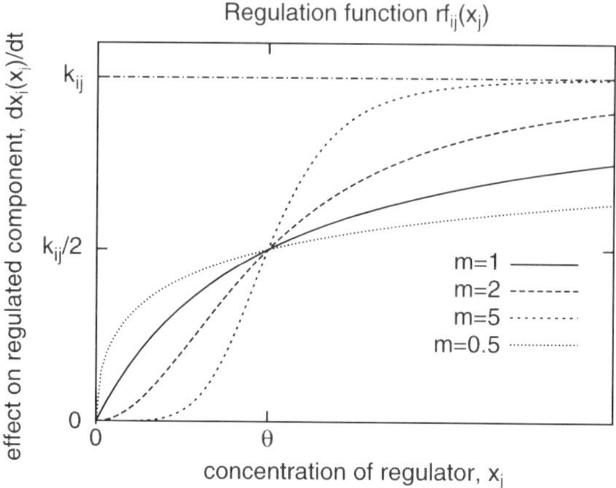

Fig. 2.4. Sigmoidal regulation function according to equation (2.25)

with *Hill coefficients* m_{ij} accounting for cooperativity between transcription factors, and with threshold values θ_{ij} which are related to the reaction constant K in equation (2.24) (see [31]). Function (2.25) is monotonically increasing or decreasing, and approaches the *regulation strength* k_{ij} for large concentrations x_j. The coefficient k_{ij} is positive when j activates i, zero when the concentration of j has no influence on i, and negative in case of an inhibition. Figure 2.4 illustrates equation (2.25) for the case of an activation with different values for the Hill coefficients m. A coefficient $m = 1$ corresponds to independent regulation (according to equation (2.24)). An exponent $m > 1$ indicates cooperative interaction between transcription factors x_j, causing a sigmoidal shape. Compared to $m = 1$, the effect on the regulated component is lower for small regulator concentrations, but increases quickly around the threshold θ, so that it exceeds the curve for $m = 1$ for concentrations $x_j > \theta$. When transcription factors influence each other negatively, for example they compete for a single binding site, this is expressed by an exponent $m < 1$, and the corresponding curve shows a steep slope for low regulator concentrations. It rapidly flattens for higher concentrations due to mutual inhibition.

To our knowledge, equation (2.25) was first proposed by Jacob and Monod in the year 1961 [49], and experiments carried out by Yagil and Yagil in 1971 supported the theory [105]. The latter estimated values of Hill coefficients and dissociation constants for different enzymes in *Escherichia coli*, one of the best studied bacterial model organisms, which is found, for example, in the human gut.

Let us stop here for a moment and reconsider the modeling approach according to equation (2.15). Although it looks rather general, it implies that there is a functional relation between the state of the system at time t, that is, the concentration vector $x(t)$ in our case, and the temporal change of this state at time t. This is a strong assumption which underlies all approaches used for network inference from expression data. For models based on chemical reaction kinetics, it implies that regulating reactions are in chemical equilibrium, otherwise there would be no unique relation between $x(t)$ and

$\dot{x}(t)$. This assumption is feasible for gene regulatory networks, when one considers the time scales in the system: Regulation via binding of a transcription factor to DNA happens at a time scale of seconds, and is thus much faster than the whole gene expression process, which lasts several minutes or even hours [6].

From a mathematical point of view, one of the main requirements on the regulation functions is that they should be bounded (concentrations should not rise to infinity) and monotone. Different parameterizations are used in the literature to guarantee these properties. Equation (2.25) is the direct result from chemical reaction kinetics, but exponents such as the Hill-coefficient m_{ij} are often hard to estimate from a numerical point of view. Thus other parameterizations such as $k_{ij}(1+e^{-x_j})^{-1}$ [22, 104] or $k_{ij}\tanh(x_j)$ [32] can be found. Several authors use step functions of the form

$$f_{ij}(x_j) = \begin{cases} 0 & \text{if } x_j \leq \theta_{ij} \\ k_{ij} & \text{otherwise} \end{cases} \tag{2.26}$$

to approximate equation (2.25) [30, 31, 34, 40, 66, 92]. This is the limit function for large Hill coefficients $m \to \infty$, and these models are known as *piecewise linear differential equations* (PLDEs). Equation (2.26) provides a partition of the state space into cuboids, separated by the threshold values θ_{ij}. Within each cuboid, the model is linear and thus analytically tractable. On the other hand, problems concerning the behavior of the system at the thresholds θ_{ij} can occur and may lead to additional steady states or limit cycles [30,31]. Note also that a step function is not differentiable at the thresholds and therefore does not satisfy the conditions in system (2.15).

In contrast to simple linear models, systems of the form (2.16) with bounded regulation functions are stable in the sense that there exists a trapping region in state space which eventually attracts all trajectories. This is an important feature in order to provide a global description of the biological system. Furthermore, monotonicity of the regulation function leads to a Jacobian matrix with constant signs (It should be noted at this point that positive self-regulation might lead to exceptions from this rule and must be treated carefully in this context – It can lead to changing signs of the Jacobian matrix depending on the location in state space, and thus statements about systems with constant J hold only for the parts of the state space in which J has constant signs). For ODE systems with positive Jacobian matrix, important statements about their dynamic behavior can be made. For example, Gouzé and Thomas emphasized the role of feedback circuits in the corresponding interaction graph [41, 93]. A positive circuit is required for multi-stationarity or hysteresis, and a negative feedback loop with at least two components is needed for periodic behavior. Thus feedback mechanisms in regulatory networks are fundamental network substructures which are related to certain dynamic behavior.

S-Systems

A further widely used class of ordinary differential equation models are *S-systems* [100], in which regulatory influences are described by power law functions:

$$\frac{dx_i(t)}{dt} = \alpha_i \prod_{j=1}^{n} x_j(t)^{g_{ij}} - \beta_i \prod_{j=1}^{n} x_j(t)^{h_{ij}} \tag{2.27}$$

The *kinetic orders* g_{ij} and $h_{ij} \in \mathbb{R}$ and the *rate constants* α_i and $\beta_i \geq 0$ have to be estimated in these models, these are $2n^2 + 2n$ parameters. The first term describes the effect of positive regulators, the second one refers to inhibitors, respectively. In contrast to additive models, here, single influences are multiplied. S-systems have been shown to capture many relevant types of biological dynamics [53]. A hyperbolic regulation such as described by equation (2.24) can be well approximated with exponents 0.5 and -0.5, respectively [95]. Steady states of (2.27) can be determined analytically, making these models attractive for network inference. Nevertheless, most of the model parameters are exponents, which are typically hard to estimate numerically. Cho *et al.* [26] and Kikuchi *et al.* [53] have used S-systems to reconstruct regulatory networks from gene expression data with genetic algorithms. Thomas *et al.* [95] developed an algorithm to estimate the rate constants from experimental data. They evaluated their approach with a simulated three gene system.

2.5.1 Network Inference

We now turn to the network inference problem for ordinary differential equation models, which is usually formulated as an optimization problem with an objective function that is minimized with respect to the network parameters ω. A common choice for this objective function is the sum of squared errors between measurements and model predictions. The corresponding optimization problem has the form

$$\min_{\omega} \left(F_1(\omega) = \sum_{t=1}^{T} \sum_{i=1}^{n} \| x_{i,model}(\omega,t) - x_{i,exp}(t) \|^2 \right). \tag{2.28}$$

Here, $x_{i,model}(\omega,t)$ denotes the model prediction for the concentration of network component i at time t, which is compared with the corresponding experimental result $x_{i,exp}(t)$. In order to minimize F_1 with respect to the parameter vector ω, numerical integration of the system is required to calculate $x_{i,model}(\omega,t)$. Usually, optimization of equation (2.28) can not be carried out analytically, and one has to apply heuristic methods such as gradient descent or genetic algorithms. This means that the numerical integration has to be carried out several times, and computing time quickly becomes the limiting factor [101]. This can be avoided by optimizing the sum of squared errors of time derivatives rather than of the concentrations directly:

$$\min_{\omega} \left(F_2(\omega) = \sum_{t=1}^{T} \sum_{i=1}^{n} \| \dot{x}_{i,model}(\omega,t) - \dot{x}_{i,exp}(t) \|^2 \right) \tag{2.29}$$

In this formula, $\dot{x}_{i,model}(\omega,t)$ is obtained from the model equations, and $\dot{x}_{i,exp}(t)$ is the corresponding slope estimate from the experimental data. Contrary to the minimization problem (2.28), solving problem (2.29) does not require numerical integration of the ordinary differential equations. Instead, one needs an appropriate method to estimate the slopes $\dot{x}_{i,exp}(t)$ from the data. For this purpose, it can be useful to smooth the data in a preprocessing step, in particular in case of high levels of noise in the data.

Since quantitative models such as ordinary differential equation models depend on many parameters, but the number of samples available for parameter estimation is usually small in comparison, the main problem in this setting is *overfitting*. This means that the model is overtuned to specific characteristics of the training data, which do not reflect actual properties of the true underlying model, but are noise. Such overfit models will show bad performance on validation data which has not been used for training.

Different algorithms have been proposed to counter overfitting. *Early stopping* divides the data into three classes. The training data are used for learning, and this process is interrupted by testing performance of the learned model on the validation set. The procedure is stopped when performance on the validation data does not improve any further. As the result depends on both, training- and validation data, a third dataset is required to validate the inferred model.

Another method, called *weight decay* in the context of neural networks, regularizes the objective function by adding a term which penalizes models with many degrees of freedom. Popular criteria used for this purpose are *Akaike's information criterion* (AIC) [1]

$$F_{AIC} = -2\ln\mathscr{L} + 2k \tag{2.30}$$

and the *Bayesian information criterion* (BIC)

$$F_{BIC} = -2\ln\mathscr{L} + k\ln(n), \tag{2.31}$$

where in both equations k is the number of free model parameters, $\mathscr{L}$ the value of the error function and n the sample size. These criteria were used as objective functions in the inference of the yeast cell cycle network in Nachman *et al.* [68] and Chen *et al.* [22], respectively.

More biologically motivated approaches restrict the search space by including biological knowledge into the learning process. This can be done by introducing constraints to the optimization problem, such as upper limits for the number of regulators for every gene, or ranges for model parameters. Alternatively, similar to the criteria introduced above, one can modify the objective function by adding a term penalizing networks with a large number of strong regulatory interactions (see, for example, [98]). In Section 2.6, we will introduce an inference method which uses this latter approach.

2.5.2 Advantages and Disadvantages of ODE Models

Continuous dynamical systems provide a very detailed quantitative description of a network's dynamic, as they are time- and state-continuous models. They can show a rich variety of dynamic behaviors, such as multi-stationarity, switch-like behavior, hysteresis or oscillations. For nonlinear systems based on chemical reaction kinetics, parameters can directly be interpreted as kinetic rates of a chemical or physical reaction, for example, as degradation rates or velocities of binding reactions [44]. Some of these rates are experimentally accessible, which provides either a possibility to evaluate the model afterwards, or to restrict the range of the corresponding parameter values prior to the inference process. For example, binding coefficients between macromolecules can often be measured in vitro, and they differ only slightly in vivo [85]. Other rate constants, such as rates of phosphorylation of a transcription factor subsequent to a stimulus, are hard to verify experimentally [15, 29, 85].

When analyzing ODEs, one can exploit the well-established theory of differential equations [42]. To examine, for example, the long-term behavior of an ODE system, methods have been developed to calculate steady states or limit cycles and to determine their basins of attraction. *Bifurcation analysis* aims at predicting parameter values for which the qualitative behavior of the system changes, because the stability of steady states or periodic solutions changes when varying parameters, or solutions appear and disappear. Many tools have been developed to conduct such an analysis numerically [35].

A drawback of differential equation models is the relatively large number of parameters which have to be estimated in the network inference process. Time courses with many time points are needed for this purpose, but such data is rarely available. Many high-throughput techniques aim at measuring a lot of components simultaneously, but good time resolution is hard to obtain. This is the main reason why inference of ODEs from experimental data is currently restricted to small networks with only few components.

Another problem lies in the quality of experimental data. Microarray data are mostly used to infer gene regulatory networks. They contain a lot of noise, and may not be appropriate to make quantitative statements. Thus, when modeling regulatory networks with differential equations, it is often inevitable to include prior biological knowledge or to make simplifying assumptions. Of course, this often makes the approach specific for a certain biological system and not ad hoc applicable to other organisms or subsystems.

2.6 Bayes Regularized Ordinary Differential Equations

We would now like to give an example from our own work, combining ordinary differential equations with a (dynamic) Bayesian network approach. The underlying model used is a system of differential equations, but we embed the differential equations in a probabilistic framework with conditional probabilities as in Bayesian networks, and use Bayes' theorem for the inference. In our approach, the differential equations are used to specify the mean of the conditional probability distributions for the genes at a given time point, conditioned on the expression pattern at a previous time point. We then estimate the parameters of the differential equations, and thus determine the conditional probability distributions and the network topology. This topology is assumed to be fully connected initially, but we will show how to drive the solution to sparse networks using a specifically designed prior distribution on the ODE parameters.

Two aspects make such a Bayesian approach attractive for the inference of gene regulatory networks from gene expression data. The stochastic approach captures the stochastic nature of biological processes within a cell and the noise due to the experimental procedure. Moreover, prior knowledge can be included into the posterior distribution by choosing appropriate prior distributions reflecting our knowledge of the system. Furthermore, the probabilistic nature of the approach makes it possible to compute confidences for model parameters and also predictions of the network [50], work on this is ongoing in our groups.

To become more concrete, we consider an additive ODE model with sigmoidal regulation functions of the form

$$\dot{x}_i(t) = s_i - \gamma_i x_i(t) + \sum_{j=1}^{n} k_{ij} \frac{x_j^{m_{ij}}}{x_j^{m_{ij}} + \theta_{ij}^{m_{ij}}}. \tag{2.32}$$

The parameters s_i and γ_i are basic *synthesis*- and *degradation* rates, they determine the dynamics of component i when all regulators of i are absent. Coupling of the differential equations is due to the sum of regulation functions, compare equation (2.25). The sum in (2.32) is over all genes in the network and reflects the influence of the j-th gene on gene i. The network is thus assumed to be fully connected, unless the corresponding parameters k_{ij} become zero. More details on this model can be found in [38, 74].

We discretize this equation with a simple *Euler discretization*, that is, we approximate the time derivatives on the left hand side by difference quotients, and we furthermore add a noise term $r_i(t)$ to the output. We then get

$$x_i(t+\Delta t) = \underbrace{x_i(t) + \Delta t \left[s_i - \gamma_i x_i(t) + \sum_{j=1}^{n} k_{ij} \frac{x_j(t)^{m_{ij}}}{x_j(t)^{m_{ij}} + \theta_{ij}^{m_{ij}}} \right]}_{h_i(\omega)} + r_i(t). \tag{2.33}$$

The noise term $r_i(t)$ is assumed to be normally distributed with mean 0 and variance $\sigma_i(t)^2$. The assumption of normally distributed noise corresponds to assuming that the noise stems from many small, independent sources, which is arguably a reasonable approximation at least for the experimental noise. Δt is a discretization parameter, the smaller the time step Δt, the better does equation (2.33) approximate the continuous system (2.32). Biological data sets usually comprise large time steps, when using differential equations models one therefore interpolates over time in order to get sufficiently small time resolution.

Assuming independence of all noise terms for every time point and all network components from one another, the likelihood $\mathscr{L}$ decomposes into a product over all time points $t_1, \ldots, t_T$ and all n network components:

$$\mathscr{L} = p(\mathscr{D} \mid \omega) = \prod_{z=1}^{T} \prod_{i=1}^{n} \frac{1}{\sqrt{2\pi\sigma_i(t_z)^2}} \exp\left[-\frac{1}{2\sigma_i(t_z)^2} \left(h_i(\omega) - x_i(t_z)\right)^2 \right] \tag{2.34}$$

Clearly, the independence and normality assumptions are a simplification. Noise on $x(t)$ will lead to correlated, non-normal noise on $x(t+\Delta t)$. Furthermore, modeling errors will accumulate over time, and are certainly correlated. On the other hand, the assumptions are probably reasonable for experimental noise, and clearly they are a trade-off between model complexity/computational tractability and a detailed and realistic model, and similar assumptions are frequently used in Bayesian learning approaches.

If no prior knowledge is included into the learning process, an optimal parameter vector ω can be computed by maximization of equation (2.34) with respect to ω. This is known as *maximum likelihood estimation* (MLE):

$$\hat{\omega}_{MLE} = \arg\max_{\omega} \mathscr{L}(\omega) \tag{2.35}$$

In case all variances are equal, that is, $\sigma_i(t) = \sigma$ for all $i = 1, \ldots, n$ and for all $t = t_1, \ldots, t_T$, $\hat{\omega}_{MLE}$ is equivalent to the result one gets when minimizing the sum of squared errors between measurements and model predictions with respect to ω. This is easily seen when taking the negative logarithm of (2.34), and dropping terms independent of ω.

To include prior knowledge into the inference process, we use Bayes' theorem to compute the posterior probability distribution, that is, the conditional distribution over the parameter vector ω, given the data:

$$p(\omega \mid \mathscr{D}) = \frac{p(\mathscr{D} \mid \omega)p(\omega)}{p(\mathscr{D})}. \tag{2.36}$$

Here, the right hand side includes a product of the likelihood $\mathscr{L}$ and the prior distribution $p(\omega)$ over the model parameters. Maximizing equation (2.36) with respect to ω once again leads to a point estimate for ω, this is known as *maximum a-posteriori* (MAP) estimation:

$$\hat{\omega}_{MAP} = \arg\max_{\omega} p(\omega \mid \mathscr{D}) \tag{2.37}$$

When no prior information about the system under consideration is available, the prior distribution $p(\omega)$ is often chosen to be an improper uniform distribution, and $\hat{\omega}_{MLE}$ then equals $\hat{\omega}_{MAP}$. In the following section, we will detail our choice of prior distribution over model parameters.

2.6.1 Prior Distributions for Network Parameters

We now need to specify prior distributions for the model parameters s_i, γ_i, k_{ij}, θ_{ij} and m_{ij}.

The parameters s_i and γ_i are basic synthesis and degradation rates for the network components. Both parameters should neither become negative nor too large. We therefore choose independent gamma distributions for these two parameters. The gamma distribution is given by

$$g(x) = \frac{a^r x^{r-1}}{\Gamma(r)} \exp[-ax]. \tag{2.38}$$

Here, $\Gamma(r)$ is the gamma function. The shape of the distribution depends on the shape parameter $r > 0$. The parameter $a > 0$ determines the scale of the distribution. The smaller a, the more spread out is the distribution. The parameters a and r must be carefully chosen depending on the numerical range of the experimental measurements, and prior knowledge on synthesis and degradation rates can be included through specific settings.

For the parameters k_{ij}, a hierarchical prior distribution is used, which has specifically been designed to favor *sparse* networks. Biologically motivated, most of the k_{ij} should be close to zero, and only few k_{ij} should differ significantly from zero – corresponding to only few significant edges in the network. This is achieved using independent mean-zero normal distributions as prior distributions on k_{ij}, with standard deviations distributed according to a gamma distribution. The idea here is that most of the normal distributions should be concentrated strongly around zero in order to keep the corresponding k_{ij} small, and should only in few cases be allowed to become wider, if the data indicates so. This expectation of narrow normal distributions is reflected by the gamma distribution on the standard deviations of the normal distributions. Combining these two distributions and marginalizing over the standard deviation s,

$$p(k) = \prod_{i=1}^{n} \int_0^\infty \mathcal{N}(k|\mu = 0, \sigma = s) g(s) ds, \tag{2.39}$$

where

$$\mathcal{N}(k|\mu,\sigma) = \frac{1}{\sqrt{2\pi}\sigma} \exp\left[-\frac{1}{2}\frac{(k-\mu)^2}{\sigma^2}\right] \tag{2.40}$$

is the normal density with mean μ and variance σ^2.

When drawing samples from the distribution $p(k)$, most of the values will be small, since they stem from normal distributions with mean zero and a small variance. Figure 2.5 shows the distribution $p(k)$ resulting from the integration (2.39) for the two-dimensional case ($n = 2$). As can be seen, this prior favors solutions were only one parameter k_i is distinct from zero over solutions where both k_1 and k_2 differ

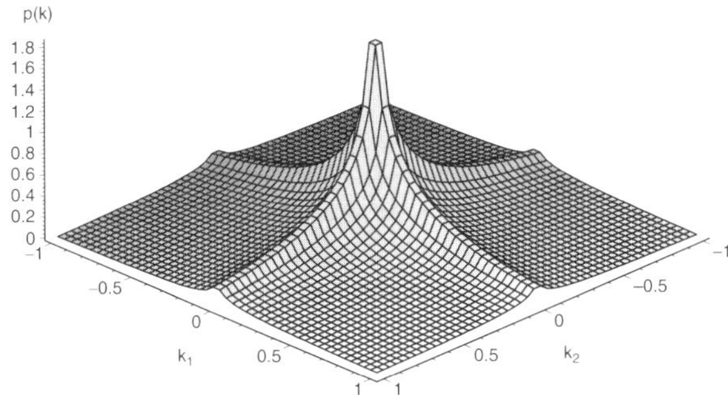

Fig. 2.5. The two dimensional hierarchical prior distribution with parameters $r = 1$ and $a = 1.0001$ for edge weights k_{ij}

significantly from zero. This is distinct from standard regularization schemes used such as the L_2 penalty, which would correspond to a Gaussian prior – and which would give the same penalty to points at equal distance from the origin, independent of the number of nonzero components. Note also that this prior is stronger than a Laplace prior on k.

At this point, we remark that the choice of prior distribution on network parameters clearly influences results of the computation, and it is not necessarily guaranteed, that this reflects biological reality. This is a classical example of the *bias-variance tradeoff*, where a stronger prior will lead to a stronger bias in learning, but less variance, and vice versa. In the setting of network learning described here, a strong prior driving the network to sparse solutions is needed to avoid overfitting of the model, this is discussed in more detail in [75], where we compare maximum likelihood and maximum a-posteriori under various settings on simulated data.

We use fixed values for the exponents m_{ij} and threshold parameters θ_{ij} for numerical reasons, this corresponds to assuming a delta distribution on these parameters. The reason for this decision is numerical instability of the optimization routine when m and θ are optimized, and insufficient experimental data to learn these parameters properly.

The negative logarithm of the posterior distribution (2.36) is then minimized using conjugate gradient descent. Alternatively, one could sample from the posterior distribution using a Markov chain Monte Carlo approach, work on this is presently ongoing and will be published elsewhere. For technical details on both approaches see [50, 51], where the same hierarchical prior distribution as the one used here on the k_{ij} is used in combination with a Cox regression model to predict survival times of cancer patients from gene expression measurements.

2.6.2 Application to the Yeast Cell Cycle Network

In this section, we will show results of an application of the Bayesian approach described above to a dataset on the yeast cell cycle. More details as well as an additional evaluation of the method on simulated data can be found in [75].

The yeast cell cycle is one of the best studied eukaryotic regulatory systems. A proper functioning of this regulatory mechanism is essential for the organism to survive. Core elements of its machinery are highly conserved in evolution among eukaryotes, making studies on a model organism such as budding yeast worthwhile, as many results can be transfered to higher organisms. Many publications on the yeast cell cycle exist, see, for example, [8].

We examined eleven genes from a publicly available dataset by Spellman *et al.* [87], these genes are known to be central to the cell cycle [57]. The dataset contains log ratios between synchronized cells and control experiments of the whole yeast genome, approximately 6000 genes were measured several times during the cell cycle, in total over 69 time points. The reference network we use for evaluation of our results is a reduction of the network described in Li *et al.* [57].

Results

Time series data of eleven genes was used, including cln1, cln2, cln3, clb5, clb6, cdc20, cdc14, clb1, clb2, mcm1 and swi5. Measurements corresponding to nodes in the reference network involving several genes were represented by the mean value of the genes, missing values were estimated by linear interpolation over time. Conjugate gradient descent was used to fit the model to the data, with prior distribution parameters $a = 0.1$ and $r = 0.01$ for the synthesis and degradation rates, and $a = 5$ and $r = 1.7$ for the prior on the k_{ij}. Fixed values of $\theta_{ij} = 1$ and $m_{ij} = 2$ were used for the threshold parameters and Hill coefficients of the ODE model. Since we expect sparse solutions, the gradient descent was started near the origin, see [75] and [50] for technical details.

To evaluate our results, we compared the inferred network structure with the reference network. Figure 2.6 shows the reference network (*left*) and the network inferred with the Bayesian approach (*right*). The 16 edges with highest weights are marked in bold, continuous bold lines indicate true positives, dashed bold lines correspond to false positives. Thin lines appear in the reference network, but are not revealed in our approach. 12 of 16 regulations are true positives, the remaining four interactions are not stated in the literature. Note that, in the latter case, it is not clear whether there is no such regulation or whether it exists but has not been described yet. The corresponding values for *specificity*, that is, the fraction of revealed true regulations, and *sensitivity*, the fraction of true negatives, are 0.55 and 0.85, respectively.

Receiver Operator Characteristics (ROC) curves can be used to assess the inferred network structure more quantitatively. By using a cutoff value c on the weights k_{ij} and including only edges with $|k_{ij}| > c$ in the network, one can compute sensitivity and $1 -$ specificity. Sensitivity and $1 -$ specificity can then be plotted against one another for different cutoff values c, assuming that the reference network is the correct underlying network. The resulting ROC curves provide a comprehensive overview over all combinations of sensitivity and specificity that can be achieved with a given model. ROC curves can further be summarized by computing the *Area Under the ROC Curve*, the AUC. The AUC is a numerical value between 0.5 and 1, where

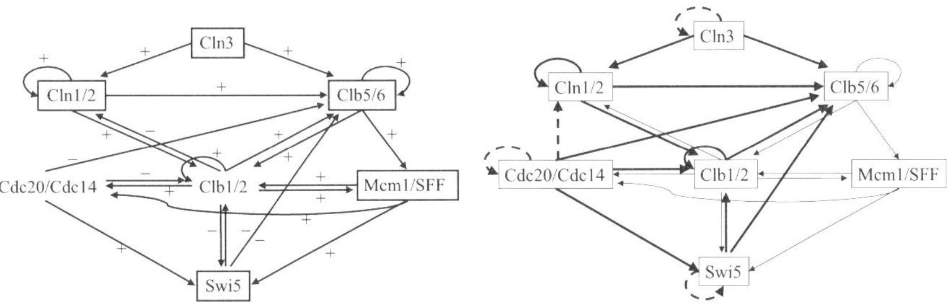

Fig. 2.6. Regulatory network of the yeast cell cycle (*left*) (see [75] and [57]) and the network inferred with the Bayesian approach (*right*). True positives are marked in bold, false positives are marked with bold dashed lines, false negatives correspond to thin lines

0.5 would be equivalent to guessing for each edge whether it is present or not, and an AUC of 1 would correspond to a prediction with perfect sensitivity and specificity. We computed AUC values for our approach, the corresponding AUC value is 0.68, indicating that the main regulatory interactions are revealed. The approach outperforms maximum likelihood estimation, which yields an AUC value of 0.61, showing that the sparsity constraint introduced through the prior distribution helps the learning process.

Computational demands of the approach are comparable to demands of other differential equations model approaches. For small networks with 5 to 10 nodes, running times are typically in the range of a few minutes, depending on the number of time points available. For large networks, the limiting factor is usually not computing time, but insufficient amounts of data to reliably estimate parameters of the differential equations.

2.7 Other Approaches

In this section, we will give an overview over models that go beyond ordinary differential equations. We will focus on three further model classes, delay differential equations (DDEs), partial differential equations (PDEs), and stochastic equations. DDEs are used to account for time delays in regulatory mechanisms, which is often necessary when the system consists of reactions taking place at different time scales. Spatial inhomogeneities within a cell are captured with PDEs, which contain derivatives of time and space, and include, for example, diffusion processes. Unfortunately, without further knowledge concerning diffusion coefficients and locations of transcription and translation, it is not possible to learn parameters for such models using only microarray data.

Stochastic equations try to model the stochastic nature of single reactions and provide the most detailed level of description. Here as well, far more information is needed than microarray expression data can provide. Thus, all three model classes are not ad hoc suitable for the inference of large scale regulatory networks from expression data, and have mostly been investigated only theoretically or used to model very specific regulatory mechanisms so far. No "standard method" exists to estimate parameters for these models, and we will therefore only point out some basic concepts and difficulties with these models rather than give a recipe on how to infer networks using them.

2.7.1 Delay Differential Equations

All modeling approaches discussed so far implicitly assume a local time dependence and spatial homogeneity. To include transport processes into the model, time-delay differential equations (DDEs) or partial differential equations (PDEs) are used (see, e.g. [23, 29, 85]). This is particularly interesting for eukaryotic organisms, where macromolecules such as mRNA have to be transported from one cell compartment into another prior to translation, or proteins are produced somewhere in a cell and become active somewhere else. How important such factors may be can be seen from recent reports indicating that the spatial distribution of proteins within a cell seems to have an effect on the embryonic development in eukaryotes [29].

Active transport processes, which require energy and can be in opposition to the concentration gradient, are modeled with DDEs of the form

$$\dot{x}(t) = f(x(t), x_{delay}(t)), \tag{2.41}$$

with

$$x_{delay}(t) = \int_{-\infty}^{0} x(t-\tau)G(x(t-\tau))d\tau. \tag{2.42}$$

In these systems, the left hand side depends on the current state $x(t)$ and the state vector $x_{delay}(t)$, which is a weighted integral over past states. The sum of the weights is normalized, that is,

$$\int_{-\infty}^{0} G(x(t-\tau))d\tau = 1. \tag{2.43}$$

In the simple case where one can assume a fixed duration τ_0 between binding of a transcription factor to a binding site within a promoter of a gene and the effect it has on the amount of protein, the distribution over weights can be modeled using a delta distribution:

$$G(x(t-\tau)) = \delta(\tau_0) \tag{2.44}$$

and hence

$$\int_{-\infty}^{0} x(t-\tau)G(x(t-\tau))d\tau = x(t-\tau_0). \tag{2.45}$$

In equation (2.41), $f : \mathbb{R}^n \times C^1 \to \mathbb{R}^n$ is a functional operator which maps n continuously differentiable functions defined on $\mathbb{R}$ onto a vector in $\mathbb{R}^n$. This makes DDEs more difficult to analyze than ODEs, in which $f : \mathbb{R}^n \to \mathbb{R}^n$ is an ordinary function which maps a vector $x(t)$ onto another vector $\dot{x}(t)$.

In order to solve equation (2.41), not only an initial state vector $x(t_0)$, but an entire interval of initial data is required. Thus, the state space is infinite dimensional. This also leads to infinitely many eigenvalues when linearizing the system in order to analyze the behavior of steady states. The characteristic equation is not a simple polynomial, but involves exponential functions. No standard method to solve such equations exists, and stability analysis of steady states can be a hard task. In general, not much is known about effects that are caused by time delays. Most work in this field examines the stability of a steady state depending on time delays for a certain system. Chen and Aihara [23], for example, consider an oscillating two-gene model and claim that time delays increase the stability region of oscillations in their model, making the oscillations robust against parameter changes. Santillán and Mackey [78] built a model of the tryptophan operon in *Escherichia coli*, one of the prototypic gene control systems. They also included time delays into their nine differential equations, and they estimated 28 parameters. Simulations were carried out numerically with a fourth order Runge-Kutta method. A lot of specific knowledge about the operon as well as steady state approximations were included into the parameter estimation processes, hence the estimation method cannot ad hoc be generalized and used for arbitrary organisms.

2.7.2 Partial Differential Equations

PDEs describe spatial inhomogeneities and diffusion processes and distinguish between different cell compartments, for example nucleus and cytoplasm [29]. The corresponding differential equations consist of a sum of functions $f_i(x(t))$, which describe the regulatory network as in equation (2.16), and a term for the diffusion process:

$$\frac{\partial x_i}{\partial t} = f_i(x(t)) + \delta_i \frac{\partial^2 x_i}{\partial l^2}, \text{ with } 0 \leq l \leq \lambda, i = 1, \ldots, n. \tag{2.46}$$

In contrast to ODEs, this equation contains derivatives with respect to both time and space. The variable δ_i is the diffusion constant, and l is the position in the cell. Boundary conditions such as

$$\frac{\partial^2}{\partial l^2} x_i(0,t) = 0 \text{ and } \frac{\partial^2}{\partial l^2} x_i(\lambda,t) = 0 \tag{2.47}$$

ensure that components stay within the cell.

The lack of appropriate analysis methods and missing experimental data providing information on transport processes make both DDEs and PDEs currently inappropriate for the inference of regulatory networks from gene expression data. Also little is known about "typical durations" of mRNA or protein transport, and data about spatial distributions of cell components is only gradually becoming available with recent developments in live-cell imaging techniques.

2.7.3 Stochastic Kinetic Approaches

Finally, a *stochastic kinetics* modeling approach provides the by far most detailed level of description [29, 44, 85], but also has the highest computational cost [44].

Probabilistic models were developed to explain the observed variety in experiments, in particular when the number of molecules is small [62]. In these models, concentrations are discrete and change according to some probability distribution. The probability of a system to be in state X at time $t + \Delta t$ is given by

$$p(X,t+\Delta t) = p(X,t)\left(1 - \sum_{j=1}^{m} \alpha_j \Delta t\right) + \sum_{j=1}^{m} \beta_j \Delta t, \tag{2.48}$$

see, for example [29]. Here, X is a discrete concentration vector and $p(X,t)$ is a probability distribution. The term $\alpha_j \Delta t$ is the probability that a reaction j takes place in the time interval Δt, and the sum runs over all m possible reactions. The second term is the probability that the system will be brought to state X from any other state via a reaction j. Taking the limit $\Delta t \to 0$ leads to the well known *Master equation*, a first order differential equation that describes the evolution of the probability to occupy a discrete set of states:

$$\frac{\partial p(X,t)}{\partial t} = \sum_{j=1}^{m} (\beta_j - \alpha_j p(X,t)) \tag{2.49}$$

Modeling gene regulatory networks with these equations requires much more information than with ordinary differential equations since every single reaction is considered. Moreover, the computational costs are very high since a large number of simulations is needed to approximate $p(X,t)$ [44]. The Master equation can in some cases be approximated with stochastic differential equations. These so called *Langevin equations* assume that internal fluctuations cancel out on average, and the system can be described by a deterministic ODE and a noise term. Numerical solutions are obtained for these equations using Monte Carlo simulations. Alternatively, a stochastic simulation approach provides information on individual behavior instead of examining the whole distribution $p(X,t)$. Gillespie developed an efficient algorithm to describe a spatially homogeneous chemical system with a stochastic approach, the *stochastic simulation algorithm* [39], which is equivalent to the spatially homogeneous Master equation. This algorithm was used by McAdams and Arkin [65], who examined the influence of statistical variations during regulatory cascades on cellular phenomena across cell populations, and by Arkin *et al.* [7], who considered the influence of fluctuations in the rate of gene expression on the choice between lytic and lysogenic growth in phage λ. The latter is the pioneering work on the role of fluctuations in gene expression.

2.8 Conclusion

In this chapter, we have attempted to give an overview over a number of different models used for gene regulatory network reconstruction. We started with simple binary models, which assume that each gene is in one of two possible states, *expressed* or *not expressed*. We then extended the scope all the way to complex quantitative models, which can capture kinetic properties of the chemical reactions underlying gene regulation. All these models have their own specific strengths and weaknesses. So, when faced with an actual biological system to be analyzed or simulated, what is the appropriate model to use?

The answer is – it depends. It depends on the biological question we are interested in, and it will also depend on the experimental data we have at our disposition or can measure on the system considered. Furthermore, it will depend on the kind of biological knowledge we already have on the system under consideration.

There are three central questions that should be considered when choosing a model. These are:

1. What do we hope to learn from the model?
2. How large is the system we need to model?
3. Do we have the right data, and is there additional knowledge we can use?

The first question asks for the ultimate objective driving our modeling attempts. Occam's razor is the principle to choose the simplest model that can explain the data we have, and very similarly, we should choose the simplest model that can answer the questions we ask. So, if our interest is in qualitative properties such as *"Does component j interact with component i in the network?"* or *"Do two components have the same regulators?"*, then qualitative models such as Boolean networks probably provide the appropriate framework. If the questions are of a quantitative nature, such as *"What happens when changing the affinity of a transcription factor to a specific binding site?"* or *"How does a change in concentration of component i affect the dynamic behavior of the system?"*, then obviously quantitative models are required.

At the same time, one should be highly alert to the complexity of the modeling task. This brings us to the second question above. Large genetic systems are extremely difficult to model, and extrapolating a detailed differential equations model for a single gene with its several kinetic parameters to larger systems will quickly render the model prohibitively complicated [16]. The sheer quantity of parameters in such models will make their application impossible to networks involving more than just a few genes. So, there also is a tradeoff here. When the complexity of the biological system modeled is low, thus single genes or only few genes are of interest, computer modeling can go into much detail and quantitative differential equation models or even stochastic molecular simulations are feasible, permitting simulations of detailed single gene dynamics and time courses of gene activity. On the other hand, when mid-size to large genetic networks are desired, models must focus on less detail and concentrate on the overall qualitative behavior of the system. This may still allow inference about the state dynamics of a system in terms of a flow pattern grasping the qualitative aspects of state transitions, but quantitative models for the entire system are usually impossible, simply because of the lack of sufficient data to estimate all parameters in those models.

In our experience, differential equations models quickly reach their limit when more than a handful of genes are modeled, and while additional constraints such as the sparsity constraint introduced in Section 2.6 can extend the feasible network size slightly, these approaches are not useful for large-scale network inference with several hundred to thousands of components. However, they provide a very detailed, quantitative model for small networks. Bayesian networks permit slightly larger network models, but here too, one needs to be cautious about overfitting and insufficient data when more than a few dozen genes are modeled. Boolean models and relevance network approaches finally permit the largest number of genes to be included in network models, and application involving thousands of genes have been reported, see, for example, [21]. It remains to be seen how reliable such large-scale networks are.

For a numerical evaluation and comparison of different approaches on simulated data see, for example, [9].

This brings us to the third question, concerning the available data for the modeling task. The large bulk of work on transcriptional network reconstruction has concentrated on deterministic, coarse-grained models. Even when quantitative models are used, the conclusions drawn from them are usually of a qualitative nature. This is mainly due to the incomplete knowledge on the chemical reactions underlying gene regulation, and the lack of detailed kinetic parameters and concentration measurements required for these models [29]. Often, the lack of suitable data is the limiting factor in network inference. However, this can sometimes be alleviated by the inclusion of additional biological knowledge in the learning process. For example, if information on transcription factor binding sites is available, this may be used to reduce the search space for model topologies. The inclusion of such prior knowledge is an ongoing research problem. If quantitative data of good quality is available, maybe supported by additional data sources such as measurements of kinetic parameters and prior biological knowledge on interactions in the network, detailed quantitative models are often feasible [7].

Even though large scale techniques such as DNA microarrays can provide genome-wide expression measurements, microarrays provide effectively only more or less qualitative data at present. In addition, they measure many genes under few different conditions or time points, whereas for network inference, one would rather have few (relevant) genes under many conditions and time points.

However, this data bottleneck can reasonably be expected to be relieved in the near future. With the advent of novel experimental techniques to measure RNA and protein concentrations, accompanied by large databases providing access to this and other published and unpublished data, quantitative models will increasingly be used in the future, bringing us closer to the ultimate goal, the simulation of whole cells [97].

References

1. H. Akaike, *A new look at the statistical model identification*, IEEE Trans. Automatic Control **19** (1974), 716–723.
2. T. Akutsu, S. Miyano, and S. Kuhara, *Identification of genetic networks from a small number of gene expression patterns under the boolean network model*, Pac Symp Biocomput **4** (1999), 17–28.
3. ———, *Algorithms for identifying boolean networks and related biological networks based on matrix multiplication and fingerprint function*, RECOMB'00: Proceedings of the fourth annual international conference on Computational molecular biology (New York, NY, USA), ACM Press, 2000, pp. 8–14.
4. R. Albert and H.G. Othmer, *The topology of the regulatory interactions predict the expression pattern of the segment polarity genes in Drosophila melanogaster*, J Theor Biol **223** (2003), 1–18.
5. B. Alberts, A. Johnson, J. Lewis, M. Raff, K. Roberts, and P. Walter (eds.), *Molecular biology of the cell*, 4 ed., Garland Publishing, New York, 2002.

6. U. Alon, *An introduction to systems biology - design principles of biological circuits*, Chapman & Hall/CRC Mathematical and Computational Biology Series, New York, 2007.
7. A. Arkin, J. Ross, and H.H. McAdams, *Stochastic kinetic analysis of developmental pathway bifurcation in phage λ - infected Escherichia coli cells*, Genetics **149** (1998), no. 4, 1633–1648.
8. J. Bähler, *Cell-cycle control of gene expression in budding and fission yeast*, Annu. Rev. Genet. **39** (2005), 69–94.
9. M. Bansal, V. Belcastro, A. Ambesi-Impiombato, and D. di Bernardo, *How to infer gene networks from expression profiles*, Molecular Systems Biology **3** (2007), 78.
10. M. Bansal, G.D. Gatta, and D. di Bernardo, *Inference of gene regulatory networks and compound mode of action from time course gene expression profiles*, Bioinformatics **22** (2006), no. 7, 815–822.
11. K. Basso, A.A. Margolin, G. Stolovitzky, U. Klein, R. Dalla-Favera, and A. Califano, *Reverse engineering of regulatory networks in human B-cells*, Nature Genetics **37** (2005), 382–390.
12. J. Beirlant, E. Dudewicz, L. Gyorfi, and E. van der Meulen, *Nonparameteric entropy estimation: An overview*, Int J Math Stat Sci **6** (1997), no. 1, 17–39.
13. P. Berg and M. Singer (eds.), *Dealing with genes*, University Science books, 1992.
14. A. Bernard and J. Hartemink, *Informative structure priors: Joint learning of dynamic regulatory networks from multiple types of data*, Pac Symp Biocomput (2005), 459–70.
15. H. Bolouri and E.H. Davidson, *Modeling transcriptional regulatory networks*, BioEssays **24** (2002), 1118–1129.
16. S. Bornholdt, *Less is more in modeling large genetic networks*, Science **310** (2005), no. 5747, 449–450.
17. E. Boros, T. Ibaraki, and K. Makino, *Error-free and best-fit extension of partially defined boolean functions*, Information and Computation **140** (1998), 254–283.
18. S. Bulashevska and R. Eils, *Inferring genetic regulatory logic from expression data*, Bioinformatics **21** (2005), no. 11, 2706–2713.
19. W. Buntine, *Theory refinement on bayesian networks*, Proceedings of the 7th Conference on Uncertainty in Artificial Intelligence (Los Angeles, CA, USA) (B. D'Ambrosio, P. Smets, and P. Bonissone, eds.), Morgan Kaufmann Publishers, 1991, pp. 52–60.
20. A. Butte and I. Kohane, *Mutual information relevance networks: functional genomic clustering using pairwise entropy measurements*, Pac Symp Biocomput, 2000, pp. 418–429.
21. A.J. Butte, P. Tamayo, D. Slonim, T.R. Golub, and I.S. Kohane, *Discovering functional relationships between rna expression and chemotherapeutic susceptibility using relevance networks*, Proc Natl Acad Sci U S A **97** (2000), no. 22, 12182–12186.
22. K.-C. Chen, T.-Y. Wang, H.-H. Tseng, C.-Y.F. Huang, and C.-Y. Kao, *A stochastic differential equation model for quantifying transcriptional regulatory network in Saccharomyces cerevisiae*, Bioinformatics **21** (2005), no. 12, 2883–2890.
23. L. Chen and K. Aihara, *A model for periodic oscillation for genetic regulatory systems*, IEEE Trans. Circuits and Systems I **49** (2002), no. 10, 1429–1436.
24. T. Chen, H.L. He, and G.M. Church, *Modeling gene expression with differential equations*, Pac Symp Biocomput, 1999, pp. 29–40.
25. D.M. Chickering, D. Geiger, and D. Heckerman, *Learning bayesian networks: Search methods and experimental results*, Proceedings of the Fifth Conference on Artificial Intelligence and Statistics (Ft. Lauderdale), Society for Artificial Intelligence and Statistics, 1995, pp. 112–128.

26. D.-Y. Cho, K.-H. Cho, and B.-T. Zhang, *Identification of biochemical networks by S-tree based genetic programming*, Bioinformatics **22** (2006), no. 13, 1631–1640.
27. J. Collado-Vides and R. Hofestädt (eds.), *Gene regulations and metabolism - postgenomic computational approaches*, MIT Press, 2002.
28. G.M. Cooper and R.E. Hausman (eds.), *The cell: A molecular approach*, 4 ed., ASM Press and Sinauer Associates, 2007.
29. H. de Jong, *Modeling and simulation of genetic regulatory systems: A literature review*, J Comput Biol **9** (2002), no. 1, 67–103.
30. H. de Jong, J.-L. Gouzé, C. Hernandez, M. Page, T. Sari, and J. Geiselmann, *Qualitative simulation of genetic regulatory networks using piecewise-linear models*, Bull Math Biol **66** (2004), no. 2, 301–340.
31. H. de Jong and M. Page, *Qualitative simulation of large and complex genetic regulatory systems*, Proceedings of the 14th European Conference on Artificial Intelligence (W. Horn, ed.), 2000, pp. 141–145.
32. P. D'Haeseler, *Reconstructing gene networks from large scale gene expression data*, Ph.D. thesis, University of New Mexico, 2000.
33. M. Drton and M.D. Perlman, *Model selection for gaussian concentration graphs*, Biometrika **91** (2004), no. 3, 591–602.
34. R. Edwards and L. Glass, *Combinatorial explosion in model gene networks*, Chaos **10** (2000), no. 3, 691–704.
35. B. Ermentrout, *Simulating, analyzing and animating dynamical systems: A guide to xppaut for researchers and students*, 1 ed., Soc. for Industrial & Applied Math., 2002.
36. N. Friedman, M. Linial, I. Nachman, and D. Pe'er, *Using bayesian networks to analyze expression data*, J Comput Biol **7** (2000), no. 3-4, 601–620.
37. N. Friedman, K. Murphy, and S. Russell, *Learning the structure of dynamical probabilistic networks*, Proceedings of the 14th Annual Conference on Uncertainty in Artificial Intelligence (San Francisco, CA, USA), Morgan Kaufmann Publishers, 1998, pp. 139–147.
38. J. Gebert and N. Radde, *Modelling procaryotic biochemical networks with differential equations*, AIP Conference Proceedings, vol. 839, 2006, pp. 526–533.
39. D.T. Gillespie, *Exact stochastic simulation of coupled chemical reactions*, J Phys Chem **81** (1977), no. 25, 2340–2361.
40. L. Glass and S.A. Kauffman, *The logical analysis of continuous, non-linear biochemical control networks*, J Theor Biol **39** (1973), 103–129.
41. J.L. Gouze, *Positive and negative circuits in dynamical systems*, J Biological Systems **6** (1998), no. 21, 11–15.
42. J. Guckenheimer and P. Holmes, *Nonlinear oscillations, dynamical systems, and bifurcations of vector fields*, Springer Series, New York, 1983.
43. M. Gustafsson, M. Hörnquist, and A. Lombardi, *Constructing and analyzing a large-scale gene-to-gene regulatory network - lasso-constrained inference and biological validation*, IEEE Transaction on Computational Biology and Bioinformatics **2** (2005), no. 3, 254–261.
44. J. Hasty, D. McMillen, F. Isaacs, and J.J. Collins, *Computational studies of gene regulatory networks: in numero molecular biology*, Nature Review Genetics **2** (2001), no. 4, 268–279.
45. D. Heckerman, *A tutorial on learning with bayesian networks*, Technical Report MSR-TR-95-06, Microsoft Research, Redmond, WA, USA, 1995.
46. D. Heckerman, D. Geiger, and D.M. Chickering, *Learning bayesian networks: The combination of knowledge and statistical data*, Machine Learning **20** (1995), 197–243.

47. D. Heckerman, A. Mamdani, and M. Wellman, *Real-world applications of bayesian networks*, Communications of the ACM **38** (1995), no. 3, 24–30.
48. S. Huang, *Gene expression profiling, genetic networks, and cellular states: an integrating concept for tumorigenesis and drug discovery*, Journal of Molecular Medicine **77** (1999), 469–480.
49. F. Jacob and J. Monod, *Genetic regulatory mechanisms in the synthesis of proteins*, J Mol Biol **3** (1961), 318–356.
50. L. Kaderali, *A hierarchical bayesian approach to regression and its application to predicting survival times in cancer*, Shaker Verlag, Aachen, 2006.
51. L. Kaderali, T. Zander, U. Faigle, J. Wolf, J.L. Schultze, and R. Schrader, *Caspar: A hierarchical bayesian approach to predict survival times in cancer from gene expression data*, Bioinformatics **22** (2006), no. 12, 1495–1502.
52. S. Kauffman, *Metabolic stability and epigenesis in randomly constructed genetic nets*, J Theor Biol **22** (1969), 437–467.
53. S. Kikuchi, D. Tominaga, M. Arita, K. Takahashi, and M. Tomita, *Dynamic modeling of genetic networks using genetic algorithm and S-systems*, Bioinformatics **19** (2003), no. 5, 643–650.
54. H. Lähdesmäki, I. Shmulevich, and O. Yli-Harja, *On learning gene regulatory networks under the boolean network model*, Machine Learning **52** (2003), 147–167.
55. W. Lam and F. Bacchus, *Using causal information and local measures to learn bayesian networks*, Proceedings of the 9th Conference on Uncertainty in Artificial Intelligence (Washington, DC, USA), Morgan Kaufmann Publishers, 1993, pp. 243–250.
56. R. Laubenbacher and B. Stigler, *A computational algebra approach to the reverse engineering of gene regulatory networks*, J Theor Biol **229** (2004), no. 4, 523–537.
57. F. Li, T. Long, Y. Lu, Q. Ouyangm, and C. Tang, *The yeast cell-cycle network is robustly designed*, Proc. Natl. Acad. Sci. U. S. A **101** (2004), 4781–4786.
58. S. Liang, S. Fuhrman, and R. Somogyi, *Reveal, a general reverse engineering algorithm for inference of genetic network architectures*, Pac Symp Biocomput **3** (1998), 18–29.
59. D. Madigan, J. Garvin, and A. Raftery, *Eliciting prior information to enhance the predictive performance of bayesian graphical models*, Communications in Statistics: Theory and Methods **24** (1995), 2271–2292.
60. J.M. Mahaffy, D.A. Jorgensen, and R.L. van der Heyden, *Oscillations in a model of repression with external control*, J Math Biol **30** (1992), 669–691.
61. J.M. Mahaffy and C.V. Pao, *Models of genetic control by repression with time delays and spatial effects*, J Math Biol **20** (1984), 39–57.
62. L. Mao and H. Resat, *Probabilistic representation of gene regulatory networks*, Bioinformatics **20** (2004), no. 14, 2258–2269.
63. A.A. Margolin, I. Nemenman, K. Basso, C. Wiggins, G. Stolovitzky, R. Dalla-Favera, and A. Califano, *Aracne: An algorithm for the reconstruction of gene regulatory networks in a mammalian cellular context*, BMC Bioinformatics **7 (Suppl 1)** (2006), S7.
64. A.A. Margolin, K. Wang, W.K. Lim, M. Kustagi, I. Nemenman, and A. Califano, *Reverse engineering cellular networks*, Nature Protocols **1** (2006), 663–672.
65. H.H. McAdams and A. Arkin, *Stochastic mechanisms in gene expression*, Proc. Natl. Acad. Sci. U. S. A. **94** (1997), 814–819.
66. T. Mestl, E. Plahte, and S.W. Omholt, *A mathematical framework for describing and analyzing gene regulatory networks*, J Theor Biol **176** (1995), no. 2, 291–300.
67. K. Murphy and S. Mian, *Modelling gene expression data using dynamic bayesian networks*, Tech. report, Computer Science Division, University of California, Berkeley, CA, USA, 1999.

68. I. Nachman, A. Regev, and N. Friedman, *Inferring quantitative models of regulatory networks from expression data*, Bioinformatics **20** (2004), no. 1, i248–i256.
69. S. Ott, S. Imoto, and S. Miyano, *Finding optimal models for small gene networks*, Pac Symp Biocomput **9** (2004), 557–567.
70. J. Pearl, *Causality: Models, reasoning and inference*, Cambridge University Press, Cambridge, 2000.
71. J. Pearl and T. Verma, *A theory of inferred causation*, Knowledge Representation and Reasoning: Proceedings of the Second International Conference (New York) (J. Allen, R. Fikes, and E. Sandewal, eds.), Morgan Kaufmann Publishers, 1991, pp. 441–452.
72. D. Pe'er, *Bayesian network analysis of signaling networks: A primer*, Science STKE **281** (2005), p 14.
73. B.-E. Perrin, L. Ralaivola, A. Mazurie, et al., *Gene networks inference using dynamic bayesian networks*, Bioinformatics **19 Suppl. II** (2003), i138–i148.
74. N. Radde, J. Gebert, and C.V. Forst, *Systematic component selection for gene network refinement*, Bioinformatics **22** (2006), 2674–2680.
75. N. Radde and L. Kaderali, *Bayesian inference of gene regulatory networks using gene expression time series data*, BIRD 2007, LNBI **4414** (2007), 1–15.
76. R.W. Robinson, *Counting labeled acyclic graphs*, New Directions in the Theory of Graphs (F. Harary, ed.), Academic Press, New York, 1973, pp. 239–273.
77. C. Sabatti and G.M. James, *Bayesian sparse hidden components analysis for transcription regulation networks*, Bioinformatics **22** (2006), no. 6, 739–746.
78. M. Santillán and M.C. Mackey, *Dynamic regulation of the tryptophan operon: A modeling study and comparison with experimental data*, Proc. Natl. Acad. Sci. U. S. A. **98** (2001), no. 4, 1364–1369.
79. M.J. Schilstra and H. Bolouri, *Modelling the regulation of gene expression in genetic regulatory networks*, Document for NetBuilder, a graphical tool for building logical representations of genetic regulatory networks.
80. C.E. Shannon and W. Weaver, *The mathematical theory of communication*, University of Illinios Press, 1963.
81. I. Shmulevich, E.R. Dougherty, and W. Zhang, *From boolean to probabilistic boolean networks as models of genetic regulatory networks*, Proceedings of the IEEE **90** (2002), no. 11, 1778–1792.
82. I. Shmulevich, A. Saarinen, O. Yli-Harja, and J. Astola, *Inference of genetic regulatory networks under the best-fit extension paradigm*, Proceedings of the IEEE EURASIP Workshop on Nonlinear Signal and Image Proc. (W. Zhang and I. Shmulevich, eds.), 2001.
83. A. Silvescu and V. Honavar, *Temporal boolean network models of genetic networks and their inference from gene expression time series*, Complex Systems **13** (1997), no. 1, 54–75.
84. P.W.F. Smith and J. Whittaker, *Edge exclusion tests for graphical gaussian models*, Learning in Graphical Models (M. Jordan, ed.), MIT Press, 1999, pp. 555–574.
85. P. Smolen, D.A. Baxter and J.H. Byrne, *Modeling transcriptional control in gene networks*, Bull Math Biol **62** (2000), 247–292.
86. R.V. Solé, B. Luque, and S.A. Kauffman, *Phase transitions in random networks with multiple states*, Technical Report 00-02-011, Santa Fe Institute, 2000.
87. P.T. Spellman, G. Sherlock, M.Q. Zhang, et al., *Comprehensive identification of cell cycle-regulated genes of the yeast Saccharomyces cerevisiae by microarray hybridization*, Mol Biol Cell **9** (1998), 3273–3297.
88. D. Spiegelhalter, A. Dawid, S. Lauritzen, and R. Cowell, *Bayesian analysis in expert systems*, Statistical Science **8** (1993), 219–282.

89. P. Sprites, C. Glymour, and R. Scheines, *Causation, prediction, and search*, Springer Verlag, New York, 1993.
90. H. Steck and T. Jaakkola, *On the dirichlet prior and bayesian regularization*, Advances in Neural Information Processing Systems 15 (Cambridge, MA, USA), MIT Press, 2002.
91. J. Suzuki, *A construction of bayesian networks from databases based on an mdl scheme*, Proceedings of the 9th Conference on Uncertainty in Artificial Intelligence (Washington, DC, USA), Morgan Kaufmann Publishers, 1993, pp. 266–273.
92. D. Thieffry and R. Thomas, *Qualitative analysis of gene networks*, Pac Symp Biocomput **3** (1998), 77–88.
93. R. Thomas, *On the relation between the logical structure of systems and their ability to generate multiple steady states or sustained oscillations*, Springer Series in Synergetics **9** (1981), 180–193.
94. R. Thomas and R. d'Ari, *Biological feedback*, CRC Press, Boca Raton, FL, USA, 1990.
95. R. Thomas, S. Mehrotra, E.T. Papoutsakis, and V. Hatzimanikatis, *A model-based optimization framework for the inference on gene regulatory networks from dna array data*, Bioinformatics **20** (2004), no. 17, 3221–3235.
96. R. Thomas, D. Thieffry, and M. Kauffman, *Dynamical behaviour of biological regulatory networks – I. biological role of feedback loops and practical use of the concept of the loop-characteristic state*, Bull Math Biol **57** (1995), 247–276.
97. M. Tomita, *Whole-cell simulation: A grand challenge for the 21st century*, Trends Biotechnol. **19** (2001), no. 6, 205–210.
98. E.P. van Someren, B.L.T. Vaes, W.T. Steegenga, A.M. Sijbers, K.J. Dechering, and J.T. Reinders, *Least absolute regression network analysis of the murine osteoblast differentiation network*, Bioinformatics **22** (2006), no. 4, 477–484.
99. E.P. van Someren, L.F.A. Wessels, and M.J.T. Reinders, *Linear modeling of genetic networks from experimental data*, ISMB 2000: Proceedings of the 8th International Conference on Intelligent Systems for Molecular Biology, 2000, pp. 355–366.
100. E.O. Voit, *Computational analysis of biochemical systems*, Cambridge University Press, 2000.
101. E.O. Voit and J. Almeida, *Decoupling dynamical systems for pathway identification from metabolic profiles*, Bioinformatics **20** (2004), no. 11, 1670–1681.
102. J. von Neumann, *The theory of self-reproducing automata*, University of Illinois Press, 1966.
103. A. Wagner, *Circuit topology and the evolution of robustness in two-gene circadian oscillators*, Proc. Natl. Acad. Sci. U. S. A. **102** (2005), 11775–11780.
104. D.C. Weaver, *Modeling regulatory networks with weight matrices*, Pac Symp Biocomput, 1999, pp. 112–123.
105. G. Yagil and E. Yagil, *On the relation between effector concentration and the rate of induced enzyme synthesis*, Biophysical Journal **11** (1971), no. 1, 11–27.
106. M. Zou and S.D. Conzen, *A new dynamic bayesian network (dbn) approach for identifying gene regulatory networks from time course microarray data*, Bioinformatics **21** (2005), no. 1, 71–79.

3

Belief Networks for Bioinformatics

Jeroen (H.H.L.M.) Donkers and Karl Tuyls

Bio- and medical Informatics Competence Center (BioMICC),
MICC, Universiteit Maastricht, PO Box 616, 6200 MD Maastricht,
The Netherlands. jeroen.donkers@educ.unimaas.nl,
k.tuyls@micc.unimaas.nl

Summary. Recent publications illustrate successful applications of belief networks[1] (BNs) and related probabilistic networks in the domain of bioinformatics. Examples are the modeling of gene regulation networks [6,14,26], the discovering of metabolic [40,83] and signalling pathways [94], sequence analysis [9, 10], protein structure [16, 28, 76], and linkage analysis [55]. Belief networks are applied broadly in health care and medicine for diagnosis and as a data mining tool [57, 60, 61]. New developments in learning belief networks from heterogeneous data sources [40, 56, 67, 80, 82, 96] show that belief networks are becoming an important tool for dealing with high-throughput data at a large scale, not only at the genetic and biochemical level, but also at the level of systems biology.

In this chapter we introduce belief networks and describe their current use within bioinformatics. The goal of the chapter is to help the reader to understand and apply belief networks in the domain of bioinformatics. To achieve this, we (1) make the reader acquainted with the basic mathematical background of belief networks, (2) introduce algorithms to learn and to query belief networks, (3) describe the current state-of-the-art by discussing several real-world applications in bioinformatics, and (4) discuss (free and commercially) available software tools.

The chapter is organized as follows. We start (in Section 3.1) with introducing the concept of belief networks. Then (in Section 3.2) we present some basic algorithms to infer on belief networks and to learn belief networks from data. Section 3.3 is dedicated to a (non-exhaustive) range of extensions to and variants of the standard belief-network concept. We continue (in Section 3.4) by discussing some techniques and guidelines to construct belief networks from domain knowledge. Section 3.5 reviews some recent applications of belief networks in the domain of bioinformatics. In Section 3.6 we discuss a range of tools that are available for constructing, querying, and learning belief networks. Finally, (in Section 3.7) we provide a brief guide to the literature on belief networks.

[1]Many names have been used in the literature for this concept: Bayesian Networks, Bayesian Probability Networks, Probability Networks, Directed Graphical Models, and Belief Networks. In this chapter we use the latter name only.

Donkers and Tuyls: *Belief Networks for Bioinformatics*, Studies in Computational Intelligence (SCI) **94**, 75–111 (2008)
www.springerlink.com

3.1 An Introduction to Belief Networks

Belief networks represent probabilistic relations between a (finite) set of variables. These variables can include observations, diagnoses, hidden causes, and so on. The variables can have binary domains (e.g., "yes", "no"), discrete domains (e.g., "A", "C", "G", "T"), or continuous domains (e.g., gene expression level). A belief network on variables $X_1,\ldots,X_n$ is a compact representation of the joint probability distribution $P(X_1,X_2,\ldots,X_n)$. Assume we observe evidence e_k for variable X_k, i.e., $X_k = e_k$, then the belief network can be used to compute the *beliefs* in unobserved variables, which is the conditional probability of the unobserved variables given the evidence: $Bel(X_i = x_i) = P(X_i = x_i|X_k = e_k)$. When additional evidence ($X_m = e_m$) is added to the network, the beliefs in variables can be updated: $Bel(X_i = x_i) = P(X_i = x_i|X_k = e_k, X_m = e_m)$.

In the general case, the update of beliefs (also called *inference*) is a computationally challenging task. Two marginal probabilities have to be determined, which in the case of discrete domains (denoted by $D(X_j)$) equals to computing the following sums of joint probabilities (we write $P(x_i)$ for $P(X_i = x_i)$):

$$P(x_i|e_k) = \frac{P(x_i,e_k)}{P(e_k)} = \frac{\sum \cdots \sum_{xj \in D(X_j), j \neq i, j \neq k} P(x_1,\ldots x_n,x_i,e_k)}{\sum \cdots \sum_{xj \in D(X_j), j \neq k} P(x_1,\ldots x_n,e_k)}. \tag{3.1}$$

This requires summing up in the order of $|D(X_1)| \times |D(X_2)| \times \ldots |D(X_n)|$ probabilities, which is exponential in the number of variables.

3.1.1 Belief Networks Defined

The crux of belief networks is that the computation of marginals can be speeded up considerably by using *conditional independencies* that might exist between variables. Variables X and Y are said to be independent *given* a set of variables Z (indicated by $X \perp\!\!\!\perp Y|Z$) if $P(X,Y|Z) = P(X|Z)P(Y|Z)$, or equivalently, $P(X|Y,Z) = P(X|Z)$.

A belief network on a set of variables $\mathscr{X} = \{X_1,\ldots,X_n\}$ consists of two parts. The first part is a directed graph G in which each variable X_i is represented by a node (also indicated by X_i) and in which the arrows represent conditional independencies as follows. If $Pa(X_i)$ is the set of parent nodes of X_i then X_i is independent of all other variables given $Pa(X_i)$:

$$\forall X_j \in \mathscr{X} \backslash X_i \cup Pa(X_i) : \quad X_i \perp\!\!\!\perp X_j|Pa(X_i). \tag{3.2}$$

These conditional independencies, the *Markov blanket*, can be used to reformulate the joint probability distribution. Given the chain rule of probability $P(X_1, X_2,\ldots,X_n) = \prod_i P(X_i|X_1,\ldots X_{i-1})$ and the definition of conditional independence, the joint probability reduces to:

$$P(X_1,\ldots X_n) = \prod_i P(X_i|Pa(X_i)), \tag{3.3}$$

provided that the graph G is acyclic. It means that instead of knowing the whole probability distribution explicitly, it is sufficient to know *conditional probabilities* of each node in the network, given its parents. In case a node X does not have any parents, these conditional probabilities are the *priors* $P(X = x)$ of variable X.

This brings us at the second part of a belief network: to every node X in a belief network, a local probability distribution $P(X|Pa(X))$ is connected. In case of discrete variables, this will be a distribution represented by a probability table. In case of continuous variables, it might be a multidimensional gaussian. The local probability distributions have two advantages. Not only is it easier to elicit or learn local probabilities than global joint probabilities, we will also see in the next section that inference algorithms can make use of local distributions to reduce computation time.

3.1.2 Belief Network Example

Figure 3.1 gives an example of a small belief network with three binary variables A, B, and C. The leftmost table gives the values of a joint probability distribution over the three variables. This distribution contains the independency $C \perp\!\!\!\perp A|B$. We can see, for instance, that $P(C = T|A = T, B = T) = \frac{0.014}{0.014+0.006} = 0.7$ is equal to $P(C = T|B = T) = \frac{0.014+0.28}{0.014+0.006+0.28+0.12} = \frac{0.294}{0.42} = 0.7$, so $P(C|B,A)$ is equal to $P(C|B)$ and therefore $P(A,B,C)$ is equal to $P(A)P(B|A)P(C|B)$. The simple graph represents the conditional independency of C and A given B by the absence of a link between nodes C and A. The three smaller tables in the figure represent the probability tables of node A, B and C respectively.

The example at the left side of Fig. 3.1 shows that the graphical structure of a belief network represents conditional independencies. However, there might exist additional conditional independencies in the distribution that are not represented by the graph structure. It means that it is possible to derive several belief networks from the same joint probability distribution that are all correct representations but use different conditional independencies. An argument to prefer one of these equivalent networks over another is to require that the arrows also have a *causal* meaning. In a causal belief network, the parents of a node are interpreted as the *direct causes* of the variable. When in the example of Fig. 3.1, A represents "haplotype", B "protein

A	B	C	$P(A,B,C)$
T	T	T	0.014
T	T	F	0.006
T	F	T	0.036
T	F	F	0.144
F	T	T	0.280
F	T	F	0.120
F	F	T	0.080
F	F	F	0.320

A	$P(A)$
T	0.2
F	0.8

A	B	$P(B\|A)$
T	T	0.1
T	F	0.9
F	T	0.5
F	F	0.5

B	C	$P(C\|B)$
T	T	0.7
T	F	0.3
F	T	0.2
F	F	0.8

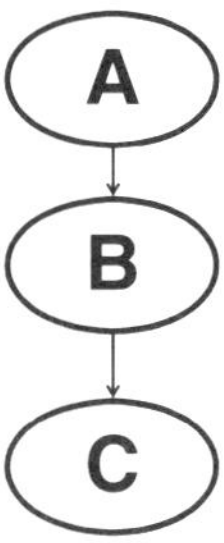

Fig. 3.1. An example probability distribution and a Belief network representing it

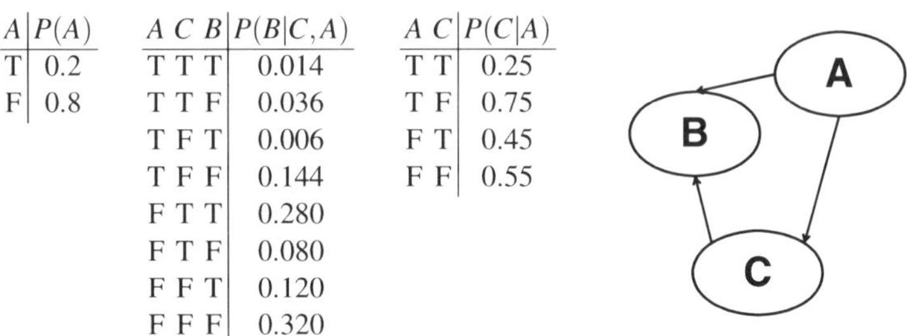

A	$P(A)$
T	0.2
F	0.8

A	C	B	$P(B\|C,A)$
T	T	T	0.014
T	T	F	0.036
T	F	T	0.006
T	F	F	0.144
F	T	T	0.280
F	T	F	0.080
F	F	T	0.120
F	F	F	0.320

A	C	$P(C\|A)$
T	T	0.25
T	F	0.75
F	T	0.45
F	F	0.55

Fig. 3.2. Another belief network representing the distribution of Fig. 3.1

type", and C "phenotype", then the network would model a causal chain between gene and body.

The probability distribution in Fig. 3.1 could alternatively be modeled in a correct way by the network in Fig. 3.2. Here, the arrow between B and C is reversed and an arrow between A and C is added. This has the effect that the network does not explicitly represent the conditional independency between A and C anymore. Moreover, the network has lost its causal meaning.

3.1.3 D-Separation

A belief network represents (many) more conditional independencies than the ones that are obvious from the child-parent relations. One can use *d-separation* to find all conditional independencies that can be deduced from the structure of a belief network. Namely, when nodes X and Y are d-separated by a set of nodes Z, then X and Y are conditionally independent given Z. D-separation is based on blocking undirected paths between X and Y. An undirected path p between X and Y is blocked by a set of nodes Z if:

1) p contains a chain $i \rightarrow z \rightarrow j$ such that $z \in Z$;
2) p contains a fork $i \leftarrow z \rightarrow j$ such that $z \in Z$;
3) p contains a collision $i \rightarrow e \leftarrow j$ such that neither e nor any of its descendants are in Z.

If all possible paths between X and Y are blocked by Z then Z d-separates X from Y and, hence, $X \perp\!\!\!\perp Y|Z$. The third demand for blocking a path is the least intuitive. It can be explained as follows. If two variables A and B influence the same third variable C, then as long as we do not know anything about the combined effect, the two variables A and B are independent. As soon as we know the effect on C, the independency between A and B is lost. If we know that two defect genes can cause the same disease, then knowing the state of one gene does not change our knowledge on the other gene, when we do not know whether the disease occurred. As soon as the disease is observed, however, evidence that the first gene is defect will lower our belief of the second gene being defect as well.

3.1.4 Local Probabilities

As stated, each node X_i in a belief network carries a conditional probability function $P(X_i|Pa(X_i))$. When a node has many parents, the number of parameters in this function can be large. In the case of discrete variables, the size of the table is exponential in the number of parents, and it might be difficult to assess all probabilities separately. There are two ways to overcome this problem. The first way is to introduce intermediate variables that might model the relation between the parent nodes on a more precise and detailed level. The intermediate nodes are inserted between parents and child (*divorcing* parents). Although this produces more nodes and more tables, the size of the tables can be much smaller, leading to a considerably lower number of probabilities to be assessed.

The second way is to restrict the parameter space of the probability function $P(X_i|Pa(X_i))$. This has the advantage of an easier assessment and more efficient inference and learning algorithms. However, the approach might lead to an over-simplified belief network. An example of parameter reduction is the *noisy OR* [69]. In this model, a binary variable D has common causes C_1, C_2, $\ldots C_n$, each represented by a binary variable. When any of the C_i's is true, then D is true, unless some inhibitor Q_i prevents it with probability q_i (so $P(D=F|C_i=T)=q_i$). The main assumption is that all inhibitors Q_i are independent: $P(D=F|C_1,\ldots,C_n)=\prod_{C_i=T} q_i$. It means that for n parents, only n parameters, q_i have to be assessed instead of 2^n parameters. Variants of this model are the *noisy AND*, and the *generalized noisy OR*, which is defined for multivalued variables.

Another example of parameter reduction is to exploit *context-specific independencies* (CSI) [11] to reduce the number of parameters. It is based on the observation that conditional independencies between X and Y given Z might occur only when some other variable C has a given value c. For example, only for females, there is a dependency between age and newborn health; for males these variables are independent (see Fig. 3.3). These context-specific independencies can be derived from a probability table and can be used to transform the table into a compact tree. The nodes in this conditional probability tree represent the parent variables, the edges represent the states of these variables, and the leaves represent the conditional probabilities. When CSIs are present, the number of leaves in the tree will be smaller than the number of entries in the original table. In Fig. 3.3 there are only 6 probabilities in the tree, whereas the full table would contain 8 probabilities.

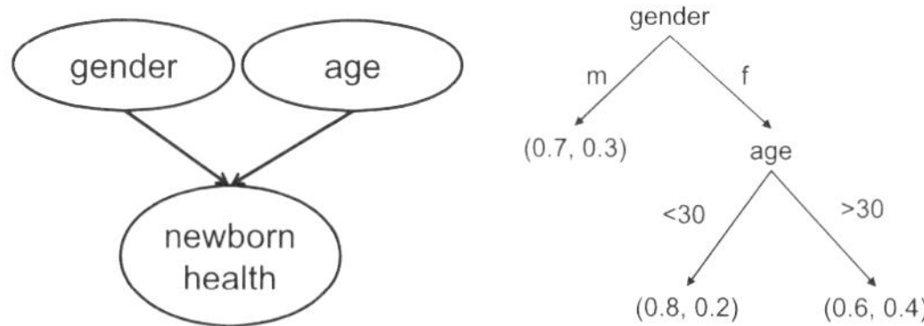

Fig. 3.3. Conditional probability tree example. The variable 'Newborn health' has values 'good' and 'bad'

3.1.5 Markov Random Fields

Belief networks are not the only way to represent a joint probability network as a graph. A closely related graphical model is the *Markov Random Field* or *Markov Network* [41]. Similar to belief networks, every variable X_i is represented by a node in the graph. But in this case, the nodes in the graph are connected by *undirected* edges. Let $adj(X_i)$ be all the nodes that are adjacent (i.e., directly connected) to X_i, then the edges in a Markov field are placed in such way that:

$$\forall X_j \in \mathscr{X} \backslash X_i \cup adj(X_i): \quad X_i \perp\!\!\!\perp X_j | adj(X_i). \tag{3.4}$$

so, $adj(X_i)$ acts as the Markov blanket of X_i, just like the parents of a node in a belief network. In contrast to belief networks, there are no conditional probability functions connected to nodes. Instead, each *clique* $k \in K$ (a fully connected subset of nodes) in the graph is provided with a *potential* $\phi_k(\cdot)$, which assigns a non-negative real value to all combinations of values of nodes $X \in k$ (indicated by $X\{k\}$). As a result of the Hammersley-Clifford theorem (see [41]), these potentials partition the joint probability as follows.

$$P(X_1, \ldots X_n) = \frac{1}{Z} \prod_{k \in K} \phi_k(X\{k\}), \tag{3.5}$$

where Z is a normalization constant.

Markov Random Fields represent different dependencies and independencies than belief networks. For instance, they can represent circular dependencies between variables. The undirected graph approach is also useful in the cases in which direction of influence has no meaning, for instance when variables represent pixels in an image or atoms in a protein molecule. In the clique-tree algorithms we describe in the next section, belief networks are transformed into equivalent Markov fields. A complication is that, in general, inference in Markov Random Fields is harder than in belief networks. These transformations, however, are performed in a careful way such that inference is in fact improved.

3.2 Basic Algorithms for Belief Networks

In Section 3.1 we explained that querying BNs (i.e., computing marginal probabilities) is in general exponential in the number of variables. The (sparse) structure of belief networks, however, allows for efficient algorithms. In Section 3.2 we discuss several *exact inference algorithms*, such as the polytree and junction-tree algorithms. In large, dense, or loopy BNs, however, these exact algorithms might be inapplicable or not sufficiently efficient. We therefore also treat *approximate inference algorithms* such as Gibbs sampling.

As will be shown below, BNs do not need to be constructed by hand. It is possible to learn them from available data. Since Belief Networks can incorporate causal relations, they can be used to discover causalities from data and to estimate the strength of causal relations. In this section we discuss two groups of learning algorithms. The first group concentrates on *learning the probabilities* for a given Belief Network structure. The second group learns the probabilities as well as the *structure* of the network.

3.2.1 Querying Belief Networks

A belief network can be used to answer several types of questions. The most well-known type of query is the *belief update*: what is the probability of nodes X given the evidence entered in nodes E? Nodes X are called the *query nodes* and E the *evidence nodes*. In the descriptions of the inference algorithms below we will emphasize this type of query.

A second type of query is to obtain the *most probable explanation (MPE)*: given evidence E, what combination of values for all other variables $X = \mathscr{X} \backslash E$ has the highest probability? In other words find: $\text{argmax}_X P(X|E)$. Since in this query, only maximum probability is requested, the inference algorithms require less computations.

When we are only interested in the maximum-probability value of a subset of nodes, it is called a *maximum aposteriori hypothesis* or *MAP* query. So find: $\text{argmax}_Y P(Y|E)$, where $Y \subset \mathscr{X} \backslash E$. This problem is in fact harder than MPE, since it is not sufficient to calculate with maximum probabilities alone. Note that a MAP query cannot be answered simply by performing an MPE query, since variables outside Y are not fixed to their maximum probability values.

3.2.2 Exact Inference by Bucket Elimination

We will discuss four different algorithms for exact inference (belief update) in belief networks. The first algorithm is called *Bucket Elimination* [27]. The basic idea of the algorithm is to eliminate one by one all nodes from the belief network, that are not involved in the query or evidence nodes. During elimination, probability tables are changed by a process called *marginalization*.

Assume we have a joint probability distribution $P(A,B)$, then $P(A)$ can be computed by marginalizing out B as follows: $P(A = a_i) = \sum_j P(a_i, b_j)$. This process can be notated as: $P(A) = \sum_B P(A,B)$. Now, every time a node X is eliminated from the belief network, all probability tables in which X appears are combined in a large table. Then from this table, X is marginalized out. At the end, only a table containing the query and evidence nodes remains, from which the required answer can be computed. For instance, when we want to compute the probability of $P(A|e)$ in the network of Fig. 3.4, we have to compute:

$$P(A,e) = \sum_{B,C,D,E=e} P(A)P(B)P(C|A,B)P(D|B)P(E|C,D). \tag{3.6}$$

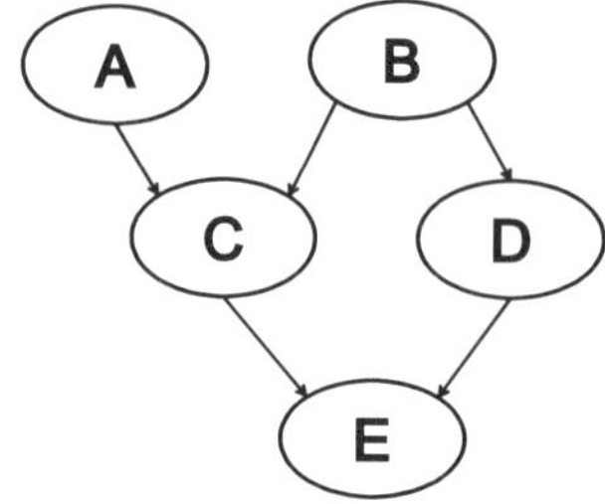

Fig. 3.4. Example belief network with five variables

When we order the variables, say alphabetically, we can rearrange the summations as follows into five *buckets*:

$$P(A,e) = P(A)\sum_{B} P(B)\sum_{C} P(C|A,B)\sum_{D} P(D|B)\sum_{E=e} P(E|C,D). \qquad (3.7)$$

First we marginalize $E = e$ out of bucket $P(E|C,D)$, which produces a new bucket, $\phi_E(C,D)$, a table of probabilities:

$$P(A,e) = P(A)\sum_{B} P(B)\sum_{C} P(C|A,B)\sum_{D} P(D|B)\phi_E(C,D). \qquad (3.8)$$

This table is multiplied with $P(D|B)$ and the resulting table is marginalized for D, resulting in bucket $\phi_D(B,C)$:

$$P(A,e) = P(A)\sum_{B} P(B)\sum_{C} P(C|A,B)\phi_D(B,C). \qquad (3.9)$$

After repeating this process for C, producing $\phi_C(A,B)$, and C, producing $\phi_C(A)$, we obtain $P(A)\phi_C(A)$, which after multiplication and normalization produces the requested probabilities for A.

The advantage of this procedure is that in the example we never dealt with tables having more than three variables at a time. The size of the tables, and therefore the complexity of the algorithm, however, depends on the *order* of the nodes at the start. Using graph-theoretical manipulations (see below), a perfect elimination ordering can be computed that minimizes the bucket sizes that occur during the elimination procedure. A disadvantage of bucket elimination is that all computations have to start over at the moment the updated beliefs of another variable have to be inspected or new evidence is entered into the network.

The well-know peeling algorithm for linkage analysis (see Subsection 3.5.4) is in fact an instance of bucket elimination.

3.2.3 Exact Inference in Polytrees by Belief Propagation

The main idea behind Pearl's belief propagation [69] is to reduce the number of recomputations needed for new queries. When evidence is entered into the belief network, the belief at all nodes is updated, so all nodes can be inspected without recomputations. In the algorithm, the nodes in a belief network act as if they are autonomous agents, each one keeping track of the own beliefs. When evidence is provided to a node's agent, which changes it beliefs, it sends messages to child and parent nodes containing information on how the belief is changed. The receiving agents adapt their beliefs using the incoming messages and propagate information further through the network. The size of the messages is bounded by the size of the local probability tables.

The algorithm can be used on any belief network, but only when the network does not contain undirected cycles, the algorithm can be proven to converge in time linear to the diameter of the graph. A belief network without undirected cycles is a *polytree*. This induces a special property: every node d-separates all its descendants from all its ascendants.

Pearl's algorithm uses two types of messages: π-messages contain evidence from the ascendant nodes and are sent from parent to child; λ-messages contain evidence from descendent nodes and are sent from child to parent. Let E^+ be the evidence in the ascendants of node X and E^- the evidence in the descendants. The incoming π-messages are used to compute $\pi(X) = P(X|E^+)$, and the incoming λ-messages to compute $\lambda(X) = P(E^-|x)$. These can be combined to compute the updated belief $P(X|E)$:

$$\begin{aligned} P(X|E) = P(X|E^+,E^-) &= \frac{P(E^+,E^-|X)P(X)}{P(E)} \qquad (3.10) \\ = \frac{P(E^+|X)P(E^-|X)P(X)}{P(E)} &= \frac{P(E^+)}{P(E)}P(E^-|X)P(X|E^+) = \alpha\pi(X)\lambda(X), \end{aligned}$$

where α is a normalization constant and is equal to $(\sum_X \pi(X)\lambda(X))^{-1}$. Formula (3.10) is based on Bayes' theorem and on the polytree structure which renders $E+$ and $E-$ independent, given X. The values for $\lambda(X)$ are computed by multiplying all incoming λ-messages, $\pi(X)$ is computed by multiplying all incoming π-messages with $P(X|Pa(X))$ and marginalizing X from the product. The value of $\pi(X)$ can only be computed when all incoming π-messages have arrived, and $\lambda(X)$ when all λ-messages have arrived. When evidence is provided to a node, both $\pi(X)$ and $\lambda(X)$ messages are equal to the evidence. Nodes without parents have $\pi(X) = P(X)$ and nodes without children have $\lambda(x_i) = 1$.

Messages are sent by a node X in two cases only: (1) $\pi(X)$ is known and all parents of X, except Y, have sent λ-messages to X. In this case a π-message is sent to Y. (2) $\lambda(X)$ is known and all children of X, except Z, have sent π-messages to X. In this case a λ-message is sent to Z. The computation of π-messages $\pi_{X\to Y}(X)$ is straightforward: multiply $\pi(X)$ with all received λ-messages and normalize. The computation of λ-messages requires more work. First, all π-messages are multiplied. Then this product is multiplied by $P(X|Pa(X))$. From the product all parents except Z are marginalized out. It is multiplied by $\lambda(X)$ and then X is marginalized out too, resulting in a function $\lambda_{X\to Z}(Z)$.

Since polytrees contain at least one leaf node (having only one adjacent node), there is always a node at the beginning that can start passing a message. Now Pearl's algorithm basically consist of the following six steps:

1. initialize $\pi(X)$ and $\lambda(X)$ for all evidence and leaf nodes.
2. for all nodes: if possible and not done already, send π or λ message.
3. for all nodes: if all π-messages have arrived, compute $\pi(X)$, if needed.
4. for all nodes: if all λ-messages have arrived, compute $\lambda(X)$, if needed.
5. proceed with step 2 until $\pi(X)$ and $\lambda(X)$ are known for all nodes.
6. Compute $P(X|E) = \alpha\pi(X)\lambda(X)$ for all nodes.

It is clear that most effort is put in the computation of $\pi(X)$ and in λ-messages. These operation require summarizations of $P(X|Pa(X))$. The efficiency of Pearl's algorithm is determined by the maximum size of these conditional probability tables.

When Pearl's message-passing algorithm is applied to general belief networks, or even belief networks that have directed cycles, the algorithm is not guaranteed to converge. However, it can be applied as an approximative inference algorithm. In the next section we will see that message-passing can be used efficiently in general belief networks too.

3.2.4 Exact Inference by Clique-Tree Belief Propagation

When we look closer at the process of bucket elimination, we can see that the buckets produced during the process in fact form a tree structure. Now instead of recomputing and destroying the buckets every time we have a new query, we could first construct the whole tree of buckets and then use a Pearl-like message-passing algorithm to propagate belief updates in this tree. The resulting algorithm is called *clique-tree* belief propagation [54, 84], but also the names *junction-tree* or *join-tree* are used frequently.

Since the computation of messages in the belief-propagation algorithm depend on the size of the probability tables, it is important to keep the buckets as small as possible. This is achieved in a series of graph-theoretical transformations, that are also known in, for instance, database theory for optimizing query graphs (hence the name "join-tree").

The first step is to transform the belief net in an equivalent Markov Random Field. This is done by *moralizing* the graph: all arrows are transformed into undirected links, and every pair of parents that are not connected directly in the belief network, are linked together. This produces the moral graph of the belief network. The second step is to *triangulize* the moral graph, which means the addition of so many links to the graph that no cycles of size four or larger exist that have no additional links (chords) between their nodes. The cliques in this triangulated graph are in fact the buckets that exchange the belief-propagation messages. Each clique C receives a potential ϕ_C such that $P(X_1,\ldots,X_n) = \prod_C \phi_C$.

A graph can be triangulized in many ways, resulting in different clique sizes. Finding a good triangulation is equivalent to finding a good node elimination order in bucket elimination. Although this problem is NP-complete, there exist algorithms that find an optimal elimination order in reasonable time. For instance, the program HUGIN uses such an algorithm [4].

When the graph is triangulized, a *clique tree* is constructed. This is a tree in which the nodes represent the cliques C_i of the graph. The cliques are connected into a tree in such a way that if there is a path from clique C_i via C_j to C_k, then $C_i \cap C_k \subseteq C_j$. Every clique C in the tree receives a potential ϕ_C. Moreover, every node X in the belief network is assigned to a clique C_X that contains both X and $Pa(X)$. The potentials ϕ_C are obtained by multiplying $P(X|Pa(X))$ for all nodes X assigned to this clique. If a clique C happens to have no node assigned to it, its potential is equal to 1.

The construction of the clique tree and computation of the potentials is called the *compilation* of the belief network. A belief network has only to be compiled once. Evidence is fed into the clique tree and belief updates are also done on the clique tree. Only when the belief network is changed by altering the structure or changing conditional probabilities, the network has to be recompiled.

Entering evidence and belief propagation in a clique tree is very similar to Pearl's algorithm. First, the tree is rooted, so that each node has a unique parent. Each node in the tree then sends λ messages to its children, and sends one π message to its (unique) parent. From the collected messages, the update belief potential ψ_C is computed at each clique C. The beliefs for the nodes X are finally computed by marginalizing ψ_C of the clique to which X was assigned.

There are several options for the order in which the messages are sent and the precise intermediate information that is stored, but the basic principle as we described above remains the same. The complexity of the algorithm is determined by the maximum clique size of the triangulated graph. In the worst case, this is the whole graph (in fact, belief propagation is NP hard [17]). Fortunately, many belief networks possess such a structure that exact belief propagation can be performed efficiently.

3.2.5 Exact Inference by Arithmetic Circuits

The final exact inference method we discuss is based on the observation that the computations that are needed to compute $P(X|E)$ can also be represented by a single polynomial [23, 24]. When we use abbreviation $\phi_{X|Pa(X)}$ for $P(X|Pa(X))$, and λ_{x_i} as evidence indicators for X, then the following formula represents the *network polynomial*:

$$f = \sum_{\mathbf{x}} \prod_{i} \lambda_{\mathbf{x}} \phi_{x_i|Pa(x_i)}. \tag{3.11}$$

The summation has to be interpreted as the sum over all possible assignments to variables X. In the case of Fig. 3.1, the polynomial is equal to:

$$f = \lambda_a \lambda_b \lambda_c \phi_a \phi_{b|a} \phi_{c|b} + \lambda_a \lambda_b \lambda_{\bar{c}} \phi_a \phi_{b|a} \phi_{\bar{c}|b} + \cdots + \lambda_{\bar{a}} \lambda_{\bar{b}} \lambda_{\bar{c}} \phi_{\bar{a}} \phi_{\bar{b}|\bar{a}} \phi_{\bar{c}|\bar{b}}, \tag{3.12}$$

where we use $\bar{a}$ for $A = F$. Queries on the belief network such as $P(X|E)$ can be translated into computations on this polynomial. The first observation is that

$P(E) = f(E)$, for evidence E. The evidence is first translated in setting the appropriate λ-values to 1 or 0 and then these values are substituted in the polynomial f. The second observation is that $P(X|E)$ can be computed as a partial derivative of f:

$$P(X = x|E) = \frac{1}{f(E)} \frac{\partial f(E)}{\partial \lambda_x}. \tag{3.13}$$

Direct computation of f and its partial derivatives is at least as complex as bucket elimination or clique-tree propagation. The advantage of the approach is that polynomials such as f can be compiled into an efficient *arithmetic circuit* on which the computations can be performed much faster. An arithmetic circuit over a set of variables V is a rooted directed acyclic graph in which the leaves contain variables in V or constants and in which the inner nodes contain either multiplication or addition operators. Evaluating an arithmetic circuit is straightforward. It consists of first fixing all variables to a value and then propagating the values upwards from the leaves to the root, via the multiplication and addition nodes. Computing a partial derivative is also possible, but it requires two passes through the network. The upward pass is linear in the number of edges in the graph, the downward pass involves more computations and its complexity depends on the maximum number of child nodes.

For every polynomial, several equivalent circuits can be constructed. Since the structure and size of the circuit determines the complexity of computations, it is important to find a minimal arithmetic circuit. It appears that such a circuit can be computed using propositional logic. It requires three steps. (1) The polynomial is encoded into a propositional theory. (2) The theory is compiled into a boolean network with special properties. (3) The boolean network is decoded into an arithmetic circuit.

The advantage of arithmetic circuits is that they can make efficient use of regularities in the structure of a belief network and inside the probability tables. Moreover, the circuits allow types of queries that are computationally demanding in message-passing algorithms, such as queries concerning sensitivity analysis. A limitation of the approach is that it can only be applied to discrete variables.

3.2.6 Approximate Inference by Sampling

When a belief network is too large, or has a too large clique size to allow exact inference, it is possible to apply methods that approximate updated beliefs.

The simplest way of approximate inference is called *forward sampling*. In this approach, the nodes are first ordered topologically (parents come before children). Then for each root node X, with probabilities $P(X = x_i)$, a random number p is drawn uniformly from the interval $[0, 1]$. If $p > \sum_{1 \leq j < i} P(X = x_j)$ and $p \leq \sum_{1 \leq j \leq i} P(X = x_j)$, X is set to value x_i. For nodes with parents, sampling takes place according to the conditional probability $P(X|Pa(X))$ and the already sampled values of the parents. When all nodes in the network are sampled this way, a counter for each of the selected values per node is increased by one. Then sampling starts all over at the root nodes. The procedure is continued until enough samples are drawn to allow an approximation at the requested level of confidence.

There are two problems with forward sampling: when probabilities are close to zero, many samples are needed for a reliable approximation. Moreover, what to do with evidence? It is possible just to skip samples that do not agree with the evidence, but that can be extremely inefficient.

The second problem can be tackled by applying *Gibbs sampling* [68]. In this approach, first one node is selected at random and then a new value for this node is sampled, based on $P(X|Pa(X))$ and the values that the parent nodes have at the moment. This is repeated for all nodes in an arbitrary order. After the samples, the counts for all values are updated and a new sample round starts. Nodes that bare evidence are not selected for re-sampling. To start off Gibbs sampling, a start value has to be provided to all nodes. When the start configuration is not representative for the belief network, it can take a long time before a reliable approximation is reached, and unfortunately, it is difficult to predict how many samples are needed exactly. Moreover, it is possible to get trapped in a local minimum.

Many techniques have been developed to increase the efficiency of sampling in belief networks, such as importance sampling, or other methods inspired on Monte-Carlo Markov Chain (MCMC) sampling (e.g., blocked Gibbs sampling [89]). In general, efficient sampling is NP-hard, but for certain restrictions on probabilities and certain network structures, optimal sampling is possible.

3.2.7 Parameter Learning

An important feature of belief networks is that they can be learned from observed data. Learning a belief network can be separated in two cases: (1) given a belief network structure, learn the conditional probabilities (this is called *parameter learning*) and (2) given a set of variables and a set of observed values, learn the structure and the parameters of the most probable belief network. The latter is called *structure learning* although parameters are learned at the same time.

We can also make a second distinction. The data from which structure and/or parameters have to be learned can be complete (meaning that for all instances the value of all variables is observed) or can involve missing data. The second case can include variables that are sometimes unobserved or hidden variables that are never observed.

The case of parameter learning with complete data is straightforward. It has been proven that the maximum-likelihood estimator $\hat{\theta}$ for the conditional probability $P(X = x_i|Y_1 = y_1, Y_2 = y_2, \ldots, Y_n = y_n)$ is equal to:

$$\hat{\theta} = \frac{\#(x_i, y_1, y_2, \ldots, y_n)}{\sum_j \#(x_j, y_1, y_2, \ldots, y_n)}. \tag{3.14}$$

So it is sufficient to count the number of occurrences in the data for each value of X for all possible configurations of X's parents. In the case that data is missing, this straightforward approach is not suitable anymore. A possible way to handle (randomly) missing data is expectation-maximization (EM) [35]. In this procedure an expectation step is alternated with a maximization step until a satisfactory result is achieved. In the E-steps, the total counts $\#(\cdot)$ for variables with missing data are estimated, based on the current $\hat{\theta}$'s and the observed data. In the M-steps, new values for all $\hat{\theta}$'s are computed, based on the new counts.

For small data sets, a bayesian learning approach might be more appropriate than maximum-likelihood estimation. It means learning the posterior probability distribution of the parameters, given the data. In practice, bayesian learning can be performed by adding nodes representing the parameters themselves to the network, and using standard belief-network inference to determine their distributions [18, 79, 87].

3.2.8 Structure Learning

In this subsection we will concisely summarize the different techniques that exist for learning a belief network structure from data. A detailed summary of the differences of those methods can be found in [22, 48]. The existing techniques are arranged around two main approaches: (1) scoring-based methods [43] and, (2) constraint-based methods [88]. Hybrid methods [1, 25], also exist but we do not further discuss them here.

In scoring-based methods the goal is to find the graph that best matches the data by first introducing a scoring function that evaluates each network with respect to the data, and then searching for the best network according to this score. A good example of such is the application of the STAGE algorithm of Boyan [12]. A neighborhood structure over graphs is defined and the searching is done via a heuristic search method such as hillclimbing, through this space of graphs. Some commonly used scores are the Bayesian scoring metric [19], and the minimum description length (MDL) principle [52].

Instead of searching in a space of directed acyclic graphs, it is also possible to look for graphs that have directed as well as undirected edges. The undirected edges represent situations in which the direction of the arrow is irrelevant for the structure since the graph is equivalent in either case. So, these graphs form *equivalence classes* of directed graphs. Using equivalence classes reduces the graph search space considerably.

In constraint-based methods one tries to match the conditional independence relations observed between variables in the data with those entailed by a graph, based on statistical independence tests. These independencies follow from the missing edges within a belief network structure. Most of these algorithms first try to come up with the *skeleton* of the network, i.e., the underlying undirected graph, after which the orientation of edges is determined. Usually, the algorithm starts with the empty graph, and in the first step, sets of variables are exhaustively sought to separate any two variables. If no such set is found, an undirected edge is added between these variables. In the next steps a maximum amount of edges is oriented, so as to respect the independencies found in the data. Finally, the result of a constraint-based method an graph equivalence class. Examples include the PC algorithm [88] and the IC algorithm [70]. For a comparison between the two techniques see [21].

Important issues when learning a structure include whether the data is complete or incomplete [36], whether there are hidden variables [32, 35], and whether the learning is on DAGs or equivalence classes [15].

3.3 Variants and Extensions

This section starts with discussing some variants of Belief Networks that superpose additional semantics on the basic BN's. In *Dynamical Belief Networks* (DBNs) [97], each variable is modeled at a series of time points. *Hidden Markov Models* (HMM) [8] are special cases of dynamical belief networks that are applied at large scale in bioinformatics. Their restricted structure allows for specialized, efficient learning and query algorithms.

Next to adding semantics on top of Belief Nets, variants are created by abstraction. We show, for instance, *Qualitative Belief Networks* (QBNs) [93] that abstract BNs by putting qualitative restrictions on conditional probabilities. These QBNs can be defined manually when little data is at hand and can be transformed into traditional BNs when more data comes available.

A recent extension of BNs that we treat in this section is the use of relational models to define frameworks for belief networks. In *Probabilistic Relational Models* (PRMs) [82], classes of objects and their relations are used to specify a framework of general probabilistic dependencies. A concrete set of instances then produces a concrete BN that can be queried using standard inference. Specialized learning algorithms are used to learn the general probabilistic framework. PRMs are useful in (large-scale) data integration.

A more general extension is the *Relational Bayesian Network* (RBN) [44], that uses a logic language to describe the probabilistic framework. From this framework and a concrete input structure, a BN can be constructed for inference. RBNs can describe both DBNs and PRMs, but can also be applied to model heredity relations in linkage analysis.

3.3.1 Dynamic Belief Networks

In this section we provide a temporal view of the static belief network. A standard belief network represents a static situation through the joint probability of a set of variables. One way to add a dynamic aspect to belief networks is to discretize time into a number of *time slices* and have *a copy* of each variable X in the belief network for each time slice $(X_{t=0}, X_{t=1}, X_{t=2}, \cdots)$. Arrows between variables in different time slices indicate the temporal dependencies that model the dynamic aspects. By introducing time slices, many options come available and are applied: e.g., (1) time slices can be equal or change, (2) arrows only run from one time slice to the next (cf. Markov property) or run over larger times (3) arrows within a slice can exist or be absent, and (4) the number of slices can be finite or infinite.

Of course, the resulting *Dynamic Belief Networks* (DBNs) are in principle just belief networks, so all standard inference and learning algorithms apply. However, the special structure of DBNs allow specialized inference algorithms that make use of the special, repetitive structure. For instance, it is not necessary to represent all time slices during inference at the same time: only a small window of a few time slices is needed for the computations. Moreover, such an approach allows handling infinite-time DBNs. For a more detailed elaboration on DBNs we refer to [64, 79, 97]. In the next subsection we go into a special case of DBNs that allows even more efficient algorithms.

3.3.2 Hidden Markov Models

Hidden Markov Models (HMM) provide a means and a representation to understand a stochastic time-series process. Good examples are sequence analysis (cf. subsection 3.5.1) and, outside bioinformatics, natural language processing and automatic speech recognition [8]. The underlying idea of HMMs is that an observed sequence of events is generated by a system that can find itself in a finite number of states. At each point in time the system makes a transition to another state, and an observed outcome is generated or emitted. The state transitions and outcomes are governed by probabilities. it is Important to note is that the system states are not visible to an external observer and are therefore called *hidden*.

We define a HMM formally as follows. A HMM is a five-tuple (S, O, A, B, Π), in which S is a finite set of system states s_i. We denote the actual system state at timestep t by q_t with $t = 1, 2, \ldots$. Π contains the initial state probabilities: $\pi_i = P(q_1 = s_i)$.

Matrix A contains the state transition probabilities $a_{ij} = P(q_t = s_i | q_{t-1} = s_j)$. So a_{ij} is the probability that the system is in state s_i at timestep t, given that the system was in state s_j at timestep $t-1$. This probability is the same at each timestep. Set O contains the symbols v_j that can be emitted by the system (the outcome). We denote the observed outcome at timestep t by o_t. Finally, B is defined as the outcome probability for each of the states s_i: $b_{ij} = P(o_t = v_j | q_t = s_i)$. In the case of continuous outcomes, continuous probability density functions are used.

The Markov property of HMMs dictates that the current state is only dependent on the previous one: $P(q_t | q_{t-1}, q_{t-2}, \ldots, q_1) = P(q_t | q_{t-1})$. This is the so-called *memory* effect. More precisely, it means that the previous state alone gives information of the future behavior of the process, i.e., knowledge of the entire history of the process does not add any new information. Moreover, for the observations we also have,

$$P(o_t | q_t, o_{t-1}, q_{t-1}, o_{t-2}, q_{t-2}, \ldots, o_1, q_1) = P(o_t | q_t), \qquad (3.15)$$

which means that observation only depends on the hidden state of the system. As opposed to Markov processes, system states in a HMM are not observable, only the observations v_i emitted by the states can be observed. A HMM is a natural extension of a Markov process.

An example clarifies these ideas (taken from [3]). Suppose we have two urns containing balls of a different color, say, red (r) and green (g). Somebody draws balls from them and shows us their color. Let q_t denote the color of the ball drawn at timestep t. Figure 3.5 shows a Markov process which models this example. Assume

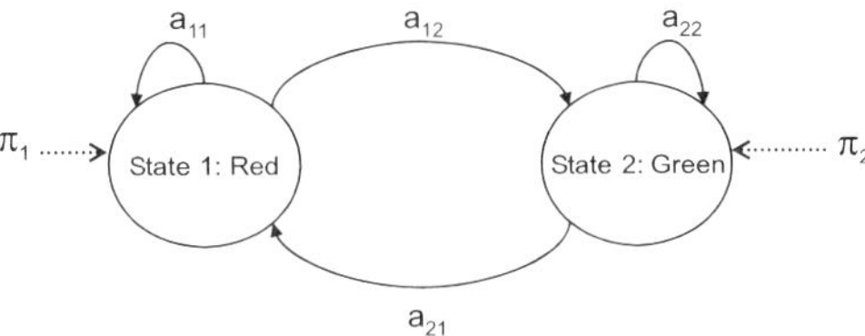

Fig. 3.5. An example of a Markov process in which state sequences are observable. More precisely, each state represents an urn containing either red balls (state 1), or green balls (state 2). It is now straightforward to calculate the probability of a sequence, given the model.

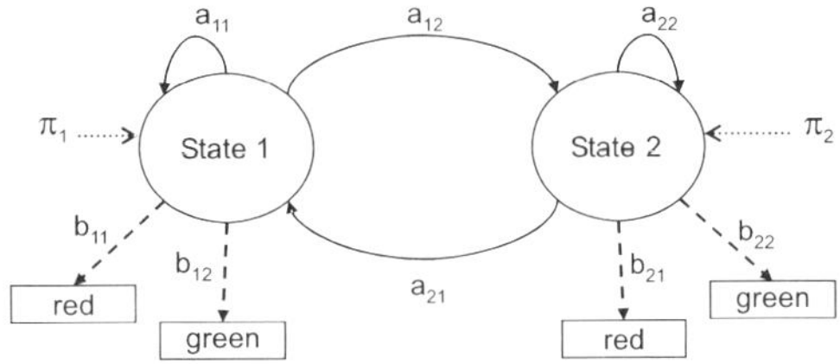

Fig. 3.6. An example of a HMM in which state sequences are not observable. More precisely, we now model the situation in which both urns can contain as well red balls as green balls. Each state has now emission probabilities for the occurrence of a green or red ball. In our example these are (b_{11}, b_{12}) and (b_{21}, b_{22}).

that a sequence "r,g,r" was generated, its probability of occurring is easily computed from π and A, because we know that the state sequence causing the outcomes was $\{s_1, s_2, s_1\}$. Assume now, that the two urns each contain balls of both colors. This situation corresponds to the *hidden* situation. More precisely, we cannot see which state sequence generated an observation sequence as for instance "r,g,r". This situation corresponds to the HMM depicted in Fig. 3.6.

There are three main tasks that can be solved with HMMs. (1) Determine the probability $P(O|H)$ of an observed sequence $O = (o_1, o_2, \ldots)$ given a HMM H. There exists an algorithm, named *forward-backward*, that can efficiently compute this probability. (2) Given a HMM H and a sequence O, find the corresponding state sequence $Q = (q_1, q_2, \ldots, q_t)$, with the highest probability of generating O. An algorithm for this is the *Viterbi algorithm*. (3) Given a training set of observed sequences $T = \{O_k\}$, determine the corresponding HMM H. An algorithm to learn a HMM from data is the *Baum-Welsh* algorithm (an EM variant). For an extensive discussion of these algorithms we refer to [3].

The main difference between HMMs and DBNs lies in the fact that a HMM only uses one discrete random variable for a hidden state, while in a DBN a set of discrete or continuous random variables can be used to represent states and observations. In order to translate a HMM into an equivalent DBN, we re-define the transitions A, the observations B and initial state distribution Π of HMMs. In a DBN, q_t and o_t are each represented by a *set* of variables, which we denote by Q_t and O_t. We define $Z_t = Q_t \cup O_t$ as the set of nodes in time slice t. We use variable $z_t^i \in Z_t$ for

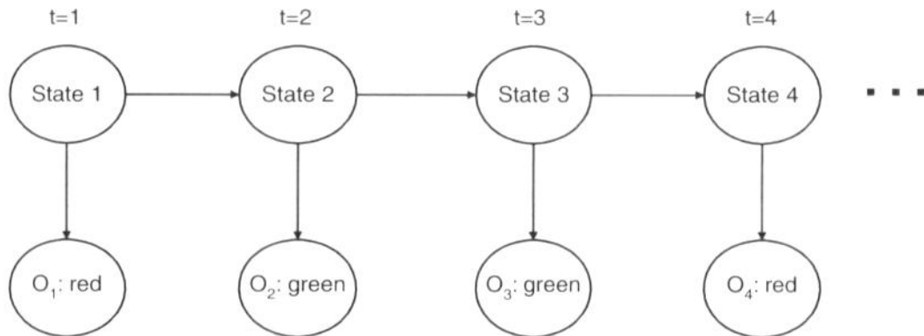

Fig. 3.7. An example of a HMM represented or "unrolled" as a DBN for the sequence $(r, g, g, r, \ldots)$

the ith node at timestep t, hidden or observed. Now we can compute $P(Z_t|Z_{t-1})$ as $\prod_i P(z_t^i|Pa(z_t^i))$. The initial state distribution $P(Z_1)$ can be represented by a regular bayesian net. Figure 3.7 provides a DBN representation of the HMM of Fig. 3.6 unrolled for four timesteps.

Since DBNs allow states and observations to be represented by an arbitrary number of variables that can show conditional independencies, DBNs often need less parameters than a HMM for the same situation.

3.3.3 Qualitative Belief Networks

It is not always possible or necessary to obtain the conditional probability distribution for a belief network. Qualitative Belief Networks (QBNs) [93] allow a kind of probabilistic belief update using qualitative probability constraints. A QBN consists of a set of variables X of which the values can be ordered from low to high. (The amount and type of values is not relevant.) The variables are connected in a DAG, expressing conditional independencies similar to the standard belief network. Instead of conditional probability distributions, qualitative influences are expressed. There are four types of influences:

- $(X,Y,+)$: for high values of x, $P(x|y_1,z) > P(x|y_2,z)$ if $y_1 > y_2$,
- $(X,Y,-)$: for high values of x, $P(x|y_1,z) < P(x|y_2,z)$ if $y_1 > y_2$,
- $(X,Y,0)$: $P(x|y_1,z) = P(x|y_2,z)$, for all x, y_1, and y_2,
- $(X,Y,?)$: covers all other cases.

The influences in a network are indicated by labeling the arcs in the DAG with either '+' or '−'. From these labels, the qualitative influence between all pairs of variables can be derived. First, influences are *symmetric*: if (X,Y,δ) then (Y,X,δ). Second, chained influences are computed by an $\otimes$ operation, and third, combined influences by an $\oplus$ operation. The $\oplus$ operation results in a '?' if '+' and '−' are combined. An efficient message-passing algorithm exists that can compute a qualitative belief update for each node given some observed evidence. The output of the algorithm is a labeling of the nodes with '0', '+', '−', or '?', meaning "unchanged", "higher probability on higher values", "lower probability on higher values", and "ambiguous". A problem with QBNs is that for larger networks, most nodes will be labeled by '?'.

When a little more information on the conditional probabilities becomes available, a QBN can be extended into a *semi-qualitative belief network*. We mention two approaches. The first one [29] is to change the '+' and '−' signs into a (real) number indicating the strength of the influence and use these numbers to construct conditional probability tables. In fact, the QBN is translated into a standard belief network on which exact inference is applied. The second approach [77] is to combine nodes with full conditional probability distributions with nodes that only have qualitative influences from their parents. An algorithm exists that can propagate *probability intervals* for this type of networks.

3.3.4 Probabilistic Relational Models and Relational Belief Networks

The variables that are represented in a belief network need not to be attributes of a single entity. In many cases, the variables are attributes of several entities that have some type of probabilistic interaction. For instance, the spatial constitution of a protein together with the occurrence of certain motives in the upstream region of a gene together determine whether the specific protein will be connected at a promotor site for the gene.

When many entities are involved in the same belief network, for instance in the case of microarray analysis or linkage analysis, it can be useful to group similar entities into entity types, to group similar relations between entities into relation types, and to define conditional probabilities in terms of attributes of these entity and relation types. The probabilistic relation, for example, between the haplotype of a child and those of its parents is equal for all children in a pedigree. A *Probabilistic Relational Model* consists of two parts. The conceptual part of a PRM contains entity types (called *classes*) with their attributes and binary relation types (called *references*), similar to an entity-relation model for relational databases. The second, probabilistic, part of a PRM defines the probabilistic relations between attributes of entities participating in a relation, possibly via a chain of relations. These probabilities are expressed in a similar way as the local conditional probability distribution in belief networks, for instance by a probability tree. The PRM can now be combined with a set of concrete entities and relations, such as an actual pedigree and actual genetic markers. It results in a network of concrete interactions between entities (the *relational skeleton*). The concrete probabilistic relations between the attributes of these entities are derived from the PRM and constitute a classical belief network. It means that a PRM together with a concrete set of entities and relations produces a belief network. This belief network can be queried using the inference algorithms that we discussed in the previous section.

Just as the parameters of belief network can be learned, it is possible to learn the probabilistic part of a PRM from data. It is not necessary to use the derived belief network for this. The parameters in the generalized probability tables can be learned directly. Since all entities in a given type are assumed to be related in a similar way, the observations of all these entities can be counted together in order to estimate the conditional probabilities.

A *Relational Belief Network* or RBN [44,45] is a somewhat more general model. RBNs also have two parts. The first part only contains a set of *relation types*. There are no separate entity types. Relations can be unary (i.e., *attributes*), binary, or have a higher degree. Since attributes are either present or absent, RBNs only allow binary variables. The second part of an RBN defines conditional probabilities as functions in terms of these relations. The language used for these functions is derived from a probabilistic logic. An example is:

$$F(v) : \text{noisy-or}\{0.3|w; edge(w,v) \wedge blue(w)\} \tag{3.16}$$

which indicates that the conditional probability table of node v is a noisy-or involving all blue nodes w connected to v. Just like in PRMs, a concrete belief network is generated from a RBN as soon as a concrete set of entities along with their relations is provided (called the *relational structure* in RBNs). However, it is possible to translate an RBN directly into arithmetic circuits for querying without explicitly generating a belief network.

The field of probabilistic logic has produced a range of additional languages that combine first-order logic with belief networks. We just want to mention the *Multi Entity Belief Networks* (MEBNs) [53] that are closely related to RBNs and PRMs. MEBNs also allow only binary variables, but they do allow the definition of *contexts* under which certain conditional probabilities are valid.

3.4 Building Belief Networks by Hand

In some cases, data is not sufficiently available to learn a reliable belief network, at least for some part of the network. In those cases belief networks have to be build manuallyfrom domain knowledge. This knowledge can originate from literature, databases, or domain experts. Once the network is constructed, learning techniques can then be used to fine-tune probabilities or improve structural elements at those parts of the network for which sufficient data is (or comes) available.

The manual construction of a (complex) belief network can be regarded as a *knowledge engineering* task [62]. This means that the task should contain a requirements-analysis phase, a model-building phase, and an implementation and evaluation phase. The requirements phase is equal to those of other knowledge-engineering tasks. It determines the general mission and goals, a specific objective for the network (e.g., diagnosis or prediction), available knowledge sources (experts, data, literature) and technical requirements (belief network tools, speed, embedding into other tools).

The model-building phase is specific for belief networks. It consists of identifying all variables and their states, fixing the belief-net structure, and the elicitation of conditional probabilities. The variables that have to be identified can be divided in three groups. The "focus" or "query" variables will be identified first; they follow directly from the network objective. The so-called "observables" will provide the evidential input for queries. These variables can be identified from the availability of observable data. Finally, "intermediates", that link observed evidence to the focus variables, have to be identified gradually, based on available knowledge of cause and effect. For each of the variables, an appropriate set of states must be defined. Special care has to be taken to achieve the correct number of states in qualitative variables and continuous variables that are to be discretized.

The second task of model building is setting up the structure of the belief network. Here a balance has to be found between parsimony (less arcs, leading to less probabilities to elicit) and model detail (more arcs). The direction of each arrow has to be decided upon, but a causal direction leads often to more conditional independencies and an easier to understand belief network. During the setting up of a structure, new intermediate variables might be introduced, or additional states can be needed for a variable.

The final task is the elicitation of probabilities. In hand-built belief networks, it is common to use approximate probability distributions such as the noisy-or gate when a node has more than one parent. Although the parameters of these gates are easier to elicit than full probability tables, they need severe assumptions on independence of common causes. The actual elicitation of probabilities often involves estimations by domain experts, which can be difficult [91]. Tools such as probability wheels, probability scales with textual anchors, and gambles exist to help the expert in this task. A different approach to probability estimation is to start with a qualitative belief network, since qualitative influences are often easier to elicit.

As in any large-scale design problem, *decomposition* might facilitate the model-building. For belief networks there are top-down and bottom-up approaches to decomposition. A top-down approach is to divide the belief network in loosely connected modules and build the modules independently. Of course the modules need to fulfill certain demands in order for the whole to be a valid belief network. Bottom-up approaches first define local dependencies between variables and then combine them into a belief network. Both object-oriented belief networks and probabilistic relational methods such as PRMs, RBNs, or MEBNs can be applied for this. We also want to mention the possible use of ontologies and other knowledge bases. For instance, PR-OWL [20] can be used to describe a probabilistic ontology, based on the MEBN language.

When the model is built, an evaluation of the network is needed [62]. First, a validation or verification can be done by means of a walk-through with domain experts of the network structure and probabilities. Next, a sensitivity analysis of the network's probabilities can be performed, for instance by studying the explanation of a belief network output (i.e., *importance analysis*). This can reveal inconsistencies or unintended side-effects. The final evaluation is the analysis of the behavior of the belief network in a representative set of well-understood cases.

The existing belief-network tools, such as NETICA and HUGIN support most parts of the knowledge-engineering process of manually constructing belief networks. Additional tools for knowledge management might be needed in cases when the elicitation process is particularly difficult.

3.5 Applications in Bioinformatics

As the proverb suggests: the proof of the pudding is in the eating. In this section we review some recent application of BNs and variants in Bioinformatics. These applications can be divided roughly into three domains: sequence analysis (discussed in Subsection 3.5.1), biological networks (Subsections 3.5.2 and 3.5.3), and linkage analysis (Subsection 3.5.4). We end the section with a general discussion on belief networks in bioinformatics (3.5.5).

3.5.1 Sequence Analysis, Protein Prediction, and Motif Finding

In the area of sequence analysis and protein prediction, belief networks are mainly used for *classification tasks*. The networks predict, given a sequence, the secondary structure of a protein, the presence of binding site, the presence of an operon, and so on.

The first known application of belief networks (or HMMs, in fact) in bioinformatics is the work by Delcher *et al.* [28] on the secondary structure prediction. Currently, HMMs are used at a regular base in bioinformatics and are described even in introductory handbooks. The Computational Biology group at the University of California in Santa Cruz, for instance, provides a range of online HMM tools (called SAM[2]) that perform multiple sequence alignment, secondary protein structure prediction, module prediction, protein prediction, and homology finding. In [76] elaborated versions of HMM-like belief networks are presented that can be used to recognize remote homologues and protein folds. These belief networks not only use a DNA or protein sequence for input, but also include secondary structure, residue accessibility. Finally, HMMs are also being used for the construction of phylogenetic models (cf. phylo-HMMs [85]).

As an example we explain the use of HMMs for multiple sequence alignment (see, for instance, profile HMMs [31]). A HMM that is trained for a set of sequences represents a kind of profile over these sequences. When sequences from the set are processed by the HMM, it will not only indicate where insertions and deletions are most likely but also will produce a score. The Hidden Markov Model for multiple alignment has three states per residue in the alignment (see Fig. 3.8 left). One state for a *match* (square), one for *insertions* (diamond) and one for a *deletion* (circle). The match state indicates whether there is a residue in the profile and in the sequence. In the belief network (Fig. 3.8 right), these states are represented as values of the hidden state variables S_i. The conditional probability tables for state transition and emission probability in the HMM's belief network result in position-dependent gap and insertion penalties, and position-dependent character distributions. This provides an advantage over other tools for multiple sequence alignment, such as profiling. Moreover, a HMM is adaptive: new sequences can be added after training to improve the model.

Below we give three other examples of the use of belief networks in sequence analysis. The first example is the prediction of operons in prokaryotes (*E. Coli* in this

[2]http://www.soe.ucsc.edu/research/compbio/HMM-apps/HMM-applications.html

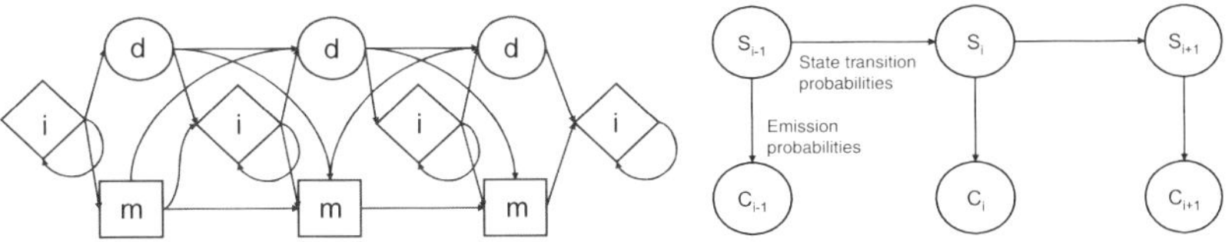

Fig. 3.8. A HMM for multiple sequence alignment and its belief network representation

case). The task of the belief network in the approach described by Bockhorst *et al.* [10], is to predict whether a group of genes forms an operon, i.e., whether they are regulated as a unit. The belief network combines evidence derived from the sequence (e.g., number of genes in the operon, within-operon spacing, and codon usage), with evidence from gene expression data. The belief network structure is hand-crafted, and contains 20 nodes. The parameters are trained using Laplace estimates. The network outperforms alternative classifiers such as Naive Bayes and C5.0.

The second example is the use of belief networks for the prediction of factor binding sites, especially in cases in which a single DNA sequence is to be investigated. In contrast to the purely sequence-based approaches, such as PSSMs (Position-specific scoring matrices, also called PWM, position-weight matrix) that score for recurrent motifs, the approach of Pudimat *et al.* [74] combines several sources of information, resulting in a large set of candidate features. These include the physical structure of the DNA near the binding site and multiple PSSMs for co-acting factors. The approach has two alternating phases: (1) feature selection, using sequential floating feature selection, which consists of extending the belief network in each step with the most informative new feature and removing non-informative other ones; and (2) structure learning, in which the belief network structure is restricted to trees in which all variables have no more than one parent (called tree-augmented networks). The approach outperforms the traditional PSSMs on sequences from MEF-2 and TRANSFAC databases. Another approach to predicting factor binding sites is to construct an extension of belief networks that allow a variable sequence context [9]. The variable-order belief network (VOBN) also outperforms PSSMs, but has not been compared to the multi-feature approach of [74].

The third example is the use of a belief network for the enhancement of ChIP-Chip data analysis in [75]. In this approach additional chromatin immunoprecipitation data is combined with the standard ChIP-chip data and further sequential data in order to enhance the method's spatial resolution of finding factor binding sites. The so-called Joint Binding Deconvolution (JBD) model is a belief network with real-valued and discrete variables. Factors such as the binding events, binding strengths and binding influences are represented in the network. Given the size and complexity of the network, the authors propose a novel approximative inference algorithm that can handle discrete and continuous variables. Experimental results show that the JBD approach indeed makes better predictions than the existing methods Rosetta, MPeak, and Ratio.

3.5.2 Gene Regulation Networks

Belief networks in the area of gene regulation networks and other biological networks are used to *model* these biological networks. Constructing the model is an act of *reverse engineering*. Once the model is constructed it can be used to *simulate* and *predict* the behavior of these networks.

[26] provides a general overview of computational techniques for discovering, modeling, and simulation gene regulation networks. The central question here is: how does the expression of one gene influence the expression of other genes? Since gene regulation involves interaction between DNA, several types of RNA, proteins and protein complexes, and other molecules, it is a very complex process. Moreover, biological data is very heterogenous with respect to availability, quantity and resolution. The computational techniques each apply a specific level of abstraction, which allows analysis only at a certain scale and with a certain quality. Boolean networks, for instance, are among the earliest techniques in the field. They use a high level of abstraction, allowing the analysis of large regulatory networks. Detailed partial differential equations, at the other hand, can model the physical interaction between the biological molecules, including spatial and dynamical aspects, allowing the analysis of only a handful of genes. Belief networks are used at several levels of abstraction in the domain of gene regulation networks, but mainly at a coarse level, trying to discover existing regulatory relations between genes. An advantage of the application of belief networks in this area is that by their probabilistic nature they can deal with noise in the data that is caused by biological diversity, lack of detail (abstraction), and measurement error.

The standard approach is to derive gene regulation networks from mRNA expression levels by using structure learning in belief networks (cf. [38]). In this approach the variables in the network represent genes and their values are the level of expression (e.g., *unexpressed*, *normal*, and *over-expressed*), as measured by micro-array analysis. The gene-regulation process is abstracted to probabilistic relations between expression levels of genes. The approach does not model, for instance, the dynamic aspects of gene regulation. The causal interpretation of the belief network is used as a semantic model for the direction of regulatory influence in gene networks. It is possible to include context variables in the network, such as experimental conditions, cell-cycle, etc. In principle, standard structure-learning techniques can be applied. The arrows that are learned this way might indicate possible direct relations between genes, but more likely they do not refer to physical interaction. It is the set of conditional independencies modeled by the structure that is more informative.

In many cases the number of involved genes is large (>1000) and the number of data points is low (<50). So the standard structure-learning techniques can hardly be applied with success. (This is of course also true for many other techniques in the domain.) While it is impossible to learn a reliable complete structure for a network when only little data points are available, it is possible to learn parts of the structure. This still can lead to the discovery of non-trivial regulatory relations between genes. Friedman, for instance, proposes to restrict the number of parents per gene. (cf. the sparse candidate algorithm [39]). Peña *et al.* [71] propose to reduce the scale by focussing on the Markov neighborhood of interesting genes. Their method asks the user to provide a seed gene S and a positive number R. It returns a belief network with genes that depend on S, but such that at most R other genes mediate in the dependency. Friedman *et al.* [38] reduce scale by looking for genes that have a large set of descendants in the graph and are therefore particular influential (or dominant), without having to determine the precise structure of the graph. They define a *dominance score* that appears to detect dominant gene in a robust way.

Another approach is to use *modules* to reduce the complexity of the structure learning task. This is done by dividing the network in modules and learning the structure per module [56]. A module can be understood as a set of functionally related and possibly co-regulated genes. These modules can be achieved automatically by using bi-clustering of expression data, by involving additional biological data such as ChIP data, promotor-sites, and data protein complexes. Modules can also be predefined using biological knowledge. Segal [81, 83] proposes a method to learn modules by modeling explicitly the common regulatory element per module. In this area, the application of probabilistic relational models (PRMs) has proven to be useful. The relation between genes, modules, expression levels and micro-arrays, for instance, lends itself nicely for a relational model (see Fig. 3.9). Moreover, the PRM approach allows introduction of additional data sources such as binding site motifs or protein-complex data (see below).

In [7] the method is extended to the discovery of *overlapping* cellular processes. The PRM in this approach (Coregulated Overlapped Process or COPR model) also models the processes as modules with a common regulatory program. To model overlapping, genes can be allocated to more than one module at the same time. To allow parameter learning with the large amount of latent variables involved, the authors propose a specialized EM algorithm. The method was successfully applied to yeast data and involved 2,034 genes, 50 processes, and 466 regulators. The modules found were validated using the Gene-Ontology annotations of the genes involved.

Recently, Chen *et al.* [14] propose an improved structure learning method for belief networks, specialized for gene networks. In their information-theoretic approach, first a loop-free undirected network is constructed based on mutual information (MI) between variables. Next they apply a node-ordering procedure to be able to apply an ordered version of the K2 algorithm. The authors applied their method to microarray

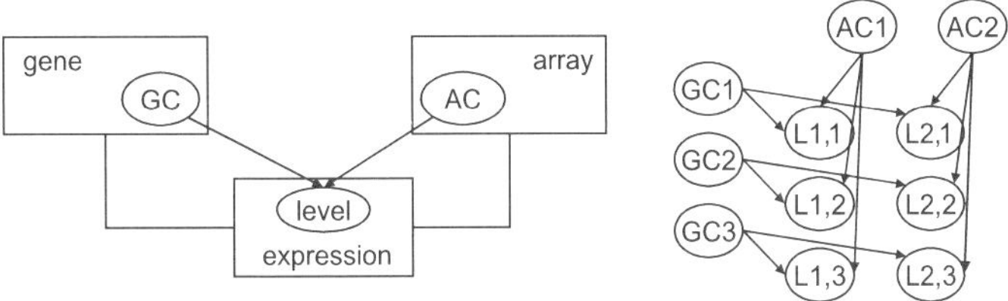

Fig. 3.9. A probabilistic relational model (left) and an instantiated belief network with three genes and two arrays (right) for bi-clustering of gene expression data. GC and AC stand for the attributes gene-cluster and array-cluster.

data (yeast cell-cycle genes) and compared the links found to those present in PathWayAssist. They only studied links in a group of 20 genes and were able to identify 65% of the experimentally known links using 77 data samples.

The MAGIC system by Troyanskaya *et al.* [90] does not apply learning, but uses a hand-built belief network that predicts a functional relation between pairs of genes based on expression data and other heterogenous inputs. The network is instantiated for every pair of genes by fixing the inputs to the evidence for the gene pair. Clustering is performed afterwards based on the belief network's prediction for each pair.

Yu *et al.* [96] propose to apply structure learning in *dynamic* belief networks to incorporate dynamic aspects of gene regulations. In their dynamic belief network, every variable represents a gene and appears in every time slice. Variables are only influenced by their parents and themselves in the *previous* time slice (i.e., first-order Markov). During their search for an optimal structure (using a genetic algorithm), arrows between parents and children are added or removed at all time slices simultaneously. The authors use an *influence score* to model the sign and magnitude of the regulatory influence. Their algorithm is tested on artificial data generated by a simulator (GeneSim). A similar approach is presented recently by Missai *et al.* [63]. Their improved method involves information-theoretic measures (MI), and allows inclusion of prior knowledge.

Perrin *et al.* [72], also describe the use of dynamic belief networks to model the dynamic behavior of a gene regulation network as well as its structure. The dynamics of their model are based on the inertial model by d'Alché. In Perrins approach, the dynamic network consists of a hidden space that represent the true (real-valued) states of the genes at each moment, and an observational space that represent the (real-valued) expression levels as measured. All gene-variables are interlinked between the time-slices of the DBN (which is in this case a Kalman filter). Structure learning is in fact performed by parameter learning (the EM algorithm). The gene network is deduced from the trained DBN by selecting only those connections between genes that show significant effect. The approach was applied to a small gene network only (E.Coli S.O.S. network).

3.5.3 Other Cellular Networks

Although belief networks seem to be employed most in gene regulation networks, they are also applied in cellular networks or pathways that involve proteins and other molecules. The general approach is similar to those of gene regulation networks, i.e., structure learning for reverse engineering. However, translating the causal interpretation of belief networks into biological relations can be more problematic. We discuss three examples below.

The first example involves the modeling of signalling protein networks on the basis of proteomic data [94]. The variables in the belief network in this approach represent the phosphorylation levels of proteins. The causal interpretation of the belief network represent the true biological causal relations that rule the signalling. E.g., phosphorylation of MEK1 *causes* phosphorylation of ERK1 and ERK2. The authors apply structure learning, restricting the network to biological possible configurations, but also restricting the maximum number of parents per node. The approach is tested on proteomics data of the MAPK/ERK signalling pathway. The structure of the obtained belief network agreed to a large extent with existing pathway knowledge. Moreover, a graphical presentation of the conditional probability tables in some of the nodes provides interesting biological insights. Finally, the belief network was used to model the effect of environmental changes in the cell such as the application of drugs.

The second example involves the discovery of molecular pathways using the combination of protein interaction data and gene expression data. In [81, 83], a partially directed PRM is used. The main assumption is that interacting proteins are more likely to be involved in the same pathway. Protein interaction, however is an undirected process, so a (directed) belief network would not be a correct model for these interactions. The authors therefore use a Markov Random Field to represent this part. Genes are used to represent the interacting proteins. It is combined with a directed model for the gene regulation aspects. The parameters from the combined PRM model are learned using an adapted EM algorithm. The results show that using the protein-interaction on top of expression data leads to an improved pathway prediction. Another approach to combining protein-interaction with expression data is provided by [95]. Their model is called a *physical model*, and is an extension of a Markov Random Field. It has potentials that incorporate decision variables which allow modeling *gene knock-out* experiments. The model incorporates binding site information (location data), protein-protein data and expression data from knock-out experiments. Moreover, additional variables such as the direction of knock-out effects are included in the model.

The third example is the generic modeling of biological networks as described by [40]. In their approach mRNA variables (gene expression), protein states in cytoplasma and nucleus, metabolics, stress and stimulator factors, are all included in one network. In the model, all biological variables have a finite set of states (e.g., expression level for genes and phosphorylation state for proteins). The variables are connected in a *regulatory dependency graph* in which each variable has a set of parents that regulate it. These regulations are modeled by conditional probability tables. Due to noise, variables may be observed through real-valued *sensor variables* that depend on the logical state of the biological variables. The network is represented by a *factor graph network* that is equivalent to a belief network when the regulatory dependency graph is acyclic. The network does not describe dynamic behavior. A special approximation algorithm is proposed that can deal with feedback loops that might occur in the network. Moreover, the authors describe learning algorithms to learn the probabilities of regulation and uncertain observations. As an example the authors study the acyclic HOG pathway and the lysine biosynthesis that has feedback loops.

3.5.4 Linkage Analysis and Genetics

The third main application area of belief networks in bioinformatics is the analysis of linkage in pedigrees. Linkage analysis is generally accepted to be a computationally hard task (cf. [73]). In this domain belief networks are used to model the probabilistic mechanisms of meiosis (segregation), crossing over, and uncertain observations (penetration) in order to determine the *most likely location* of a (disease-related) gene on a chromosome relative to a given set of genetic markers. Exact inference on the belief network can be used to compute the probabilities needed exactly [55]. Although probabilistic modeling is not new in the area (e.g., the well-known peeling algorithm is similar to variable-elimination in belief networks [42]), belief networks appear to be a rather useful tool. Figure 3.10 shows a small pedigree (one child and its parents) and the corresponding belief network for segregation and phenotype modeling of a single gene. The variables G represent the alleles of the gene on each chromosome, P models the resulting phenotype and S models the segregation process.

The most well-known tool for linkage analysis that uses belief nets is SUPERLINK [33]. This tool is also available in an online, parallelized version using the Condor computation grid [86]. The computation of exact probabilities in large cases requires specialized optimization algorithms that make use of the specific structure of belief networks that represent pedigrees. These include the computation of optimal node-elimination orders [34].

When pedigrees are too large for exact computations, even in the parallelized form, efficient approximation algorithms are available. For instance, the specialized blocking Gibbs sampling [47, 89], the improved MCMC method of [92], and the cluster variation method [2, 58] exploit the special structure of linkage belief nets to enhance computational speed and convergence behavior of approximations.

Belief networks for linkage analysis tend to grow very large. They have, however, a rather regular structure, reflecting the pedigree relations, and the spatial relations of markers. Both relational belief networks (RBN), probabilistic relational models (PRM), and multi-entity belief networks (MEBN) can use this structure to achieve a compact and generic representation of belief networks. In [46] an example application to linkage analysis is provided. In principle, RPMs and RBNs are translated

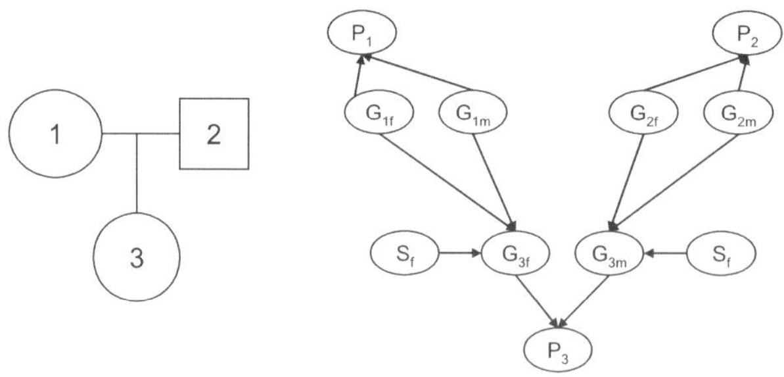

Fig. 3.10. A small pedigree and the corresponding belief network

into belief nets before probabilistic inference can take place. Recent developments, however, show that RBNs can be compiled directly into fast arithmetic circuits for efficient inference in linkage analysis [13].

3.5.5 Discussion

As this section reveals, belief networks are applied in broad areas of bioinformatics. They are used for classification, network modeling, and probabilistic inference. In linkage analysis, belief networks are constructed by hand, in sequence analysis, the parameters are learned from data and in network modeling, the structure is learned from data. It is interesting to observe that the domain of bioinformatics has led and will continue to lead to the development of specialized representations and algorithms.

The description of the different kinds of belief networks and variants in this chapter, together with their possible applications (even) in bioinformatics, makes it clear that this rather broad field is a very active and promising area of research. However, it also stresses that the field is very much in flux, and that it is rather hard to unravel who is contributing exactly what, how, and to which agenda. This leads to visible effects even within the domain of bioinformatics: algorithms that were developed for linkage analysis, did not yet find their way to the area of gene regulation networks, for instance. Therefore, we believe it would be very beneficial for the field to rethink the problems which are trying to be solved and cast them into explicit research agendas or a taxonomy of issues in belief networks and related graphical models.

In the field of bioinformatics, there are still areas in which belief networks (and related concepts) are not being applied to their full extent. As an example, belief networks can be of large help in text-mining for bioinformatics. But also in areas in which the application of belief networks is well studied, new and existing results within belief networks still need to be considered. As an example hereof we mention the possible use of (semi-)qualitative belief networks for the coarse modeling of biological networks, or the application of arithmetic circuits in belief networks for linkage analysis.

3.6 Software Tools

Since the successful application of any technique depends on the availability and quality of tools, we review a series of belief network tools in this section. On the internet a number of lists of belief-network tools are available. Not all of them are well-maintained, unfortunately. We mention the pages by Murphy[3] and Aha[4], the Google directory on belief network software[5], the Wikipedia page on belief

[3] http://bnt.sourceforge.net/bnsoft.html

[4] http://home.earthlink.net/~dwaha/research/machine-learning.html

[5] http://directory.google.com/Top/Computers/Artificial_Intelligence/Belief_Networks/Software

ACE	http://reasoning.cs.ucla.edu/ace/
ANALYTICA	http://www.lumina.com/
BAYESIA	http://www.bayesia.com/
BNARRAY	http://www.cls.zju.edu.cn/binfo/BNArray/
BNJ (Belief nets in Java)	http://bnj.sourceforge.net/
BNT (Bayes nets for Matlab)	http://bnt.sourceforge.net/
BUGS, WINBUGS, etc.	http://www.mrc-bsu.cam.ac.uk/bugs/
DEAL	http://www.math.aau.dk/novo/deal/
GENIE and SMILE	http://genie.sis.pitt.edu/
GENOMICA	http://genomica.weizmann.ac.il/
GR	http://www.ci.tuwien.ac.at/gR/
HUGIN	http://www.hugin.com/
JAVABAYES	http://www.cs.cmu.edu/~javabayes/Home/
LIBB	http://compbio.cs.huji.ac.il/LibB/
MIM	http://www.hypergraph.dk/
MAGIC	http://function.cs.princeton.edu/magic/
NETICA	http://www.norsys.com/
PRIMULA	http://www.cs.auc.dk/~jaeger/Primula/
QUIDDITY SUITE	http://www.iet.com/quiddity.html
SAMIAM	http://reasoning.cs.ucla.edu/samiam/
SUPERLINK	http://cbl-fog.cs.technion.ac.il/superlink/

Fig. 3.11. Belief network tools mentioned in the text

networks, and Kersting's page on tools for probabilistic logic[6]. A good overview is also provided in the appendix of [50].

The belief network tools can be divided into (1) integrated tools and suites that offer at least a graphical interface to build belief networks and inference algorithms for belief updates, (2) querying and other partial tools, and (3) toolkits, APIs and libraries that allow a programmer to incorporate belief networks into own developments. For convenience of the reader, Fig. 3.11 lists the urls of the tools mentioned in this section.

3.6.1 Integrated Tools and Suites

The Murphy list contains 24 packages that offer a graphical user interface for building and querying belief networks, but even more exist. The packages include commercially available suites such as HUGIN, NETICA, and ANALYTICA, but also freeware and open source programs such as GENIE, JAVABAYES and SAMIAM. (Commercial tools often have a light-weight free version or offer academic licences.) Although some tools (e.g., HUGIN) allow continuous nodes, most tools concentrate on classical belief networks with discrete probability distributions only. Especially the way in which the conditional probabilities are entered by hand differs much, and is sometimes awkward.

[6]http://www.informatik.uni-freiburg.de/~kersting/plmr/PLMR_repository.html

The packages offer a range of different inference algorithms, but the clique-tree algorithm is the most often used. The extensive (commercial) programs have a selection of inference algorithms available that can be selected and parameterized by the user. Some of the commercial and free programs allow parameter learning and/or structure learning. Again, the extensive programs offer a choice of algorithms and settings.

Some tools allow the use of belief network extensions. Decision theoretic extensions (*influence diagrams*) are provided by quite a number of tools. Other extensions are more restricted. HUGIN supports object-oriented belief networks (a predecessor of PRMs), PRIMULA in cooperation with SAMIAM allow RBNs, and the commercial QUIDDITY SUITE, for instance, allows construction of and inference in MEBNs.

It seems to be problematic to exchange belief networks between applications. A special version of XML has been proposed by Microsoft to support exchange (XBN[7]), but most often (older) HUGIN or NETICA formats are used, unfortunately leading to incompatibilities.

3.6.2 Partial Tools

Some tools only offer a part of the functionality of the above packages. The popular package BUGS offers Gibbs sampling for approximate inference in graphical models, including belief networks. It allows a large range of continues distributions for local probabilities. The program is available in several versions, including OPENBUGS and WINBUGS. Other tools for inference in graphical models are COCO, MIM and TETRAD.

The program ACE offers translation of belief networks into arithmetic circuits and querying. LIBB1.2 by Friedman is dedicated to structure learning of belief networks.

3.6.3 APIs, Libraries, and Toolkits

A number of the integrated tools discussed above, such as HUGIN and NETICA, also provide an application programming interface (API) that allows incorporation of belief networks in any application. These APIs offer the same inference algorithms as the complete tools. The belief networks are represented inside the API, hiding implementation details from the programmer.

For several programming languages, libraries exist that allow the definition and manipulation of belief networks. For instance, SMILE for C++ (with wrappers for Java, .NET and other platfoms) and for Java BNJ, JAVABAYES, and SAMIAM.

Since belief network inference involves so many mathematical computations, it makes sense to use existing mathematical frameworks such as MATLAB and R. For MATLAB the free BAYES NET TOOLKIT offers discrete and continuous variables, exact and approximate inference, parameter learning and structure learning, and dynamic belief networks. A disadvantage is that no proper graphical user interface is

[7] http://research.microsoft.com/dtas/bnformat/

available while the format in which belief networks have to be specified requires manual topological sorting of networks nodes.

For the open-source statistical framework R, an initiative GR (graphical Models in R) is going on. It resulted, for instance, in the package DEAL for learning belief networks. (This package saves belief networks in HUGIN format.) Some specialized bioinformatics R-packages exists that use belief networks such as the BNARRAY package for constructing gene-regulatory networks, which uses the DEAL package.

3.6.4 Belief Network Tools for Bioinformatics

Despite the range of available belief-network tools described above, many of the bioinformatics applications of belief networks are based on customized software. In some cases the software can be downloaded from accompanying publication websites and sometimes the belief network tools are incorporated in large (commercial) applications such as SUPERLINK and GENOMICA. The main reason is that the bioinformatics applications often require specialized inference or learning algorithms that are not (yet) available in the general tools for belief networks. An exception is perhaps the online version of the MAGIC tool for gene clustering.

3.7 Guide to the Literature

The basics of belief networks are described in a number of handbooks, starting with the inventor's one [69]. A very gentle introduction to belief network inference and learning is provided in [79]. More complete general introductions are provided by [22, 48, 50, 66]. The book by Korb and Nicholson [50] is the most gentle one, including many practical issues, examples, and tips for working with belief network tools. Jensen and Cowel *et al.* [22, 48] provide a more in-depth introduction to the algorithms for inference and learning. Learning in belief networks is treated in more detail in [3, 49, 65, 88], and in the recent Ph.D. thesis of Riggelsen [78].

Online introductions to belief networks are available on wikipedia, the UAI website[8] (including a Wiki), online tutorials such as Niedermayer's[9], and in the online manuals of several of the belief network tools.

A number of bioinformatics books explain (certain types of) belief networks. For instance [5] treats HMMs and graphical models, including belief networks. [51] emphasizes the use of HMMs in bioinformatics and [30] is restricted to sequential applications of HMMs. A standard bioinformatics handbook such as [59] covers HMM models for multiple sequence alignment.

A concise overview of literature on belief networks in bioinformatics is available at the website of Kasif[10]. An overview of belief networks in gene regulation networks is provided in [26, 37, 38]. A very good introduction in belief networks, structure

[8] http://www.auai.org

[9] http://www.niedermayer.ca/papers/bayesian/bayes.html

[10] http://sullivan.bu.edu/kasif/bayes-net.html

learning and the application in linkage analysis is provided in [55]. An overview of the use of PRMs in bioinformatics is given in [67], and in the Ph.D. thesis of Segal [80].

3.8 Conclusion

In this chapter we introduced belief networks and described their current use within bioinformatics. We discussed examples of belief networks applied on the modeling of gene regulation networks, the discovering of metabolic and signalling pathways, sequence analysis, protein structure, and linkage analysis. New developments in learning belief networks from heterogeneous data sources show that belief networks are becoming an important tool for dealing with high-throughput data at a large scale, not only at the genetic and biochemical level, but also at the level of systems biology.

References

1. Silvia Acid and Luis M. de Campos. A hybrid methodology for learning belief networks. *Int. J. Approx. Reasoning*, 27(3):235–262, 2001.
2. Cornelis A. Albers, Martijn A.R. Leisink, and Hilbert J. Kappen. The cluster variation method for efficient linkage analysis on extended pedigrees. *BMC Bioinformatics*, 7 suppl 1:xx, 2006.
3. Ethem Alpaydin. *Introduction to Machine Learning*. The MIT Press, 2004.
4. S.K. Andersen, K.G. Olesen, F.V. Jensen, and F. Jensen. Hugin - a shell for building belief universes for expert systems. In *Proceedings IJCAI*, pages 1080–1085, 1989.
5. Pierre Baldi and Søren Brunak. *Bioinformatics the Machine Learning Approach, 2nd edition*. MIT Press, Cambridge, MA, 2001.
6. Katia Basso, Adam A. Margolin, Gustavo Stolovitzky, Ulf Klein, Ricardo Dalla-Favera, and Andrea Califano. Reverse engineering of regulatory networks in humann B cells. *Nature Genetics*, 37(4):382–390, 2005.
7. Alexis Battle, Eran Segal, and Daphe Koller. Probabilistic discovery of overlapping cellular processes and their regulation. *Computational Biology*, 12(7):909–927, 2005.
8. L.E. Baum, T. Petrie, G. Soules, and N. Weiss. A maximization technique occurring in the statistical analysis of probabilistic functions of markov chains. *Ann. Math. Statist*, 41(1):164–171, 1970.
9. I. Ben-Gal, A. Shani, A. Gohr, J. Grau, S. Arviv, A. Shmilovici, S. Posch, and I. Grosse. Identification of transcription factor binding sites with variable-order bayesian networks. *Bioinformatics*, 21(11):2657–2666, 2005.
10. Joseph Bockhorst, Mark Craven, David Page, Jude Shavlik, and Jeremy Glasner. A bayesian network approach to operon prediction. *Bioinformatics*, 19(10):1227–1235, 2003.
11. Craig Boutilier, Nir Friedman, Mosies Goldszmidt, and Daphne Koller. Context-specific independence in bayesian networks. In *UAI96*, pages 115–123, 1996.
12. J. Boyan. *Learning evaluation functions for global optimization*. PhD thesis, Carnegie Mellon University, Pittsburgh, PA, 1998.
13. Mark Chavira, Adnan Darwiche, and Manfred Jaeger. Compiling relational bayesian networks for exact inference. *Int. Journal of Approximate Reasoning*, 42:4–20, 2006.

14. Xue-Wen Chen, Gopalakrishna Anantha, and Xinhun Wang. An effective structure learning method for constructing gene networks. *Bioinformatics*, 22(11):1367–1374, 2006.
15. David Maxwell Chickering. Learning equivalence classes of bayesian network structures. *Journal of Machine Learning Research*, 2:445–498, 2002.
16. Wei Chu and Zoubin Ghahramani. Protein secondary structure prediction using sigmoid belief networks to parameterize segmental semi-markov fields. In *Proceedings of the European Symposium on Artificial Neural Networks ESANN'2004*, pages 81–86, Bruges, Belgium, 2004.
17. Gregory F. Cooper. Probabilistic inference using belief networks is NP-hard. *Artificial Intelligence*, 42:393–405, 1990.
18. Gregory F. Cooper. A bayesian method for learning belief networks that contain hidden variables. *Journal of Intelligent Information Systems*, 4:71–88, 1995.
19. Gregory F. Cooper and E. Herskovits. A bayesian method for the induction of probabilistc networks from data. *Machine Learning*, 9:309–347, 1992.
20. Paulo C.G. Costa, Kathryn B. Laskey, and Kenneth J. Laskey. Pr-owl: A framework for probabilistic ontologies. In *Proceedings of the Fourth International Conference on Formal Ontology in Information Systems*, 2006.
21. Robert G. Cowell. Conditions under which conditional independence and scoring methods lead to identical selection of bayesian network models. In *UAI01*, 2001.
22. Robert G. Cowell, A. Philip Dawid, S.L. Lauritzen, and D.J. Spiegelhalter. *Probabilistic Networks and Expert Systems*. Springer Verlag, Berlin, New York, 2003.
23. Adnan Darwiche. A differential approach to inference in bayesian networks. In *UAI2000*, pages 123–132, 2000.
24. Adnan Darwiche. New advances in compiling cnf to decomposable negation normal form. In *ECAI 2004*, 2004.
25. Denver Dash and Marek J. Druzdzel. A hybrid anytime algorithm for the construction of causal models from sparse data. In *UAI99*, 1999.
26. Hidde de Jong. Modeling and simulation of genetic regulatory systems: A literature overview. *Computational Biology*, 9(1):67–103, 2002.
27. R. Dechter. Bucket elemination: a unifying framework for probabilistic inference. In *UAI96*, 1996.
28. Arthur L. Delcher, Simon Kasif, Harry R. Goldberg, and William H. Hsu. Protein secondary structure modelling with probabilistic networks. In *Proceedings of the International Conference on Intelligent Systems and Molecular Biology*, pages 109–117, 1993. (extended abstract).
29. H.H.L.M. Donkers, A.W. Werten, J.W.H.M. Uiterwijk, and H.J. van den Herik. Sequapro: A tool for semi-qualitative decision making. Technical Report CS 01-06, Department of Computer Science, Universiteit Maastricht, 2001.
30. Richard Durbin, Sean R. Eddy, Anders Krogh, and Graeme Mitchison. *Biological Sequence Analysis: Probabilistic Models of Proteins and Nucleic Acids*. Cambridge University Press, Cambridge, UK, 1999.
31. Sean R. Eddy. Profile hidden markov models. *Bioinformatics*, 14(9):755–763, 1998.
32. G. Elidan, N. Lotner, N. Friedman, and D. Koller. Discovering hidden variables: A structure-based approach. In *Advances in Neural Information Processing Systems (NIPS)*, 2000.
33. M. Fichelson and D. Geiger. Exact genetic linkage computations for general pedigrees. *Bioinformatics*, 18 Suppl. 1:S189–S198, 2002.
34. Ma'ayan Fichelson and Dan Geiger. Optimizing exact genetic link computations. *Journal of Computational Biology*, 11(2–3):263–275, 2004.

35. Nir Friedman. Learning belief networks in the presence of missing values and hidden variables. In *Proceedings of the Fourteenth International Conference on Machine Learning*, 1997.
36. Nir Friedman. The bayesian structural em algorithm. In *UAI98*, 1998.
37. Nir Friedman. Inferring cellular networks using probabilistic graphical networks. *Science*, 303:799–805, 2004.
38. Nir Friedman, Michal Linial, Iftach Nachman, and Dana Pe-er. Using bayesian networks to analyze expression data. *Computational Biology*, 7(3/4):601–620, 2000.
39. Nir Friedman, Iftach Nachman, and Dana Pe'er. Learning bayesian network structure from massive datasets: the "sparse candidate" algorithm. In *Proc. Fifteenth Conf. on Uncertainty in Artificial Intelligence (UAI)*, 1999.
40. Irit Gat-Viks, Amos Tanay, Daniela Raijman, and Ron Shamir. A probabilistic methodology for integrating knowledge and experiments on biological networks. *Computational Biology*, 13(2):165–181, 2006.
41. D. Griffeath. Introduction to markov random fields. In Kemeny, Knapp, and Snell, editors, *Denumerable Markov Chains*. Springer, 1976. 2nd edition.
42. Chris Harbron. Heuristic algorithms for finding inexpensive elimination schemes. *Statistics and Computing*, 5:275–287, 1995.
43. David Heckerman. A tutorial on learning with bayesian networks. Technical report, Microsoft Research, 1995.
44. Manfred Jaeger. Relational bayesian nets. In D. Geiger and P.P. Shenoy, editors, *Uncerrtainty in Artificial Intelligence (UAI97)*, pages 266–273, San Fransisco, CA, 1997. Morgan Kaufman Publishers.
45. Manfred Jaeger. Relational bayesian networks: a survey. *Electronic Transactions in Artificial Intelligence*, 6:xx, 2002.
46. Manfred Jeager. *The Primula System: user's guide*, 2006. http://www.cs.aau.dk/˜jeager/primula.
47. Claus Skaanning Jensen and Augustine Kong. Blocking gibbs sampling for linkage analysis in large pedigrees with many loops. *American Journal of Human Genetics*, 65:885–901, 1999.
48. Finn V. Jensen. *Bayesian Networks and Decision Graphs*. Springer Verlag, New York, Berlin, 2001.
49. M.I. Jordan. *Learning in Graphical Models*. MIT Press, Cambridge, MA, 1998.
50. Kevin B. Korb and Ann E. Nicholson. *Bayesian Artificial Intelligence*. Chapman & Hall/CRC, Boca Raton, FL, 2004.
51. Timo Koski. *Hidden Markov Models of Bioinformatics*. Springer Verlag, Berlin, New York, 2001.
52. W. Lam and F. Bacchus. Learning bayesian belief networks: An approach based on the mdl principle. *Computational Intelligence*, 10:269–293, 1994.
53. Kathryn B. Laskey. MEBN: A logic for open-world probabilistic reasoning. Technical Report C4I06-01, George Mason University C4I Center, 2006.
54. S. L. Lauritzen and D. J. Spiegelhalter. Local computations with probabilities on graphical structures and their application to expert systems. *Journal of the Royal Statistical Society, Series B*, 50(2):157–224, 1988.
55. Steffen L. Lauritzen and Nuala Sheehan. Graphical models for genetic analysis. *Statistical Science*, 18(4):489–514, 2003.
56. Phil Hyoun Lee and Doheon Lee. Modularized learning of genetic interaction networks from biological annotations and mRNA expression data. *Bioinformatics*, 21(11):2739–2747, 2005.

57. Sun-Mi Lee and Patricia A. Abbott. Bayesian networks for knowledge discovery in large datasets: basics for nurse researchers. *Journal of Biomedical Informatics*, 36(4–5):389–399, 2003.
58. M. Leisink, H.J. Kappen, and H.G. Brunner. Linkage analysis: A bayesian approach. In *ICANN 2002*, number 2415 in LNCS, pages 595–600, 2002.
59. Arthur M. Lesk. *Introduction to Bioinformatics*. Oxford University Press, New York, NY, 2005.
60. Peter J.F. Lucas. Bayesian analysis, pattern analysis and data mining in health care. *Current Opinion in Critical Care*, 10:399–403, 2004.
61. Peter J.F. Lucas. Bayesian networks in biomedicine and health-care. *Artificial Intelligence in Medicine*, 30:201–214, 2004.
62. Suzanne M. Mahony and Kathryn B. Laskey. Network engineering for complex belief networks. In *UAI97*, 1997.
63. Kristin Missal, Michael A. Cross, and Dirk Drasdo. Gene network inference from incomplete expression data: Transcriptional control of hematopoietic commitment. *Bioinformatics*, 22(6):731–738, 2006.
64. Kevin Murphy. *Dynamic Bayesian Networks: Representation, Inference and Learning*. PhD thesis, UC Berkeley, Computer Science Division, 2002.
65. R.E. Neapolitan. *Learning Bayesian Networks*. Prentice Hall, Pearson Education, Upper Saddle River, NJ, 2003.
66. R.M. Oliver and J.Q. Smith, editors. *Influence Diagrams, Belief Nets and Decision Analysis*. John Wiley & Sons, Chichester, UK, 1990.
67. David Page and Mark Craven. Biological applications of multi-relational data mining. *SIGKDD Explorations*, 5-1:69–79, 2003.
68. Judea Pearl. Evidential reasoning using stochastic simulation of causal models. *Artificial Intelligence*, 32:245–257, 1987.
69. Judea Pearl. *Probabilistic Reasoning in Intelligent Systems: Networks of Plausible Inference*. Morgan Kaufman Publishers, San Mateo, CA, 1988.
70. Judea Pearl and T. Verma. A theory of inferred causation. In *Proceedings of Principles of Knowledge Representation and Reasoning*, 1991.
71. J.M. Peña, J. Björkegen, and J. Tegnér. Growing bayesian network models of gene networks from seed genes. *Bioinformatics*, 21(Suppl. 2):ii224–ii229, 2005.
72. Bruno-Eduoard Perrin, Liva Ralaivola, Aurélien Mazurie, Samuele Bottani, Jacques Mallet, and Florence d'Alché Buc. Gene networks inference using dynamic bayesian networks. *Bioinformatics*, 19(Suppl. 2):ii138–ii148, 2003.
73. Antonio Piccolboni and Dan Gusfield. On the complexity of fundamental computational problems in pedigree analysis. *Journal of Computational Biology*, 10(5):763–773, 2003.
74. Rainer Pudimat, Ernst-Günter Schukat-Talamazzini, and Rolf Backofen. A multiple-feature framework for modelling and predicting transcription factor binding sites. *Bioinformatics*, 21(14):3082–3088, 2005.
75. Yuan Qi, Alex Rolfe, Kenzie D. MacIsaac, Georg K. Gerber, Dmitry Pokholok, Julia Zeitlinger, Timothy Danford, Robin D. Dowell, Ernest Fraenkel, Tommi S. Jaakkola, Richard A. Young, and David K. Gifford. High-resolution computational models of genome binding events. *Nature Biotechnology*, 24:963–970, 2006.
76. A. Raval, Z. Ghahramani, and D.L. Wild. A bayesian network model for protein fold and remote homologue recognition. *Bioinformatics*, 18(6):788–801, 2002.
77. Silja Renooij and Linda C. van der Gaag. From qualitative to quantitative probabilistic networks. In *UAI02*, pages 442–429, 2002.
78. Carsten Riggelsen. *Approximation Methods for Efficient Learning of Bayesian Networks*. PhD thesis, Universiteit Utrecht, 2006.

79. Stuart Russel and Peter Norvig. *Artificial Intelligence, a modern approach*. Prentice Hall, Pearson Education, Upper Saddle River, NJ, 2nd edition, 2003.
80. Eran Segal. *Rich Probabilistic Models for Genomic Data*. PhD thesis, Stanford University, 2004.
81. Eran Segal, Nir Friedman, Daphne Koller, and Aviv Regev. A module map showing conditional activity of expression modules in cancer. *Nature Genetics*, 36(10):1090–1098, 2004.
82. Eran Segal, Ben Taskar, Audrey Gash, Nir Friedman, and Daphne Koller. Rich probabilistic models for gene expression. *Bioinformatics*, 17(Suppl. 1):s234–s252, 2001.
83. Eran Segal, H. Wang, and Daphe Koller. Discovering molecular pathways from protein interaction and gene expression data. *Bioinformatics*, 19(Suppl. 1):i264–i272, 2003.
84. G. Shafer and P. Shenoy. Probability propagation. *Annals of Mathematics and Artificial Intelligence*, 2:327–352, 1990.
85. A Siepel and D. Haussler. Phylogenetic hidden markov models. In R. Nielsen, editor, *Statistical Methods in Molecular Evolution*, pages 325–351, New York, 2005. Springer.
86. M. Silberstein, A. Tzemach, N. Dovgolevsky, M. Fichelson, A Shuster, and D. Geiger. Online system for faster multipoint linkage analysis via parallel execution on thousands of personal computers. *American Journal of Human Genetics*, 78:992–935, 2006.
87. D.J. Speigelhalter, P. Dawid, S. Lauritzen, and R. Cowell. Bayesian analysis in expert systems. *Statistical Science*, 8:219–282, 1993.
88. P. Spirtes, C. Glymour, and R. Scheines. *Causation, Prediction and Search*. MIT Press, Cambridge, MA, 2000.
89. Alun Thomas, Alexander Gutin, Victor Abkevich, and Aruna Bansal. Multilocas linkage analysis by blocked gibbs sampling. *Statistics and Computing*, 10:259–269, 2000.
90. Olga G. Troyanskaya, Kara Dolinski, Art B. Owen, Russ B. Altman, and David Botstein. A bayesian framework for combining heterogeneous data sources for gene function prediction in saccharomyces cerevisiae. In *PNAS*, volume 100, pages 8348–8353, 2003.
91. Linda C. van der Gaag, Silja Renooij, Cees L.M. Witteman, B.M.P. Aleman, and B.G. Taal. How to elecit many probabilities. In *UAI99*, 1999.
92. Claudio J. Verzilli, Nigel Stallard, and John C. Whittaker. Bayeisan graphical models for genomewide association studies. *American Journal of Human Genetics*, 79:100–112, 2006.
93. Michael P. Wellman. Fundamental concepts of qualitative probabilistic networks. *Artificial Intelligence*, 44(3):257–303, 1990.
94. Peter J. Woolf, Wendy Prudhomme, Laurence Daheron, George Q. Daley, and Douglas A. Lauffenburger. Bayesian analysis of signalling networks governing embryonic stem cell fate decisions. *Bioinformatics*, 21(6):741–753, 2005.
95. Chen-hsiang Yeang and Tommi Jaakkola. Physical networks and multi-source data integration. In *Proceedings of the 7th annual International Conference on Research in Computational Molecular Biology*, 2003.
96. Jing Yu, V. Anne Smith, Paul P. Wang, Alexander J. Hartemink, and Erich D. Jarvis. Advances to bayesian network inference for generating causal networks from observational biological data. *Bioinformatics*, 20(18):3594–3603, 2004.
97. Geoffrey Zweig and Stuart J. Russell. Speech recognition with dynamic bayesian networks. In *AAAI/IAAI*, pages 173–180, 1998.

4

Swarm Intelligence Algorithms in Bioinformatics

Swagatam Das, Ajith Abraham*, and Amit Konar

Department of Electronics and Telecommunication Engineering,
Jadavpur University, Kolkata 700032, India
*Center of Excellence for Quantifiable Quality of Service,
Norwegian University of Science and Technology, Trondheim, Norway
ajith.abraham@ieee.org

Summary. Research in bioinformatics necessitates the use of advanced computing tools for processing huge amounts of ambiguous and uncertain biological data. Swarm Intelligence (SI) has recently emerged as a family of nature inspired algorithms, especially known for their ability to produce low cost, fast and reasonably accurate solutions to complex search problems. In this chapter, we explore the role of SI algorithms in certain bioinformatics tasks like microarray data clustering, multiple sequence alignment, protein structure prediction and molecular docking. The chapter begins with an overview of the basic concepts of bioinformatics along with their biological basis. It also gives an introduction to swarm intelligence with special emphasis on two specific SI algorithms well-known as Particle Swarm Optimization (PSO) and Ant Colony Systems (ACS). It then provides a detailed survey of the state of the art research centered around the applications of SI algorithms in bioinformatics. The chapter concludes with a discussion on how SI algorithms can be used for solving a few open ended problems in bioinformatics.

4.1 Introduction

The past few decades have seen a massive growth in biological information gathered by the related scientific communities. A deluge of such information coming in the form of genomes, protein sequences, gene expression data and so on have led to the absolute need for effective and efficient computational tools to store, analyze and interpret the multifaceted data.

The term *bioinformatics* literally means the science of informatics as applied to biological research. Informatics on the other hand is the management and analysis of data using various advanced computing techniques. Hence, in other words, bioinformatics can be described as the application of computational methods to make biological discoveries [1]. It presents a symbiosis of several different areas of science including biology, computer science, mathematics and statistics. The ultimate

S. Das et al.: *Swarm Intelligence Algorithms in Bioinformatics*, Studies in Computational Intelligence (SCI) **94**, 113–147 (2008)
www.springerlink.com

attempt of the field is to develop new insights into the science of life as well as creating a global perspective, from which the unifying principles of biology can be derived [2]. Three major objectives of bioinformatics can be put forward as:

- To develop algorithms and mathematical models for probing the relationships among the members of a large biological dataset.
- To analyze and interpret the heterogeneous kind of data including nucleotide and amino acid sequences, protein domains and protein structures.
- To implement tools that enable efficient storage, retrieval and management of high-volume biological databases.

Biologically inspired computing has been given importance for its immense parallelism and simplicity in computation. In recent times, quite a large number of biologically motivated algorithms have been invented, and are being used for handling many complex problems of the real world. For instance, neural computing [3] attempts to mimic the biological nervous systems of the living creatures to ensure a significant amount of parallel and distributed processing in computation. Genetic algorithms [4], [5] imitate the Darwinian evolutionary process through cross-over and mutation of biological chromosomes. They have successfully been used in many bioinformatics tasks that need intelligent search, optimization and machine learning approaches. Mitra and Hayashi [6] provides a comprehensive survey of the research in this direction.

Recently, a family of nature inspired algorithms known as Swarm Intelligence (SI) [7], [8], [9] has attracted the attention of researchers working on bioinformatics related problems all over the world. Algorithms belonging to this field are motivated by the collective behavior of a group of social insects (like bees, termites and wasps). These insects with very limited individual capability can jointly (cooperatively) perform many complex tasks necessary for their survival. For the past few years there has been a slow but steady increase of research papers reporting the success of SI based search, clustering and data mining methods applied to the field of computational biology.

This Chapter provides a detailed review of the role of SI algorithms in different aspects of bioinformatics mainly involving optimization, pattern recognition and data mining tasks. The rest of the chapter is organized as follows. Section 4.2 briefly describes the preliminary ideas of bioinformatics. In section 4.3, we have introduced the paradigm of Swarm Intelligence and outlined the technical details of two popular SI algorithms known as Particle Swarm Optimization (PSO) [10] and Ant Colony Systems (ACS) [11], [12]. We also discuss the relevance of SI in bioinformatics under this Section. Section 4.4 reviews a number of SI based methods available in the literature to address many difficult tasks in bioinformatics. A few open ended research problems as well as how these can be solved with SI algorithms have been discussed in Section 4.5. Finally the Chapter is concluded in Section 4.6.

4.2 Fundamental Concepts in Bioinformatics

In this section, we outline a few preliminary biological concepts which are essential for understanding several research problems that have been discussed in the subsequent Sections.

4.2.1 DNA

The complete set of instructions for making an organism is called its genome. It contains the master blueprint for all cellular structures and activities for the lifetime of the cell or organism. Found in every nucleus of a person's many trillions of cells, the human genome consists of tightly coiled threads of deoxyribonucleic acid or DNA and associated protein molecules, organized into structures called chromosomes. DNA plays a fundamental role in the different bio-chemical processes of living organisms in two respects:

- Firstly, it contains the information the cell requires to synthesize protein and to replicate itself. To be short, it is the storage repository for the information that is required for any cell to function [13].
- Secondly, it acts as a medium for transmitting the hereditary information (namely the synthesis plans for proteins) from generation to generation.

In humans, as in other higher organisms, a DNA molecule consists of two strands that wrap around each other to resemble a twisted ladder whose sides, made of sugar and phosphate molecules are connected by rungs of nitrogen- containing chemicals called bases. Each strand is a linear arrangement of repeating similar units called nucleotides, which are each composed of one sugar, one phosphate, and a nitrogenous base. Four different bases are present in DNA: adenine (A), thymine (T), cytosine (C), and guanine (G). The particular order of the bases arranged along the sugar-phosphate backbone is called the DNA sequence; the sequence specifies the exact genetic instructions required to create a particular organism with its own unique traits. In normal DNA, the bases form pairs: A to T and G to C. This is called complementarity. The pair of complementary strands then forms the double helix, which was first suggested by Watson and Crick in 1953. Figure 4.1 illustrates the double helix of the DNA sequence with a gene in the sequence delimited. Genes are specific sequences of bases that encode instructions on how to make proteins. We highlight them in the next subsection.

4.2.2 The Gene

Each DNA molecule contains many genes – the basic physical and functional units of heredity. A gene is a specific sequence of nucleotide bases, whose sequences carry the information required for constructing proteins, which provide the structural components of cells and tissues as well as enzymes for essential biochemical reactions. The human genome is estimated to comprise more than 30,000 genes. Human genes

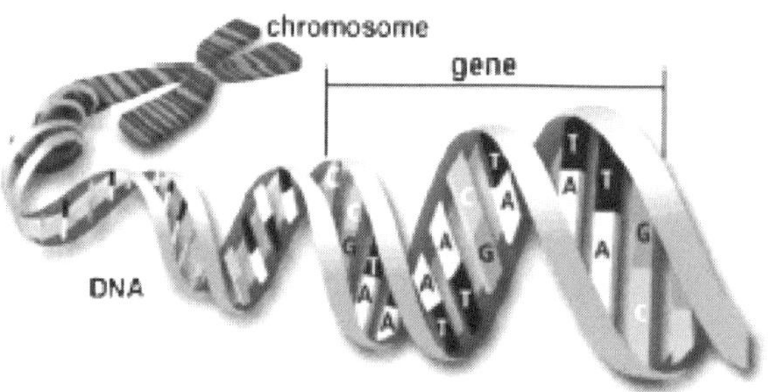

Fig. 4.1. The DNA double helix and a gene sequence

vary widely in length, often extending over thousands of bases, but only about 10% of the genome is known to include the protein- coding sequences (exons) of genes. Interspersed within many genes are intron sequences, which have no coding function.

The gene's sequence is like language that instructs cell to manufacture a particular protein. At first a gene is *transcribed* to produce messenger ribonucleic acid (m-RNA) which is next *translated* to produce proteins. It is the protein that determines the traits of an organism and is called *central dogma of life*. The m-RNA is single-stranded and has a ribose sugar molecule. There exist *promoter* and *termination* sites in a gene, responsible for initiation and termination of the transcription process. Translation consists of mapping from triplets (codons) of four bases to the twenty amino acids that serve as the building blocks of protein. There are sequences of nucleotides within the DNA that are spliced out progressively in the process of transcription and translation. Wu and Lindsay [14] provides a comprehensive survey of the research undertaken so far, in this direction.

Apart from the genes, DNA consists of three types of non coding regions:

1. Intergenic regions: Regions between genes that are ignored during the process of transcription.
2. Intragenic regions (Introns): Regions within the genes that will be spliced out after transcription, but before the RNA is used.
3. Pseudogenes: These are defunct relatives of known genes that have lost their protein-coding ability or are otherwise no longer expressed in the cell [13]. Although they may have some gene-like features (such as Promoters, CpG islands, and splice sites), they are nonetheless considered nonfunctional, due to their lack of protein-coding ability.

Figure 4.2 illustrates the different parts of a gene.

4.2.3 Proteins

An amino acid is an organic molecule that consists of an amine (NH) and a carboxylic (CO) acid group (backbone) together with a side-chain that differentiates between them. Proteins are large organic compounds made of amino acids arranged in a linear

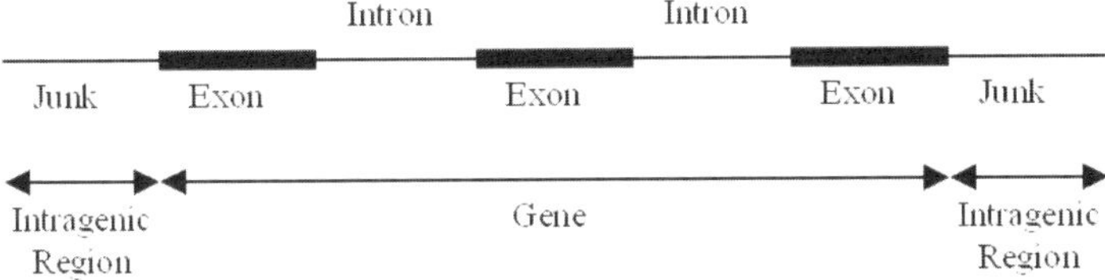

Fig. 4.2. Schematic outline of a gene

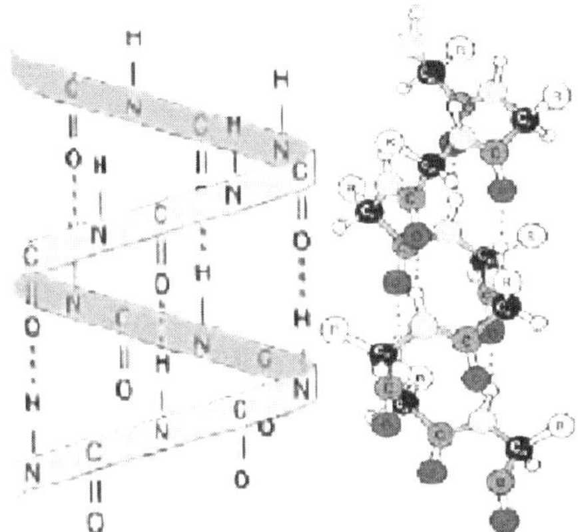

Fig. 4.3. Spiral configuration of the α helix structure. Hydrogen bonds between the CO group of one amino acids and the NH group of another amino acid holds the α helices together (adapted from [22])

chain and joined together between the carboxyl atom of one amino acid and the amine nitrogen of the other [15]. This bond is called a peptide bond. The sequence of amino acids in a protein is defined by a gene and encoded in the genetic code. Although this genetic code specifies 20 "standard" amino acids, the residues in a protein are often chemically altered in post-translational modification: either before the protein can function in the cell, or as part of control mechanisms. Proteins can also work together to achieve a particular function, and they often associate to form stable complexes.

In order to carry out their function, each protein must take a particular shape, known as its fold. When a protein is put into a solvent, within a very short time it takes a particular 3D shape. This self assembling process is called folding. More than half a century ago, Linus Pauling (Nobel prize, 1954) discovered that a major part of most proteins' folded structure consists of two regular, highly periodic arrangements of amino acids, designated "a" and "b". The key to both structures is the hydrogen bond that stabilizes the structures. The "a" structure is now called α helix (Figure 4.3). It is a spiral configuration of a polypeptide chain stabilized by hydrogen bonds between the CO group of one amino acid at position n and the NH group of the amino acid which is four residues away $(n+4)$. The "b" structure is now called β sheet (Figure 4.4). It is an essentially a flat two dimensional structure of parallel

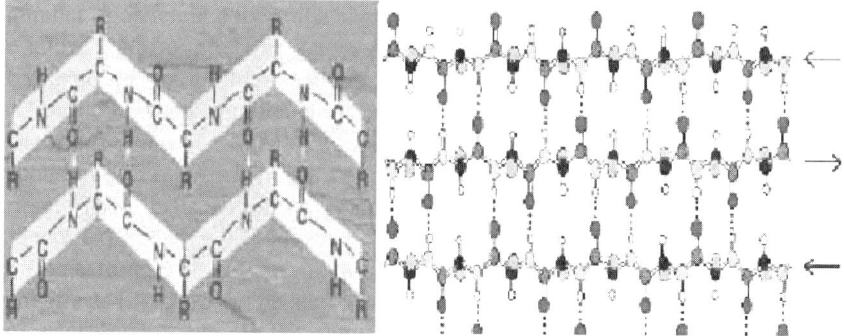

Fig. 4.4. β pleated sheets structure is stabilized by hydrogen bonds between nitrogen atoms (of the NH group of one amino acid) and oxygen atoms (of the CO group of another amino acid) of two adjacent chains (adapted from [22])

or anti-parallel β strands; each β strand consists of two polypeptide chains that are (almost) fully extended and hydrogen bonded to each other. All other local arrangements that are neither α helix nor β sheet are described as random coil: they are random in the sense that they are not periodic.

Proteins have multiple levels of structure [1]:

1. **Primary structure**: Linear structure determined solely by the number, sequence, and type of amino acid residues (R).
2. **Secondary structure**: Local structure determined by hydrogen bonding between amino acids and non-polar interactions between hydrophobic regions. These interactions produce, in general, three secondary structures: α helix (Figure 4.2), β sheet (Figure 4.3), and random coil.
3. **Tertiary structure**: It results from various interactions (mainly hydrophobic attractions, hydrogen bonding, and disulfide bonding) of the amino acids side chains (R) that pack together the elements of the secondary structure. The result is a 3D configuration of proteins.
4. **Quaternary structure**: It is characterized by the interaction of two or more individual polypeptides (often via disulfide bonds) and the result is a larger functional molecule.

4.2.4 DNA Microarray

A DNA microarray (also commonly known as DNA chip or gene array) is a collection of microscopic DNA spots attached to a solid surface, such as glass, plastic or silicon chip forming an array for the purpose of expression profiling, monitoring expression levels for thousands of genes simultaneously. Figure 4.5 illustrates a simple DNA chip [16].

Microarrays provide a powerful basis to monitor the expression of thousands of genes, in order to identify mechanisms that govern the activation of genes in an organism. Short DNA patterns (or binding sites near the genes) serve as switches

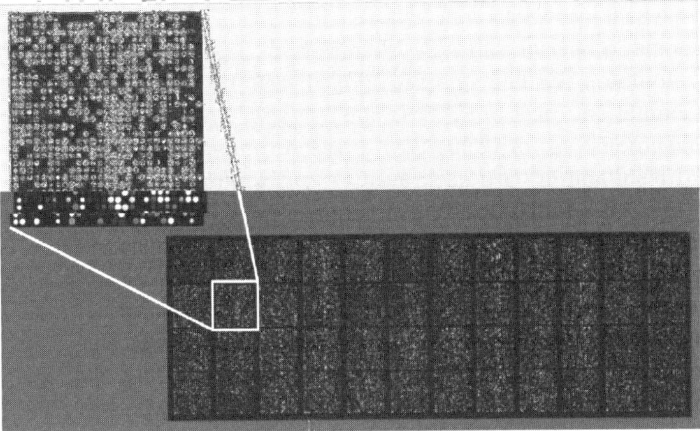

Fig. 4.5. Example of an approximately 37,500 probe spotted oligo microarray with enlarged inset to show detail (adapted from [16])

that control gene expression. Therefore, similar patterns of expression correspond to similar binding site patterns. A major cause of coexpression of genes is their sharing of the regulation mechanism (coregulation) at the sequence level. Clustering of coexpressed genes, into biologically meaningful groups, helps in inferring the biological role of an unknown gene that is coexpressed with a known gene(s). Cluster validation is essential, from both the biological and statistical perspectives, in order to biologically validate and objectively compare the results generated by different clustering algorithms.

4.3 Swarm Intelligence - an Overview and Relevance to Bioinformatics

The behavior of a single ant, bee, termite and wasp often is too simple, but their collective and social behavior is of paramount significance. A look at National Geographic TV Channel also reveals that advanced mammals including lions also enjoy social lives, perhaps for their self-existence at old age and in particular when they are wounded. The collective and social behavior of living creatures motivated researchers to undertake the study of swarm intelligence. Historically, the phrase Swarm Intelligence (SI) was coined by Beny & Wang in late 1980s [9] in the context of cellular robotics. A group of researchers in different parts of the world started working almost at the same time to study the versatile behavior of different living creatures. SI systems are typically made up of a population of simple agents (an entity capable of performing/executing certain operations) interacting locally with one another and with their environment. Although there is normally no centralized control structure dictating how individual agents should behave, local interactions between such agents often lead to the emergence of global behavior. Many biological

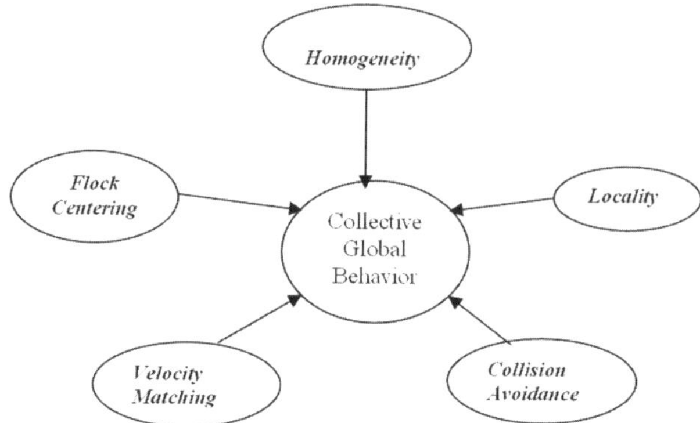

Fig. 4.6. Main traits of the collective behavior

creatures such as fish schools and bird flocks clearly display structural order, with the behavior of the organisms so integrated that even though they may change shape and direction, they appear to move as a single coherent entity [17]. The main properties of the collective behavior can be given below and is illustrated in Figure 4.6:

- **Homogeneity**: every bird in flock has the same behavioral model. The flock moves without a leader, even though temporary leaders seem to appear.
- **Locality**: its nearest flock-mates only influence the motion of each bird. Vision is considered to be the most important senses for flock organization.
- **Collision avoidance**: avoid colliding with nearby flock mates.
- **Velocity matching**: attempt to match velocity with nearby flock mates.
- **Flock centering**: attempt to stay close to nearby flock mates.

Individuals attempt to maintain a minimum distance between themselves and others at all times. This rule is given the highest priority and corresponds to a frequently observed behavior of animals in nature [18]. If individuals are not performing, an avoidance maneuver they tend to be attracted towards other individuals (to avoid being isolated) and to align themselves with neighbors [19], [20]. Couzin et al. [17] identified four collective dynamical behaviors as illustrated in Figure 4.7:

- **Swarm**: an aggregate with cohesion, but a low level of polarization (parallel alignment) among members.
- **Torus**: individuals perpetually rotate around an empty core (milling). The direction of rotation is random.
- **Dynamic parallel group**: the individuals are polarized and move as a coherent group, but individuals can move throughout the group and density and group form can fluctuate [19], [21].
- **Highly parallel group**: much more static in terms of exchange of spatial positions within the group than the dynamic parallel group and the variation in density and form is minimal.

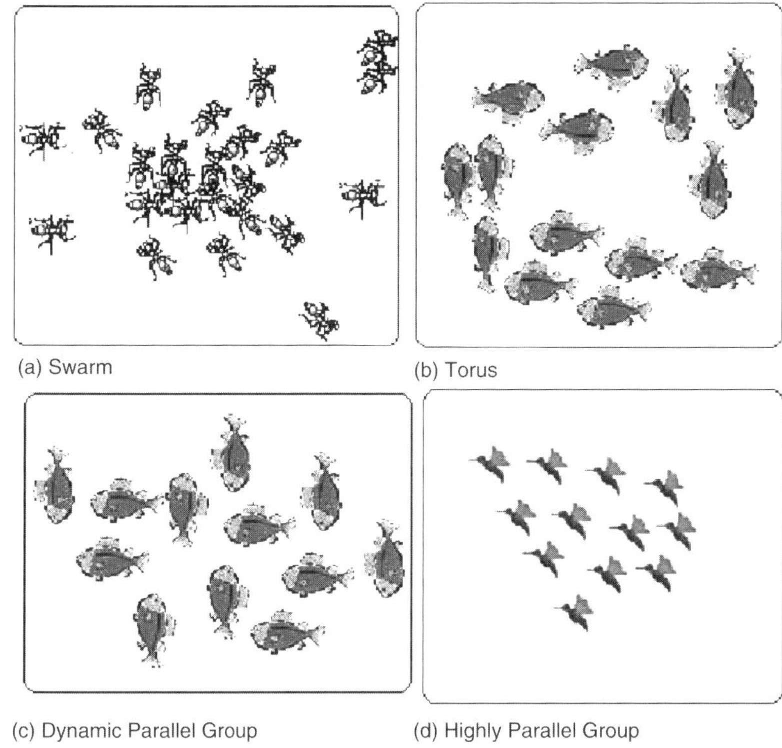

(a) Swarm (b) Torus

(c) Dynamic Parallel Group (d) Highly Parallel Group

Fig. 4.7. Different models of collective behavior (adapted from [23])

A swarm can be viewed as a group of agents cooperating to achieve some purposeful behavior and achieve some goal (see Figure 4.7) [23] . This collective intelligence seems to emerge from what are often large groups.

According to Milonas, five basic principles define the SI paradigm [24]. First is the proximity principle: the swarm should be able to carry out simple space and time computations. Second is the quality principle: the swarm should be able to respond to quality factors in the environment. Third is the principle of diverse response: the swarm should not commit its activities along excessively narrow channels. Fourth is the principle of stability: the swarm should not change its mode of behavior every time the environment changes. Fifth is the principle of adaptability: the swarm must be able to change behavior mote when it is worth the computational price. Note that principles four and five are the opposite sides of the same coin.

As it appears, 'Self-organization' is one of the fundamental features of any SI system. However, it is not a simple term to define. In general, it refers to the various mechanisms by which pattern, structure and order emerge spontaneously in complex systems. Examples of such structures and patterns include the stripes of zebras, the pattern of sand ripples in a dune, the coordinated movements of flocks of birds or schools of fish, the intricate earthen nests of termites, the patterns on seashells, the whorls of our fingerprints, the colorful patterns of fish and even the spatial pattern of stars in a spiral galaxy. Bonabeau *et al.* have tried to define self-organization using the following words [7]:

Self-organization is a set of dynamical mechanisms whereby structures appear at the global level of a system from interactions of its lower-level components.

Serra and Zanarini [25] describes the concept of self-organization generally as "highly organized behavior even in the absence of a pre-ordained design". They go on to further describe examples such as the resonance phenomenon in lasers, and in cellular automata where "unexpected and complex behaviours can be considered as self-organized." Self-organization was originally introduced in the context of physics and chemistry to describe how microscopic processes give rise to macroscopic structures in out-of-equilibrium systems. Recent research, however, suggests that it provides a concise description of a wide rage of collective phenomena in animals, especially in social insects. This description does not rely on individual complexity to account for complex spatial-temporal features, which emerge at the colony level, but rather assumes that interactions among simple individuals can produce highly structured collective behaviors. There are four main features that govern the self-organization in insect colonies:

- Positive feedback (amplification)
- Negative feedback (for counter-balance and stabilization)
- Amplification of fluctuations (randomness, errors, random walks)
- Multiple interactions

At a high-level, a swarm can be viewed as a group of agents cooperating to achieve some purposeful behavior and achieve some goal. This collective intelligence seems to emerge from what are often large groups of relatively simple agents. The agents use simple local rules to govern their actions and via the interactions of the entire group, the swarm achieves its objectives. A type of self-organization emerges from the collection of actions of the group.

An autonomous agent is a subsystem that interacts with its environment, which probably consists of other agents, but acts relatively independently from all other agents. The autonomous agent does not follow commands from a leader, or some global plan [26]. For example, for a bird to participate in a flock, it only adjusts its movements to coordinate with the movements of its flock mates, typically its neighbors that are close to it in the flock. A bird in a flock simply tries to stay close to its neighbors, but avoid collisions with them. Each bird does not take commands from any leader bird since there is no lead bird. Any bird can in the front, center and back of the swarm. Swarm behavior helps birds take advantage of several things including protection from predators (especially for birds in the middle of the flock), and searching for food (essentially each bird is exploiting the eyes of every other bird).

Below we discuss in details two algorithms from SI domain, which have gained huge popularity in a relatively short span of time all over the world. One of these algorithms, known as Ant Colony Optimization (ACO) mimics the behavior of group of real ants in multi-agent cooperative search problems. The latter one is referred to as Particle Swarm Optimization (PSO), which draws inspiration from the behavior of particles, the boids method of Craig Reynolds and socio-cognition [27].

4.3.1 Ant Colony Systems

Insects like ants, bees, wasps and termites are quite social. They live in colonies and follow their own routine of tasks independent of each other. However, when acting as a community, these insects even with very limited individual capability can jointly (cooperatively) perform many complex tasks necessary for their survival [7]. Problems like finding and storing foods, selecting and picking up materials for future usage require a detailed planning, and are solved by insect colonies without any kind of supervisor or controller.

It is a natural observation that a group of 'almost blind' ants can figure out the shortest route between a cube of sugar and their nest without any visual information. They are capable of adapting to the changes in the environment as well [28]. It is interesting to note that ants while crawling deposit trails of a chemical substance known as pheromone to help other members of their team to follow its trace. The resulting collective behavior can be described as a loop of positive feedback, where the probability of an ant's choosing a path increases as the count of ants that already passed by that path increases [12], [28].

The basic idea of a real ant system is illustrated in Figure 4.8. In the left picture, the ants move in a straight line to the food. The middle picture illustrates the situation soon after an obstacle is inserted between the nest and the food. To avoid the obstacle, initially each ant chooses to turn left or right at random. Let us assume that ants move at the same speed depositing pheromone in the trail uniformly. However, the ants that, by chance, choose to turn left will reach the food sooner, whereas the ants that go around the obstacle turning right will follow a longer path, and so will take longer time to circumvent the obstacle. As a result, pheromone accumulates faster in the shorter path around the obstacle. Since ants prefer to follow trails with larger amounts of pheromone, eventually all the ants converge to the shorter path around the obstacle, as shown in Figure 4.8.

An artificial Ant Colony System (ACS) is an agent-based system, which simulates the natural behavior of ants and develops mechanisms of cooperation and learning. ACS was proposed by Dorigo et al. [29] as a new heuristic to solve combinatorial optimization problems. This new heuristic, called Ant Colony Optimization

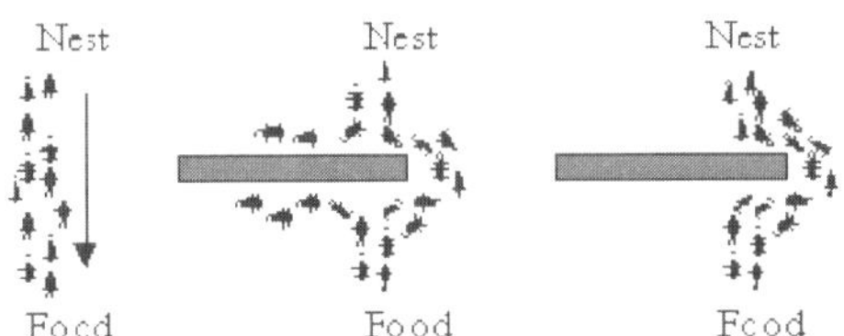

Fig. 4.8. Illustrating the behavior of real ant movements

(ACO) has been found to be both robust and versatile in handling a wide range of combinatorial optimization problems.

4.3.2 The ACO Algorithm

The main idea of ACO is to model a problem as the search for a minimum cost path in a graph. Artificial ants as if walk on this graph, looking for cheaper paths. Each ant has a rather simple behavior capable of finding relatively costlier paths. Cheaper paths are found as the emergent result of the global cooperation among ants in the colony. The behavior of artificial ants is inspired from real ants: they lay pheromone trails (obviously in a mathematical form) on the graph edges and choose their path with respect to probabilities that depend on pheromone trails. These pheromone trails progressively decrease by evaporation. In addition, artificial ants have some extra features not seen in their counterpart in real ants. In particular, they live in a discrete world (a graph) and their moves consist of transitions from nodes to nodes.

Pheromone placed on the edges acts like a distributed long term memory [29]. The memory, instead of being stored locally within individual ants, remains distributed on the edges of the graph. This indirectly provides a means of communication among the ants called *stigmergy* [30]. In most cases, pheromone trails are updated only after having constructed a complete path and not during the walk, and the amount of pheromone deposited is usually a function of the quality of the path. Finally, the probability for an artificial ant to choose an edge, not only depends on pheromones deposited on that edge in the past, but also on some problem dependent local heuristic functions.

We illustrate the use of ACO in finding the optimal tour in the classical Traveling Salesman Problem (TSP). Given a set of n cities and a set of distances between them, the problem is to determine a minimum traversal of the cities and return to the home-station at the end. It is indeed important to note that the traversal should in no way include a city more than once. Let $r(C_x, C_y)$ be a measure of cost for traversal from city C_x to C_y. Naturally, the total cost of traversing n cities indexed by i_1, i_2, $i_3, .., i_n$ in order is given by the following expression:

$$Cost(i_1, i_2, \ldots, i_n) = \sum_{j=1}^{n-1} r(C_{i_j}, C_{i_{j+1}}) + r(C_{i_n}, C_{i_1}) \tag{4.1}$$

The ACO algorithm is employed to find an optimal order of traversal of the cities. Let τ be a mathematical entity modeling the pheromone and $\tau_{ij} = 1/r_{i,j}$ is a local heuristic. Also let $allowed_k(t)$ be the set of cities that are yet to be visited by ant k located in city i. Then according to the classical ant system [11] the probability that ant k in city i visits city j is given by:

$$\begin{aligned} p_{ij}^k(t) &= \frac{[\tau_{ij}(t)]^\alpha \bullet [\eta_{ij}]^\beta}{\sum_{h \in allowed_k(t)} [\tau_{ih}(t)]^\alpha \bullet [\eta_{ih}]^\beta}, \quad if \quad j \in \quad allowed_k(t) \\ &= 0, \ otherwise \end{aligned} \tag{4.2}$$

In Equation 4.2, shorter edges with greater amount of pheromone are favored by multiplying the pheromone on edge (i, j) by the corresponding heuristic value $\eta(i, j)$. Parameters $\alpha(>0)$ and $\beta(>0)$ determine the relative importance of pheromone versus cost. Now in ant system pheromone trails are updated as follows. Let D_k be the length of the tour performed by ant k,$\Delta D_k(i, j) = 1/D_k$ if $(i, j)\varepsilon$ tour done by ant k and = 0 otherwise and finally let $\rho\varepsilon[0, 1]$ be a pheromone decay parameter which takes care of the occasional evaporation of the pheromone from the visited edges. Then once all ants have built their tours, pheromone is updated on all the edges as follows:

$$\tau(i, j) = (1-\rho)\cdot\tau(i, j) + \sum_{k=1}^{m} \Delta\tau_k(i, j) \tag{4.3}$$

From Equation 4.3, we can guess that pheromone updating attempts to accumulate greater amount of pheromone to shorter tours (which corresponds to high value of the second term in (4.3), so as to compensate for any loss of pheromone due to the first term). This conceptually resembles a reinforcement-learning scheme, where better solutions receive a higher reinforcement.

The ACS differs from the classical ant system in the sense that here the pheromone trails are updated in two ways. Firstly, when ants construct a tour they locally change the amount of pheromone on the visited edges by a local updating rule. Now if we let γ to be a decay parameter and $\Delta\tau(i, j) = \tau_0$ such that τ_0 is the initial pheromone level, then the local rule may be stated as:

$$\tau(i, j) = (1-\rho)\cdot\tau(i, j) + \gamma\cdot\Delta\tau(i, j) \tag{4.4}$$

Secondly, after all the ants have built their individual tours, a global updating rule is applied to modify the pheromone level on the edges that belong to the best ant tour found so far. If $\hat{E}$ be the usual pheromone evaporation constant, D_{gb} be the length of the globally best tour from the beginning of the trial and $\Delta\tau' = 1/D_{gb}$ only when the edge (i, j) belongs to global-best-tour and zero otherwise, then we may express the global rule as:

$$\tau(i, j) = (1-k)\cdot\tau(i, j) + k\cdot\Delta\tau'(i, j) \tag{4.5}$$

The main steps of ACS algorithm are presented as Algorithm 1. The first loop (iteration) starts with m ants being placed in n cities chosen according to some initialization rule (e.g. randomly). In the embedded loop (step) each ant builds a tour (i.e., an acceptable solution to the TSP) by repeatedly applying a stochastic state transition rule. While building its tour, the ant can modify the pheromone level on the visited edges by applying the local updating rule given by (4.4). Once all the ants have terminated their tour, the pheromone trails are modified again by the global updating rule given in (4.5). Figure 4.9 illustrates the computer simulation of the ACO technique working on a 10 city TSP problem.

Algorithm 1 Ant colony system algorithm

```
Begin
  Initialize pheromone trails;
  Repeat
  Begin /* at this stage each loop is called an iteration */
      Each ant is positioned on a starting node;
      Repeat
      Begin /* at this level each loop is called a step */
          Each ant applies a state transition rule like rule (2) to incrementally build a solution
and a local pheromone-updating rule like rule (4.4);
      Until all ants have built a complete solution;
      A global pheromone-updating rule like rule (4.5) is applied.
  Until terminating condition is reached;
End
```

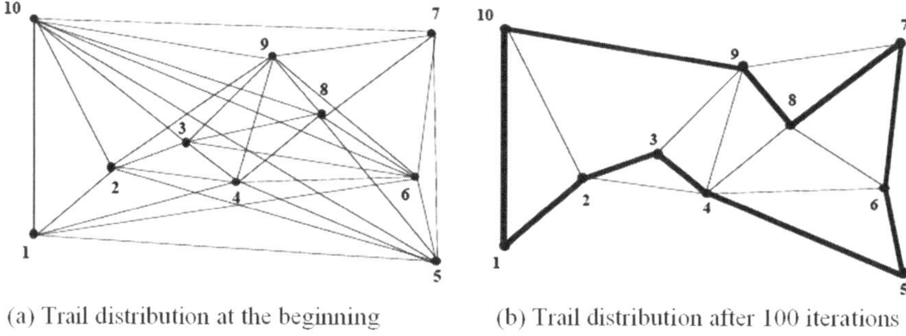

(a) Trail distribution at the beginning (b) Trail distribution after 100 iterations

Fig. 4.9. Solving the TSP problem with ACO algorithm (adapted from [29])

4.3.3 The Particle Swarm Optimisation (PSO)

The concept of particle swarms, although initially introduced for simulating human social behaviors, has become very popular these days as an efficient search and optimization technique. The Particle Swarm Optimization (PSO) [10], as it is called now, does not require any gradient information of the function to be optimized, uses only primitive mathematical operators and is conceptually very simple. Since its advent in 1995, PSO has attracted the attention of a lot of researchers all over the world resulting into a huge number of variants of the basic algorithm as well as many parameter automation strategies. PSO [27] is in principle such a multi-agent parallel search technique. Particles are conceptual entities which fly through the multi-dimensional search space. At any particular instant, each particle has a position and a velocity. The position vector of a particle with respect to the origin of the search space represents a trial solution of the search problem. At the beginning a population of particles is initialized with random positions marked by vectors $\mathbf{X}_i$ and random velocities $\mathbf{V}_i$. The population of such particles is called a 'swarm' S. A neighborhood relation N is defined in the swarm. N determines for any two particles Z_i and Z_j whether they are neighbors or not. Thus for any particle Z, a neighborhood can be assigned as $N(Z)$, containing all the neighbors of that particle.

Different neighborhood topologies and their effect on the swarm performance have been discussed in [32]. In the basic PSO, each particle P has two state variables:

1. Its current position $\mathbf{X}_i(t)$.
2. Its current velocity $\mathbf{V}_i(t)$.

And also a small memory comprising,

1. Its previous best position $\mathbf{P}_i(t)$ i.e. personal best experience in terms of the objective function value $f(\mathbf{P}_i(t))$.
2. The best $\mathbf{P}(t)$ of all $Z\varepsilon N(Z)$: i.e. the best position found so far in the neighborhood of the particle.

The PSO scheme has the following algorithmic parameters:

1. $\mathbf{V}_{max}$ or maximum velocity which restricts $\mathbf{V}_i(t)$ within the interval $[-V_{max}, V_{max}]$
2. An inertial weight factor ω.
3. Two uniformly distributed random numbers φ_1 and φ_2 which respectively determine the influence of $\mathbf{P}(t)$ and $\mathbf{g}(t)$ on the velocity update formula.
4. Two constant multiplier terms C_1 and C_2 known as "self confidence" and "swarm confidence" respectively.

Initially the settings for $\mathbf{P}(t)$ and $\mathbf{g}(t)$ are $\mathbf{P}(0) = \mathbf{g}(0) = \mathbf{x}(0)$ for all particles. Once the particles are initialized, the iterative optimization process begins where the positions and velocities of all the particles are altered by the following recursive equations. The equations are presented for the d-th dimension of the position and velocity of the $i-th$ particle.

$$\begin{aligned} V_{id}(t+1) &= \omega V_{id}(t) + C_1\phi_1.(P_d(t) - X_{id}(t)) + C_2\phi_2.(g_d(t) - X_{id}(t)) \\ X_{id}(t+1) &= X_{id}(t) + V_{id}(t+1) \end{aligned} \tag{4.6}$$

The first term in the velocity updating formula represents the inertial velocity of the particle. The second term involving $\mathbf{P}(t)$ represents the personal experience of each particle and is referred to as "cognitive part". The last term of the same relation is interpreted as the "social term" which represents how an individual particle is influenced by the other members of its society. Typically, this process is iterated for a certain number of time steps, or until some acceptable solution has been found by the algorithm or until an upper limit of CPU usage has been reached. Once the iterations are terminated, most of the particles are expected to converge to a small radius surrounding the global optima of the search space. The velocity updating scheme is illustrated in Figure 4.10 with a humanoid particle. A pseudo code for the PSO algorithm is depicted as Algorithm 2.

The PSO algorithm can be seen as a set of vectors whose trajectories oscillate around a region defined by each individual previous best position and the best position of some other individuals [27]. There are different neighborhood topologies used to identify which particles from the swarm can influence the individuals. The most common ones are known as the *gbest* and *lbest*. In the *gbest* swarm; the trajectory of each individual (particle) is influenced by the best individual found in the entire

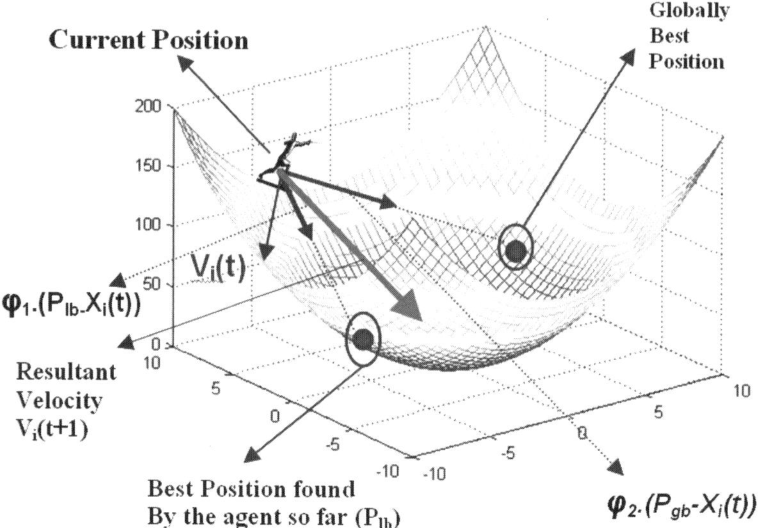

Fig. 4.10. Illustrating the velocity updating scheme of basic PSO

Algorithm 2 Particle swarm optimization algorithm

PSO Algorithm input: Randomly initialized position and velocity of the particles: $\mathbf{X}_i(0)$ and $\mathbf{V}_i(0)$
Output: Position of the approximate global optima
Begin
  While terminating condition is not reached do
    Begin
      for $i = 1$ to number of particles
        Evaluate the fitness: $=f(\mathbf{X}_i(t))$;
        Update $\mathbf{P}(t)$ and $\mathbf{g}(t)$;
        Adapt velocity of the particle using equation 4.6;
        Update the position of the particle;
        increase;
    end while;
  end;

swarm. It is assumed that *gbest* swarms converge fast, as all the particles are attracted simultaneously to the best part of the search space. However, if the global optimum is not close to the best particle, it may be impossible for the swarm to explore other areas and, consequently, the swarm can be trapped in local optima [33]. In the *lbest* swarm, each individual is influenced by a smaller number of its neighbors (which are seen as adjacent members of the swarm array). Typically, *lbest* neighborhoods comprise of two neighbors: one on the right side and one on the left side (a ring lattice). This type of swarm will converge slower but can locate the global optimum with a greater chance. *lbest* swarm is able to flow around local optima, sub-swarms being

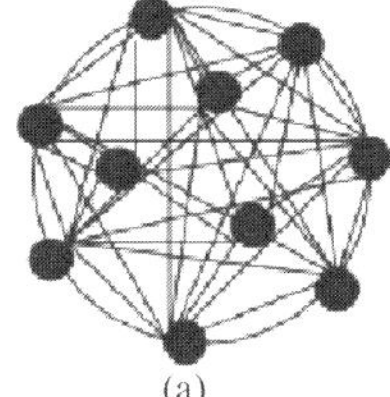

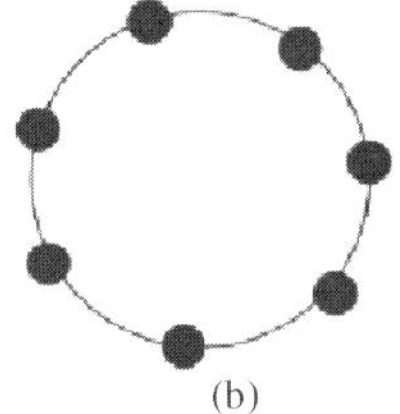

Fig. 4.11. Graphical Representation of (a) *gbest* Swarm (b) lbest Swarm (adapted from [33])

able to explore different optima [32]. A graphical representation of a *gbest* swarm and an *lbest* swarm respectively is depicted in Figure 4.11.

Watts [34] introduced the small-world network model, which allows interpolating between regular low-dimensional lattices and random networks, by introducing a certain amount of random long-range connections into an initially regular network [35]. Starting from here, several models have been developed: icing model [36], spreading of epidemics [37], evolution of random walks [38] are some of them.

4.3.4 Relevance of SI Algorithms in Bioinformatics

From the discussion of the previous subsections, we see that the SI algorithms are mainly stochastic search and optimization techniques, guided by the principles of collective behaviour and self organization of insect swarms. They are efficient, adaptive and robust search methods producing near optimal solutions and have a large amount of implicit parallelism. On the other hand, several tasks in bioinformatics involve optimization of different criteria (like energy, alignment score, overlap strength and so on); thereby making the application of SI tools more obvious and appropriate. For example, most of the ordering problems in bioinformatics, such as the sequence alignment, fragment assembly problem (FAP) and gene mapping (GM), are quite similar to the TSP (one of the most difficult ordering problems till date) with notable differences [39]. We have already discussed how TSP can be solved efficiently with the ant systems in Section 4.3.2. Thus, ACO can be tried on many of these problems and the results can be compared with the classical methods used in these contexts up to now.

The problems of bioinformatics seldom need the exact optimum solution; rather what they need are robust, fast and near optimal solutions, which SI algorithms like PSO are known to produce efficiently. Moreover, the laboratory operations on DNA inherently involve errors. These are more tolerable in executing the SI algorithms than in executing deterministic algorithms. To some extent, these errors may be regarded as contributing to population diversity, a desirable property for the convergence of the SI algorithms. The problem of integrating SI in bioinformatics, in this way, can develop a new research area.

4.4 A Review of the Present State of the Art

In this section, we provide a substantial review of the state of the art research, which focuses on the application of swarm intelligence to different bioinformatics related problems. The number of published papers reporting the applications of PSO or ACO in bioinformatics is currently smaller as compared to the huge amount of work reported for other evolutionary computing methods like GA etc in the same context. Nevertheless, we believe that in near future SI will serve as an indispensable computing methodology in the field of bioinformatics, keeping in mind the reported success of the SI algorithms over classical evolutionary algorithms in many cases [40], [41], [42]. We describe each research problem first and then illustrate how SI algorithms can be used to solve them.

4.4.1 Clustering of Gene Expression Data

Gene expression refers to a process through which the coded information of a gene is converted into structures operating in the cell. It provides the physical evidence that a gene has been "turned on" or activated. Expressed genes include those that are transcribed into m-RNA and then translated into protein and those that are transcribed into RNA but not translated into protein (e.g., transfer and ribosomal RNAs) [43].

The expression levels of thousands of genes can be measured at the same time using the modern microarray technology [44], [45]. DNA microarrays usually consist of thin glass or nylon substrates containing specific DNA gene samples spotted in an array by a robotic printing device. Researchers spread fluorescently labeled m-RNA from an experimental condition onto the DNA gene samples in the array. This m-RNA binds (hybridizes) strongly with some DNA gene samples and weakly with others, depending on the inherent double helical characteristics. A laser scans the array and sensors to detect the fluorescence levels (using red and green dyes), indicating the strength with which the sample expresses each gene. The logarithmic ratio between the two intensities of each dye is used as the gene expression data.

Proper selection, analysis and interpretation of the microarray data can lead us to the answers of many important problems in experimental biology. In the field of pattern recognition, clustering [46] refers to the process of partitioning a dataset into a finite number of groups according to some similarity measure. Currently it has become a widely used process in microarray engineering for understanding the functional relationship between groups of genes. Clustering was used, for example, to understand the functional differences in cultured primary epatocytes relative to the intact liver [47]. In another study, clustering techniques were used on gene expression data for tumor and normal colon tissue probed by oligonucleotide arrays [48].

To cluster the microarray dataset, the first thing we need a suitable similarity measure among the gene profiles. Euclidean distance serves the purpose when the objective is to partition genes displaying similar level of expression. Let $gene_i(x_{i1},$

$x_{i2},...,x_{in}$) denote the expression pattern of the $i-th$ gene. Then the Euclidean distance between the $i-th$ and the $j-th$ gene is given by:

$$d_{i,j} = \sqrt{\sum_{k=1}^{n} (x_{ik} - x_{jk})^2} \quad (4.7)$$

Another popular similarity measure used in this context is the Pearson Correlation Coefficient [49] given by

$$r = \frac{\sum_{k=1}^{n}((x_{ik} - \hat{x}_i)(x_{jk} - \hat{x}_j))/n}{\sigma_{x_i} * \sigma_{x_j}} \quad (4.8)$$

A number of standard clustering algorithms such as hierarchical clustering [50], [51], principle component analysis (PCA) [52] [53], genetic algorithms [54], and artificial neural networks [55] [56] [57], have been used to cluster gene expression data. However, in 2003, Xiao *et al.* [58] used a new approach based on the synergism of the PSO and the Self Organizing Maps (SOM) for clustering them. Authors achieved promising results by applying the hybrid SOM-PSO algorithm over the gene expression data of Yeast and Rat Hepatocytes. We will briefly discuss their approach in the following paragraphs.

The idea of the SOM [59] stems from the orderly mapping of information in the cerebral cortex. With SOM, high dimensional datasets are projected onto a one- or two- dimensional space. Typically, a SOM has a two dimensional lattice of neurons and each neuron represents a cluster. The learning process of SOM is unsupervised. All neurons compete for each input pattern; the neuron that is chosen for the input pattern wins it.

Xiao et al. [58] used PSO to evolve the weights for SOM. In the first stage of the hybrid SOM/PSO algorithm, SOM is used to cluster the dataset. Authors used a SOM with conscience at this step. Conscience directs each component that takes part in competitive learning toward having the same probability to win. Conscience is added to SOM by assigning each output neuron a bias. The output neuron must overcome its own bias to win. The objective is to obtain a better approx. of pattern distribution. The SOM normally runs for 100 iterations and generates a group of weights. In the second stage, PSO is initialized with the weights produced by SOM in the first stage. Then a *gbest* PSO is used to refine the clustering process. Each particle consists of a complete set of weights for SOM. The dimension of each particle is the number of input neurons of SOM times the number of output neurons of SOM. The objective of PSO is to improve the clustering result by evolving the population of particles.

4.4.2 The Molecular Docking Problem

Formally, the protein-ligand docking problem may be described as: We are given a geometric and chemical description of a protein and an arbitrary small organic molecule. We want to determine computationally whether the small molecule will

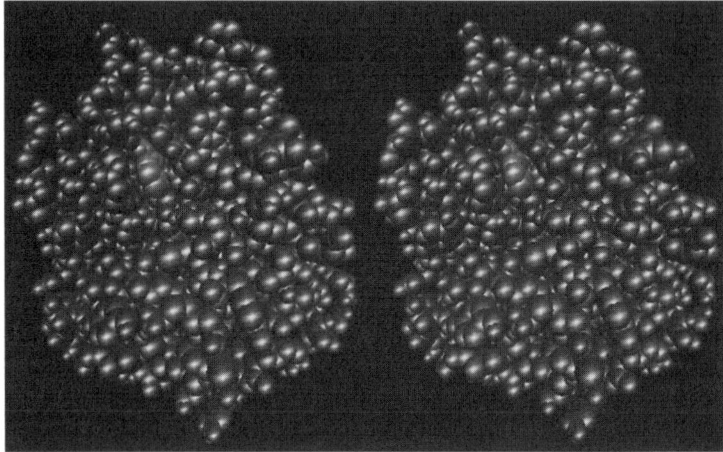

Fig. 4.12. Stereo view of benzamidine docked in the active site of trypsin

bind to the protein, and if so, we would like to estimate the geometry of the bound complex, as well as the affinity of the binding. Figure 4.12 illustrates how benzamidine, a trypsin inhibitor, docks into the active site of trypsin, a protease involved in digestion (adapted from [60]).

Liu *et al.* [61] proposed a novel PSO based docking technique, which they called SODOCK (Swarm Optimization for molecular DOCKing). After comparing with a number of state of the art docking techniques like GOLD 1.2 [62], AutoDock 3.05 [63], DOCK 4.0 [64] etc., they found promising results for SODOCK in terms of robustness, accuracy and the speed of convergence. In SODOCK, three kinds of parameters are optimized using the PSO:

- **Translation**: three parameters in this category specify the translation of the center of the ligand with respect to the center of a 3D grid box that encloses the binding site of the protein.
- **Orientation**: There are four parameters n_x,n_y,n_z and α where n_x,n_y,n_z $\varepsilon[0,1]$ specify the normal vector of the ligand whereas $\alpha\varepsilon[-\pi,\pi]$ represent the angle of self rotation along the normal vector.
- **Torsions**: These are torsion angles $tor_i\varepsilon[-\pi,\pi]$ associated with the rotating bonds, $i = 1,2,\ldots,T$.

Thus, the PSO algorithm is used to evolve a total of $N = 7 + \mathrm{T}$ parameters such that the following docking energy function is minimized:

$$E_{tot} = E_{vdw} + E_{H-bond} + E_{pot} + E_{intern} \tag{4.9}$$

The first three terms in the above expression, correspond to the intermolecular energies: van der Waals force, hydrogen bonding, and electronic potential. The last term represents the internal energy of the ligand, which also consists of the three elements.

The fitness landscape of the energy function shown in 4.9 is usually riddled with multiple local minima. In order to tackle these local peaks efficiently, Liu *et al.* integrated a local search strategy (a variant of the Solis and Wet local search [63]) in SODOCK. A generation of SODOCK has four stages: update of velocity, move of particle, local search, and update of local and global best positions. The local search may be applied to the particle according to a predefined probability P_{ls}. Finally, the local and global best positions of particles are updated if their energies are improved. The particles having the smallest energy correspond to a solution to the flexible docking problem.

4.4.3 The Multiple Sequence Alignment Problems (MSA)

Sequence alignment refers to the process of arranging the primary sequences of DNA, RNA, or protein to identify regions of similarity that may be a consequence of functional, structural, or evolutionary relationships between the sequences. Given two sequences X and Y, a pair-wise alignment indicates positions of each sequence that are considered to be functionally or evolutionarily related. From a family $S = (S_0, S_1, \ldots, S_{N-1})$ of N sequences, we would like to find out the common patterns of this family. Since aligning each pair of sequences from S separately often does not reveal the common information, it is necessary to perform multiple sequence alignment (MSA). A multiple sequence alignment (MSA) is a sequence alignment of three or more biological sequences, generally protein, DNA, or RNA. In general, the input set of query sequences are assumed to have an evolutionary relationship by which they share a linkage and are descended from a common ancestor. An example of multiple alignments of five sequences is illustrated in Figure 4.13.

To evaluate the quality of an alignment, a popular choice is to use the SP (sum of pairs) score method [65]. The SP score basically sums the substitution scores of all possible pair-wise combinations of sequence characters in one column of a multiple sequence alignment. Assuming c_i representing the $i-th$ character of a given column in the sequence matrix and match (c_i, c_j) denoting the comparing score between characters c_i and c_j, the score of a column may be computed using the formula:

$$SP = (c_1, c_2, \ldots, c_N) = \sum_{i=1}^{N-1} \sum_{j=i+1}^{N} match(c_i, c_j) \tag{4.10}$$

Progressive alignment is a widely used heuristic MSA method that does not guarantee any level of optimality [66]. ClustalW [67] is another widely popular program

```
     -----FKQCCWNSLP-----RGLSNVALVYQEFMAKCRGESENLQLVTALVINLPSMA
    --------SMFRQCIWNSLS-----HGLPETAPIYQPLKARCRGVSENLQLVTEIIINLPTLC
   -------SLWCQCIKASLPLKVIRGTPEVAPLYDQLEQVCRSENQ------VSEIVAKFASLC
------TMFKMCLWNALP-------RGLPEVAPVYRPLKARCRGDSENLQLCAERLVNLPELC
  ---------AILRSCIWNLLP---------RGLPEAAPIYEPLKARLRGESENYKLVTEIIMTLPSLC
```

Fig. 4.13. An example of multiple sequence alignments

that improved the algorithm presented by Feng and Doolittle [66]. The main shortcoming of ClustalW is that once a sequence has been aligned, that alignment can never be modified even if it conflicts with sequences added later.

Recently, Chen et al. [68] took a serious attempt to solve the classical MSA problem by using a partitioning approach coupled with the ACO algorithm. Authors algorithm consists of three stages. At first a genetic algorithm is employed to find out the near optimal cut-off points in the original sequences from where they must be partitioned vertically. In this way a partitioning method is continued recursively to reduce the original problem to multiple smaller MSA problems until the lengths of the subsequences are all less than an acceptable threshold. Next, an ant colony system is used to align each small subsection derived from the previous step. The ant system consists of N ants each of which represents a solution of alignment. Each ant searches for an alignment by moving on the sequences to choose the matching characters. Let the N sequences be $S = S_0, S_1, \ldots, S_{N-1}$. In that case an artificial ant starts from $S_0[0]$, the first character of S_0, and selects one character from each of the sequences of $S_1, \ldots, S_{N-1}$ matching with $S_0[0]$. From the sequence $S_i, i = 1, 2, \ldots, n_1$, the ant selects a character $S_i[j]$ by a probability determined by the matching score with $S_0[0]$, deviation of its location from $S_0[0]$ and pheromones trail on the logical edge between $S_i[j]$ and $S_0[0]$. In addition, an ant may choose to insert an empty space according to a predetermined probability. Next, the ant starts from $S_0[1]$, selects the characters of $S_1, \ldots, S_{N-1}$ matching with $S_0[1]$ to form the second path. Similarly, starting from $S_0[2], \ldots, S_0[|S_0| - 1]$, the ant can form other paths. Here $|S_0|$ indicates the number of characters in the sequence $|S_0|$. All these $|S_0|$ paths forming an alignment solution is reproduced in Figure 4.14.

To evaluate an alignment represented by a set of paths, the positions of characters not selected by the ants are calculated first by aligning them to the right and adding gaps to the left. Next their SP (sum-of-pairs) score is using relation 4.9. Finally, a solution to the MSA is obtained by concatenating the results from smaller sub-alignments. Chen *et al.* showed that the Divide-Ant-MSA algorithm outperforms the SAGA [69] a leading MSA program based on genetic algorithm (GA) in terms of both speed and accuracy especially for longer sequences.

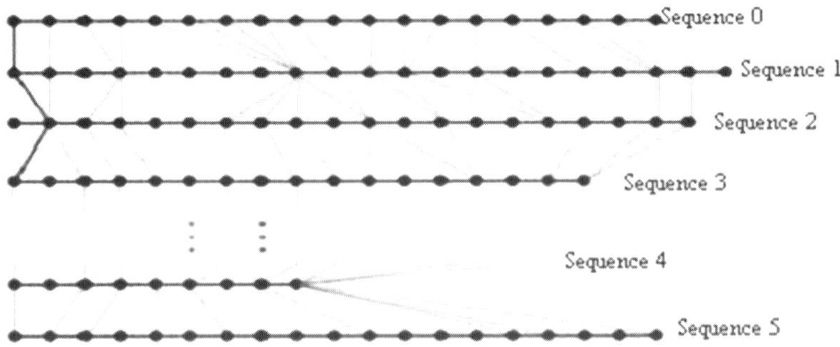

Fig. 4.14. An example alignment as presented by the paths traced out by the ants

Rasmussen and Krink in [70] focussed on a new PSO based training method for Hidden Markov models (HMMs) in order to solve the MSA problem. The authors demonstrated how the combination of PSO and evolutionary algorithms can generate better protein sequence alignments than with more traditional HMM training methods, such as Baum-Welch [71] and simulated annealing [72].

4.4.4 The Construction of Phylogenetic Trees

Every species on earth undergo slow change of their hereditary traits in course of evolution. The phylogenetic or evolutionary trees are schematic binary trees showing the evolutionary interrelationships among various species that are believed to have a common ancestor [15]. The leaves of such a tree represent the present day species while the interior nodes represent the hypothesized ancestors. Phylogenetic trees may be rooted or un-rooted (Figure 4.15). An un-rooted tree simply represents phylogenies but does not provide an evolutionary path. In case of a rooted tree, one of the internal nodes is used as an out-group, and, in essence, becomes the common ancestor of all the other external nodes. The out-group therefore enables the root of a tree to be located and the correct evolutionary pathway to be identified.

In a phylogenetic tree, the phylogenies are reconstructed based on comparisons among the present-day objects. The term object is used to denote the units for which one wants to reconstruct the phylogeny. The input data essential for constructing phylogeny are of two types [15].

- Discrete characters, such as beak shape, number of fingers, presence or absence of a molecular restriction site. Each character can have a finite number of states. The data relative to these characters are placed in an objects character matrix called character state matrix.
- Comparative numerical data, called distances between objects. The resulting matrix is called distance matrix.

Given data (character state matrix or distance matrix) for n taxa (object), the phylogenetic tree reconstruction problem is to find the particular permutation of taxa

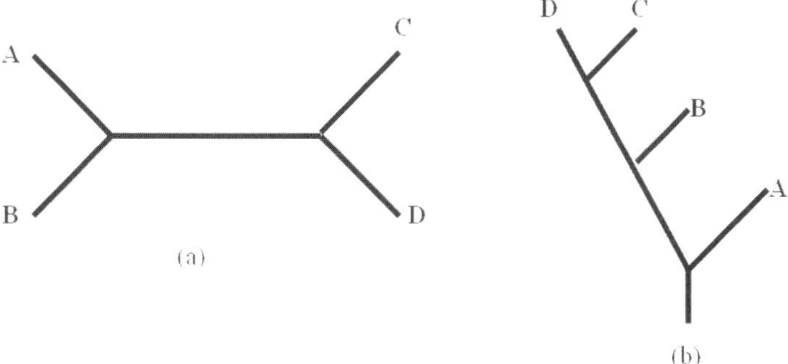

Fig. 4.15. Topologies of phylogenetic trees (a) un-rooted (b) rooted

that optimize the criteria (distance). Felenstein has shown that considering n species, it is possible to construct N number of trees where, N is given by the following equations for unrooted trees and rooted trees [73]:

$$N = \frac{(2n-3)!}{2^{n-2}(n-2)!} \tag{4.11}$$

$$N = \frac{(2n-5)!}{2^{n-3}(n-3)!} \tag{4.12}$$

The problem poses severe computational challenge before us, example, if we would like to find the best tree using the method of maximum similarity for (only) 15 species, we should try 213, 458, 046, 676, 875 trees.

The phylogenetic tree construction problem bears close resemblance to a standard TSP, (Traveling Salesman Problem) described earlier in Section 4.3.2. One can simply associate one imaginary city to each taxa, and define as the distance between two cities the data obtained from the data matrix for the corresponding pair of taxas. This kind of formulation of the problem paves the path for the application of heuristic algorithms like GA [74], [75] and ACO. Perretto *et al.* [76] proposed a slightly modified artificial ant colony based algorithm for the construction of phylogenetic trees. Their approach starts with building a two-dimensional fully-connected graph using the distance matrix among the species. In this graph, nodes represent the species and edges represent the evolutionary distances between species. An example of such a graph is provided in Figure 4.16. The ants start from a randomly selected node and continue traveling across the structured graph. At each node a transition function similar in form to equation 4.2, determines its direction.

The method described in [76], differs from the classical ant systems based TSP in only one respect. In case of the former algorithm, moves are made between nodes, but here, the ant system creates an intermediary node between the two previously selected ones. This node will represent the ancestral species of the other two, and it will not be in the list of nodes (species) to be set in the tree. Using such an intermediary node, distances to the remaining nodes (species) are recomputed.

The ants initially start from a randomly selected node and continue traveling across the structured graph. At each node a transition function similar in form to

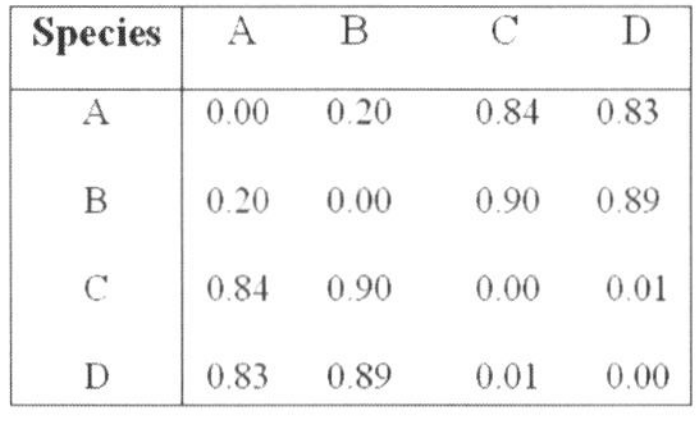

Species	A	B	C	D
A	0.00	0.20	0.84	0.83
B	0.20	0.00	0.90	0.89
C	0.84	0.90	0.00	0.01
D	0.83	0.89	0.01	0.00

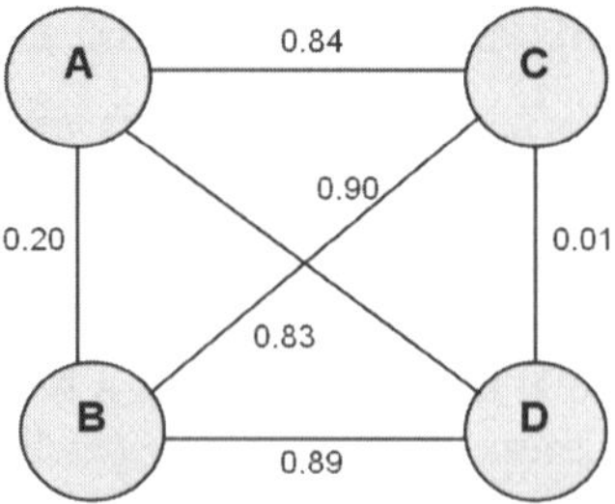

Fig. 4.16. Distance matrix for four species and the corresponding two-dimensional graph

equation 4.2 determines its direction. In original ACO based algorithm for TSP, moves are made between nodes. But here, the ant system creates an intermediary node between the two previously selected ones. This node will represent the ancestral species of the other two, and it will not be in the list of nodes (species) to be set in the tree. Using such an intermediary node, distances to the remaining nodes (species) are recomputed. This procedure is repeated until all nodes belong to the list of already visited nodes, and then a path is constructed. The score of this path is given by the sum of the transition probabilities of the adjacent nodes of the path. Paths constructed by the ants are then used for updating the pheromone trail. An increment of the pheromone trail is made at all nodes belonging to at least one path, created in an execution cycle. This key point helps to avoid trapping in a local maximum. In this way, following an algorithm very close in spirit to the ant colony algorithm for solving the TSP, the phylogenetic trees may be reconstructed efficiently.

Ando and Iba [77] proposed an ant algorithm for the construction of evolutionary trees from a given DNA sequence. Authors algorithm searches for a tree structure that minimizes the score for a given set of DNA sequences. It uses the mutual distance matrix of the leaves as the input. While the ACO for TSP visits the respective cities once to construct a round trip, ants in tree constructing algorithm visit leaves and vertices of the tree to construct a suffix representation of the bifurcating tree. The algorithm is shown to compete with conventional methods of the exhaustive search or the sequential insertion method, taken by the most popular methods.

4.4.5 The RNA Secondary Structure Prediction

Ribonucleic acid (RNA) is a nucleic acid polymer (like DNA) consisting of nucleotide monomers. Unlike deoxyribonucleic acid (DNA), which contains deoxyribose and thymine, RNA nucleotides contain ribose rings and uracil. As pointed out in Section 4.2.2, RNA serves as the template for translation of genes into proteins, transferring amino acids to the ribosome to form proteins, and also translating the transcript into proteins.

Like protein secondary structure (discussed in Section 4.2.3), RNA secondary structure may be conveniently viewed as an intermediate step in the formation of a three dimensional structure [13]. RNA secondary structure is composed primarily of double-stranded RNA regions formed by folding the single-stranded molecule on itself. To produce such double-stranded regions, a downstream sequence of the bases in RNA must be complementary to another upstream sequence so that Watson-Crick base pairing can occur between the complementary nucleotides G-C and A-U (analogous to the G-C and A-T base pairs in DNA). Among the several recognizable "domains" of secondary structure three well known ones are hairpin loops, bulges and internal loops. Figure 4.17 shows the folding of a single stranded RNA molecule into a hairpin structure.

Secondary structure of RNA molecules can be predicted computationally by calculating the minimum free energy (MFE) structure for all different combinations of hydrogen bonds and domains. Neethling and Engelbrecht [78] attempted to optimize the structure of RNA molecules using a modified PSO algorithm. The SetPSO, which

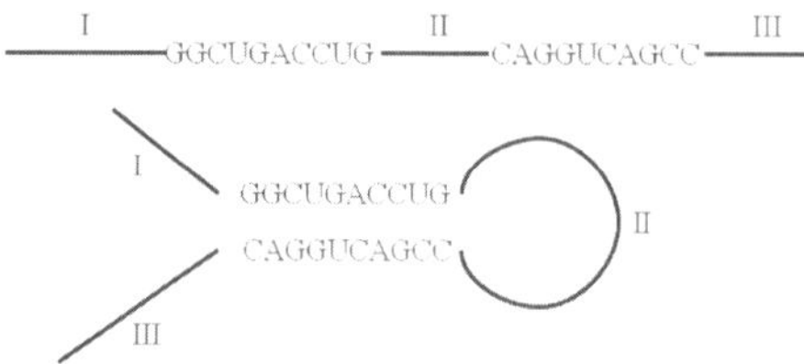

Fig. 4.17. Illustrating the formation of a hair-pin RNA secondary structure

the authors proposed for this purpose, can operate on mathematical sets in order to solve set-based combinatorial optimization problems in discrete search spaces. In SetPSO, the position of each particle denotes a set and this necessitates the redefinition of the addition and subtraction operators suitably. The addition of the position vectors of two particles here essentially means the union of the two sets, which they represent (i.e. $A+B$ now represents the set of all elements which belong to both A and B). On the other hand the subtraction operation is basically the set-theoretic difference between two sets A and B (i.e. $A-B$ denotes a set of all elements which belong to A but not to B).

In the course of folding back of RNA, the process of binding of the adjacent complementary bases is known as stacking. A stack or stem representing a valid RNA secondary structure should satisfy a few constraints like each base can pair with only one canonical base, no pseudo knots should be allowed etc. The collection of all feasible stems (i.e. those obeying the constraints) forms a universal set U. Each particle of the SetPSO is then initialized as a randomly chosen subset of U. Positions and velocities of these particles are updated using the modified addition and subtraction operators with a view to minimizing the thermodynamic free energy function defined for the RNA structure [79]. Although the SetPSO based algorithm yielded near-optimal configuration for RNA molecules in a number of benchmarks, further research is necessary to select a more robust energy function, which can eliminate the formation of pseudo-knots in the predicted structure.

4.4.6 Protein Secondary Structure Prediction

Protein secondary structures have already been introduced in Section 4.2.3. Protein structures are primarily determined by techniques such as NMRI (nuclear-magnetic resonance imaging) and X-ray crystallography, which are expensive in terms of equipment-cost, computation and time. In addition, they require isolation, purification and crystallization of the target protein. Computational approaches to protein structure prediction are therefore very attractive and cost effective. Since the processes involved in the folding of proteins are very complex and only partially understood simplified models like Dill's Hydrophobic-Polar (HP) have become one of the major tools for studying proteins [80]. The HP model is based on the observation that hydrophobic interaction is the driving force for protein folding and the hydrophobicity of amino acids is the main force for development of a native conformation of small globular proteins [80], [81]. In the HP model, each amino acid can be either of

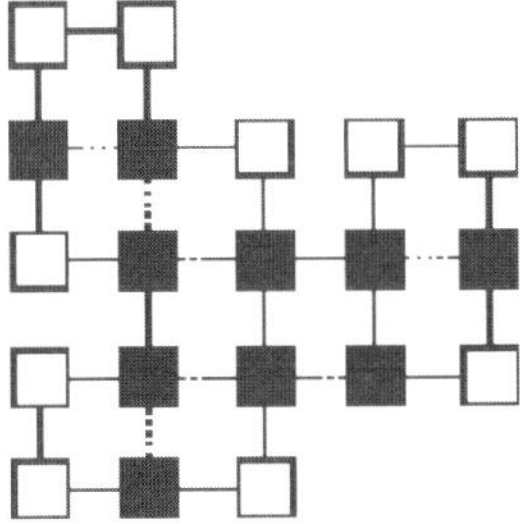

Fig. 4.18. A sample protein conformation in the 2D HP model. The underlying protein sequence (Sequence 1 from Table 1) is HPHPPHHPHPPHPHHPPHPH; black squares represent hydrophobic amino-acids, while white squares symbolize polar amino-acids

two types: *H* (hydrophobic, i.e., non-polar) or *P* (hydrophilic, i.e., polar). For simplicity, we denote *H* by "1" (black) and *P* by "0" (white). The protein conformations of this sequence are restricted to self-avoiding paths on a lattice. An example for a protein conformation under a 2D HP lattice model is illustrated in Figure 4.18.

But, finding the optimal folds even in case of the simplest two-dimensional HP model is computationally hard and knows no polynomial time solution [82]. Shmygelska and Hoos proposed a modified version of the ACO for solving this NP-hard problem [83], [84]. The ants, in their method, first randomly select a starting point within the given protein sequence. From this starting point, the given protein sequence is folded in both directions, adding one amino-acid symbol at a time. In this way, the tours of these ants construct candidate conformation for a given HP protein sequence, apply local search to achieve further improvements, and finally update the pheromone trails based on the quality of the solutions found. The ant system incorporates a local search element as a means of by-passing local minima and preventing the algorithm from premature convergence.

Chu *et al.* [85] extended the 2-D solutions of the HP protein folding problems to the 3-D case by using a parallel ant colony system. They proposed a Multi Ant Colony Optimization (MACOS) algorithm for optimizing the 3-D HP lattice configuration. The MACOS utilizes multiple colonies of artificial ants. It employs separate pheromone matrices for each colony and allows limited cooperation between different colonies.

4.4.7 Fragment Assembly Problem (FAP)

The fragment assembly problem (FAP) deals with the sequencing of DNA. Currently strands of DNA, longer than approximately 500 base pairs, cannot be sequenced very accurately. As a consequence, in order to sequence larger strands of DNA, they are first broken into smaller pieces. The FAP is then to reconstruct the original molecule's sequence from the smaller fragment sequences. FAP is basically a permutation problem, similar in spirit to the TSP, but with some important differences (circular tours, noise, and special relationships between entities) [15]. It is also NP-complete in nature. Meksangsouy and Chaiyaratana [86] attempted to solve the DNA fragment reordering problem with the ant colony systems. The authors investigated two types of assembly problems: single-contig and multiple-contig problems. The simulation results indicate that in single-contig problems, the ant colony system algorithm outperforms the nearest neighbor heuristic algorithm when multiple-contig problems are considered.

4.5 A Few Open-ended Problems and Future Research Directions

In the last section, we discussed the research works already undertaken for making an efficient use of SI tools in bioinformatics. The papers published in this context, may be small in volume, but are of immense significance to the researchers of tomorrow. We note that the SI algorithms are yet to be applied to a huge lot of NP-hard problems from computational biology, for which no universally acceptable solution is known till date. In the present Section, we address a few research problems of this kind and provide hints on their possible solution through the application of SI algorithms.

4.5.1 Identifying Gene Regulatory Networks

A Gene Regulatory Network (GRN) may be defined as a collection of genes in a cell that interact with one another, governing the rate of transcription [87]. Inferring the network from gene expression data obtained through DNA microarray constitutes one of the most challenging problems in the field of bioinformatics. Genes can be viewed as nodes in a complex network, with input being proteins such as transcription factors, and outputs being the level of gene expression. The node itself can also be viewed as a function which can be obtained by combining basic functions upon the inputs (in the Boolean network described below these are Boolean functions or gates computed using the basic AND, OR and NOT gates in electronics). These functions have been interpreted as performing a kind of information processing within the cell which determines cellular behavior.

PSO can be utilized very effectively to solve the GRN identification problem. Each particle may represent the real valued expression levels of all the genes. Each gene has a specific expression level for another gene; thus a total of N genes correspond to N^2 expression levels. Fitness of the particles may be computed from the absolute error with generated expression pattern (sum of all expressions) from the target expression pattern. Investigations of the same problem with evolutionary algorithms can be found in [88], [89], [90], [91].

4.5.2 Protein Tertiary Structure Prediction and Folding

Once a protein sequence has been determined, deducing its unique 3-D native structure is a daunting task. Experimental methods to determine detailed protein structure, such as x-ray diffraction studies and nuclear magnetic resonance (NMR) analysis, are highly labor intensive. Since it was discovered that proteins are capable of folding into their unique functional 3D structures without any additional genetic mechanisms, over 25 years of effort has been expended into the prediction of 3D structure from sequence. Despite the large amount of effort expended, the protein folding or protein structure prediction problem, as it has come to be known, remains largely unsolved [92].

Since PSO is known as a fast and accurate global optimization method, it may be integrated in the *ab initio* approach to protein tertiary structure prediction [93], [94], [95]. The *ab initio* approach is a mixture of science and engineering. The science is in understanding how the three-dimensional structure of a protein is attained. The engineering portion is in finding the 3-Dstructure from a given the sequence. The *ab initio* folding process can be broken down into two components: devising a scoring function that can distinguish between correct/good (native or native like) structures from incorrect (non-native) ones, and a search method to explore the conformational space. The PSO may be used in the searching phase in order to enhance the performance of the process as a whole.

4.5.3 Characterization of Metabolic Pathways between Different Genomes

In biochemistry, a metabolic pathway is a series of chemical reactions occurring within a cell, catalyzed by enzymes, resulting in either the formation of a metabolic product to be used or stored by the cell, or the initiation of another metabolic pathway (then called a flux generating step). Many pathways are elaborate, and involve a step by step modification of the initial substance to shape it into the product with the exact chemical structure desired [96].

The goal of characterizing the metabolic pathways is to estimate the "best" set of parameter values, which minimizes the error between the process data and the model metabolic network response. This parameter estimation problem can be formulated as a non-convex, nonlinear optimization problem and can therefore be solved using global optimization techniques. This feature makes the problem ideal for the application of algorithms like PSO.

4.5.4 Characterization of Metabolic Pathways between Different Genomes

One of the most promising applications of bioinformatics appears in computer-aided molecular design (CAMD). In pharmaceutical development, this effort is focused on modeling the drugs and the biological receptors that the drugs bind to so that better binding, and therefore, more potent or precise drugs can be developed [97], [98]. SI algorithms like PSO may find important applications for the design of a ligand molecule, which can bind to the active site of a target protein.

Unlike the GA based methods [99], [100], PSO or ACO have not been applied to the molecular design problem till date. The formulation of the drug design problem with PSO requires the control of a fitness function. The fitness function must be capable of determining which of two arbitrary molecules is better for a specific task. The algorithm may begin by generating a population of particles each representing one randomly oriented molecule. The individual molecules in a population are then evolved towards greater fitness by using the velocity updating schemes of the PSO or its variants. However, finding a suitable representation of the particles (thereby enabling them to contain information about the each random molecule) constitutes a major research issue in this direction.

4.6 Conclusions

With an explosive growth of the annotated genomic sequences in available form, bioinformatics has emerged as a challenging and fascinating field of science. It presents the perfect harmony of statistics, biology and computational intelligence methods for analyzing and processing biological information in the form of gene, DNA, RNA and proteins. SI algorithms on the other hand, have recently gained wide popularity among the researchers, for their amazing ability in finding near optimal solutions to a number of NP hard, real world search problems. A survey of the bioinformatics literature reveals that the field has a plethora of problems that need fast and robust search mechanisms. Problems belonging to this category include (but are not limited to) the multiple sequence alignment (MSA), protein secondary and tertiary structure prediction, protein ligand docking, promoter identification and the reconstruction of evolutionary trees. Classical deterministic search algorithms and the derivative based optimization techniques are of no use for them as the search space may be enormously large and discontinuous at several points. SI presents a collection of multi-agent parallel search techniques which can be very effective for solving bioinformatics related tasks of this kind. We fervently hope that the SI community will make significant contribution to the emerging research area of modern computational biology in near future.

This article surveyed several important applications of SI tools in bioinformatics. We also illustrated a few open-ended research problems of computational biology, where the SI algorithms like PSO and ACO may find very good use. Even though the current approaches in biological computing with SI algorithms are very helpful in identifying patterns and functions of proteins, genes etc., the final results are far from being perfect.

There are a few general issues which should be addressed by the researchers in future in order to exploit the SI algorithms to their full potential in bioinformatics: firstly, the basic velocity updating scheme in PSO or the pheromone trail updating mechanism in ACO are common to all applications; research should now focus on the design of problem specific operators to get better results. Secondly, the parameters of the ACO or PSO require extensive experimentation so that the appropriate range of values can be identified for different bioinformatics tasks. Finally, algorithms like PSO or ACO and their variants involve a large degree of randomness and different runs of the same program may yield different results; so it is necessary to incorporate problem specific domain knowledge in the SI tools to reduce randomness and computational time and current research should progress in this direction also.

References

1. Baldi P and Brunak S (1998) Bioinformatics: The Machine Learning Approach, MIT Press, Cambridge, MA.
2. Altman RB, Valencia A, Miyano S and Ranganathan, S (2001) Challenges for intelligent systems in biology, IEEE Intelligent Systems, vol. 16, no. 6, pp. 14–20.
3. Haykin S. (1999) Neural Networks: A Comprehensive Foundation, Prentice Hall.
4. Holland JH (1975) Adaptation in Natural and Artificial Systems, University of Michigan Press, Ann Arbor.
5. Goldberg DE (1975) Genetic Algorithms in Search, Optimization and Machine Learning, Addison-Wesley, Reading, MA.
6. Mitra S and Hayashi Y (2006) Bioinformatics with soft computing, IEEE Transactions on Systems, Man, and Cybernetics, Part C: Applications and Reviews, Vol. 36, pp. 616–635.
7. Bonabeau E, Dorigo M, Theraulaz G (2001) Swarm intelligence: From natural to artificial systems. Journal of Artificial Societies and Social Simulation, 4(1).
8. Engelbrecht AP (2005) Fundamentals of Computational Swarm Intelligence. Wiley.
9. Beni G and Wang U (1989) Swarm intelligence in cellular robotic systems. In NATO Advanced Workshop on Robots and Biological Systems, Il Ciocco, Tuscany, Italy.
10. Kennedy J, Eberhart R (1995) Particle swarm optimization, In Proceedings of IEEE International conference on Neural Networks. 1942–1948.
11. Dorigo M (1992) Optimization, learning, and natural algorithms, Ph.D. dissertation (in Italian), Dipartimento di Elettronica, Politecnico di Milano, Milano, Italy.
12. Dorigo M, Di Caro G and Gambardella L (1999) Ant colony optimization: A new meta-heuristic. In PJ Angeline, Z Michalewicz, M Schoenauer, X Yao, and A Zalzala, (eds), Proceedings of the Congress on Evolutionary Computation, IEEE Press, Vol. 2, pp. 1470–1477.
13. Lewin B (1995) Genes VII. Oxford University Press, New York, NY.
14. Wu AS and Lindsay RK (1996) A Survey of Intron Research in Genetics, In Proc. 4th Conf. of on Parallel Problem Solving from Nature, pp. 101–110.
15. Setubal J and Meidanis J (1999) Introduction to Computational Molecular Biology, International Thomson Publishing, 20 park plaza, Boston, MA 02116.
16. http://en.wikipedia.org/wiki/DNA_microarray
17. Couzin ID, Krause J, James R, Ruxton GD, Franks NR (2002) Collective Memory and Spatial Sorting in Animal Groups, Journal of Theoretical Biology, 218, pp. 1–11.
18. Krause J and Ruxton GD (2002) Living in Groups. Oxford: Oxford University Press.
19. Partridge BL, Pitcher TJ (1980) The sensory basis of fish schools: relative role of lateral line and vision. Journal of Comparative Physiology, 135, pp. 315–325.
20. Partridge BL (1982) The structure and function of fish schools. Science American, 245, pp. 90–99.
21. Major PF, Dill LM (1978) The three-dimensional structure of airborne bird flocks. Behavioral Ecology and Sociobiology, 4, pp. 111–122.
22. Branden CI and Tooze J (1999) Introduction to Protein Structure: 2nd edition. Garland Publishing, New York, 2nd edition.
23. Grosan C, Abraham A and Monica C (2006) Swarm Intelligence in Data Mining, in Swarm Intelligence in Data Mining, Abraham A, Grosan C and Ramos V (Eds), Springer, pp. 1–16.
24. Milonas MM (1994) Swarms, phase transitions, and collective intelligence, In Langton CG Ed., Artificial Life III, Addison Wesley, Reading, MA.

25. Serra R and Zanarini G (1990) Complex Systems and Cognitive Processes. New York, NY: Springer-Verlag.
26. Flake G (1999) The Computational Beauty of Nature. Cambridge, MA: MIT Press.
27. Kennedy J, Eberhart R and Shi Y (2001) Swarm Intelligence, Morgan Kaufmann Academic Press.
28. Dorigo M and Gambardella LM (1996) A Study of Some Properties of Ant Q, in Proc. PPSN IV - 4th Int. Conf. Parallel Problem Solving From Nature, Berlin, Germany: Springer-Verlag, pp. 656–665.
29. Dorigo M and Gambardella LM (1997) Ant colony system: A cooperative learning approach to the traveling salesman problem, IEEE Trans. Evol. Comput., vol. 1, pp. 53–66.
30. Deneubourg JL (1997) Application de l'ordre par fluctuations? la descriptio de certaines? tapes de la construction dun id chez les termites, Insect Sociaux, vol. 24, pp. 117–130.
31. Dorigo M, Maniezzo V and Colorni A (1996) The ant system: Optimization by a colony of cooperating agents, IEEE Trans. Syst. Man Cybern. B, vol. 26.
32. Kennedy J (1999) Small Worlds and Mega-Minds: Effects of Neighborhood Topology on Particle Swarm Performance, Proceedings of the 1999 Congress of Evolutionary Computation, vol. 3, IEEE Press, pp. 1931–1938.
33. Kennedy J and Mendes R (2002) Population structure and particle swarm performance. In Proceedings of the IEEE Congress on Evolutionary Computation (CEC), IEEE Press, pp. 1671–1676.
34. Watts DJ and Strogatz SH (1998) Collective dynamics of small-world networks. Nature, 393, 440–442.
35. Dall'Asta L, Baronchelli A, Barrat A and Loreto V (2006) Agreement dynamics on small-world networks. Europhysics Letters.
36. Barrat A and Weight M (2000) On the properties of small-world network models. The European Physical Journal, 13, pp. 547–560.
37. Moore C and Newman MEJ (2000) Epidemics and percolation in small-world networks. Physics. Review. E 61, 5678–5682.
38. Jasch F and Blumen A (2001) Trapping of random walks on small-world networks. Physical Review E 64, 066104.
39. Chen J, Antipov E, Lemieux B, Cedeno W, and Wood DH (1999) DNA computing implementing genetic algorithms, Evolution as Computation, Springer Verlag, New York, pp. 39–49.
40. Vesterstrom J and Thomsen R (2004) A comparative study of differential evolution, particle swarm optimization, and evolutionary algorithms on numerical benchmark problems, In Proceedings of the IEEE Congress on Evolutionary Computation (CEC 04), IEEE Press, pp. 1980–1987.
41. Das S, Konar A, Chakraborti UK (2005) A New Evolutionary Algorithm Applied to the Design of Two-dimensional IIR Filters in ACM-SIGEVO Proceedings of Genetic and Evolutionary Computation Conference (GECCO-2005), Washington DC.
42. Hassan R, Cohanim B and de Weck O (2005) Comparison of Particle Swarm Optimization and the Genetic Algorithm, AIAA-2005-1897, 46th AIAA/ASME/ASCE/AHS/ASC Structures, Structural Dynamics & Materials Conference.
43. Luscombe NM, Greenbaum D and Gerstein M (2001) What is Bioinformatics? A Proposed Definition and Overview of the Field, Yearbook of Medical Informatics, pp. 83–100.
44. Quackenbush J (2001) Computational analysis of microarray data, National Review of Genetics, vol. 2, pp. 418–427.

45. Special Issue on Bioinformatics, IEEE Computer, vol. 35, July 2002.
46. Jain AK, Murty MN and Flynn, PJ (1999) Data clustering: a review, ACM Computing Surveys, vol. 31, no. 3, pp. 264–323.
47. Baker TK, Carfagna MA, Gao H, Dow ER, Li O, Searfoss GH, and Ryan TP (2001) Temporal Gene Expression Analysis of Monolayer Cultured Rat Hepatocytes, Chem. Res. Toxicol., Vol. 14, No. 9.
48. Alon U, Barkai N, Notterman DA, Gish K, Ybarra S, Mack D and Levine AJ (1999) Broad Patterns of Gene Expression Revealed by Clustering Analysis of Tumor and Normal Colon Tissues Probed by Oligonucleotide Arrays, Proc. Natl. Acad. Sci. USA, Cell Biology, Vol. 96, pp. 6745–6750.
49. Maurice G and Kendall M (1961) The Advanced Theory of Statistics, Vol. 2, Charles Griffin and Company Limited.
50. Wen X, Fuhrman S, Michaels GS, Carr DB, Smith S, Barker JL, and Somogyi R (1998) Large-scale temporal gene expression mapping of central nervous system development, Proc. Natl. Acad. Sci. USA, Neurobiology, Vol. 95, pp. 334–339.
51. Spellman EM, Brown PL, Brown D (1998) Cluster Analysis and Display of Genome-wide expression patterns, Proc. Natl. Acad. Sci. USA 95: 14863–14868.
52. Yeung KY, Ruzzo WL (2001) Principal Component Analysis for Clustering Gene Expression Data, Bioinformatics, 17, pp. 763–774.
53. Raychaudhuri S, Stuart JM and Altman RB (2000) Principal Components Analysis to Summarize Microarray Experiments: Application to Sporulation Time Series, Pacific Symposium on Biocomputing 2000, Honolulu, Hawaii, pp. 452–463.
54. Li L, Weinberg CR, Darden TA and Pedersen LG (2001) Gene Selection for Sample Classification Based on Gene Expression Data: Study of Sensitivity to Choice of Parameters of the GA/KNN Method, Bioinformatics, 17, pp. 1131–1142.
55. Herrero J, Valencia A and Dopazo J (2001) A hierarchical unsupervised growing neural network for clustering gene expression patterns, Bioinformatics, 17, pp. 126–136.
56. Tamayo P, Slonim D, Mesirov J, Zhu Q, Kitareewan S, Dmitrovsky E, Lander ES and Golub TR (1999) Interpreting patterns of gene expression with self organizing maps: Methods and applications to hematopoietic differentiation. PNAS, 96, pp. 2907–2912.
57. Toronen P, Kolehmainen M, Wong G, Castren E (1999) Analysis of Gene Expression Data Using Self-organizing Maps, FEBS letters 451, pp. 142–146.
58. Xiao X, Dow ER, Eberhart RC, Miled ZB and Oppelt RJ (2003) Gene Clustering Using Self-Organizing Maps and Particle Swarm Optimization, Proc of the 17th International Symposium on Parallel and Distributed Processing (PDPS '03), IEEE Computer Society, Washington DC.
59. Kohonen T (1995) Self-organizing Maps, 2nd ed., Springer-Verlag, Berlin.
60. http://cnx.org/content/m11456/latest/
61. Liu BF, Chen HM, Huang HL, Hwang SF and Ho SY (2005) Flexible protein-ligand docking using particle swarm optimization, in Proc. of Congress on Evolutionary Computation (CEC 2005), IEEE Press, Washinton DC.
62. Jones G, Willett P, Glen RC, Leach AR and Taylor R (1997) Development and validation of a genetic algorithm for flexible docking. Journal of Molecular Biology, 267(3): pp. 727–748.
63. Morris GM, Goodsell DS, Halliday RS, Huey R, Hart WE, Belew RK and Olson AJ (1998) Automated docking using a lamarckian genetic algorithm and an empirical binding free energy function. Journal of Computational Chemistry, 19(14): pp. 1639–1662.

64. Ewing TJA, Makino S, Skillman AG and Kuntz ID (2001) Dock 4.0: Search strategies for automated molecular docking of flexible molecule databases. Journal of Computer-Aided Molecular Design, 15(5): pp. 411–428.
65. Lipman DJ, Altschul SF and Kececioglu JD (1989). A tool for multiple sequence alignment. Proc. Natl. Acad. Sci. USA, 86: pp. 4412–4415.
66. Feng DF, Doolittle RF (1987) Progressive sequence alignment as a prerequisite to correct phylogenetic trees. J. Mol. Evol. 25, pp. 351–360.
67. Thompson JD, Higgins DG and Gibson TJ (1994) CLUSTAL W: improving the sensitivity of progressive multiple sequence alignment through sequence weighting, position specific gap penalties and weight matrix choice. Nucleic Acids Research, vol. 22, No. 22, pp. 4673–4680.
68. Chen Y, Pan Y, Chen L, Chen J (2006) Partitioned optimization algorithms for multiple sequence alignment, Proc. of the 20th International Conference on Advanced Information Networking and Applications - (AINA'06), IEEE Computer Society Press, Washington DC., Volume 02, pp. 618–622.
69. Notredame C and Higgins DG, SAGA: sequence alignment by genetic algorithm, Nucleic Acids Research, vol. 24, no. 8, pp. 1515–1524.
70. Rasmussen TK and Krink T (2003) Improved hidden Markov model training for multiple sequence alignment by a particle swarm optimization-evolutionary algorithm hybrid, BioSystems 72 (2003).
71. Stolcke A and Omohundro S (1993) Hidden Markov Model induction by Bayesian model merging. In NIPS 5, pp. 11–18.
72. Hamam Y and Al-Ani T (1996). Simulated annealing approach for Hidden Markov Models. 4th WG-7.6 Working Conference on Optimization-Based Computer-Aided Modeling and Design, ESIEE, France.
73. Felsenstein J (1973). Maximum likelihood estimation of evolutionary trees from continuous characters.Am. J. Hum. Gen. 25: 471–492.
74. Lewis PO (1998), A genetic algorithm for maximum likelihood phylogeny inference using nucleotide sequence data, Molecular Biology and Evolution, vol. 15, no. 3, pp. 277–283.
75. Lemmon AR and Milinkovitch MC (2002) The metapopulation genetic algorithm: An efficient solution for the problem of large phylogeny estimation, Proc. Natl Acad Sci U S A., vol. 99, no. 16, pp. 10516–10521.
76. Perretto M and Lopes HS (2005) Reconstruction of phylogenetic trees using the ant colony optimization paradigm, Genetic and Molecular Research 4 (3), pp. 581–589.
77. Ando S and Iba H (2002) Ant algorithm for construction of evolutionary tree, in Proc. of Congress on Evolutionary Computation (CEC 2002), IEEE Press, USA.
78. Neethling M and Engelbrecht AP (2006) Determining RNA Secondary Structure using Set-based Particle Swarm Optimization, in Proc. of Congress on Evolutionary Computation (CEC 2006), IEEE Press, USA.
79. Hofacker IL (2003) Vienna rna secondary structure server, Nucleic Acids Research, vol. 31:13, pp. 3429–3431.
80. Lau KF and Dill KA (1989) A lattice statistical mechanics model of the conformation and sequence space of proteins. Macromolecules 22, pp. 3986–3997.
81. Richards FM (1977) Areas, volumes, packing, and protein structures. Annu. Rev. Biophys. Bioeng. 6, pp. 151–176.
82. Krasnogor N, Hart WE, Smith J and Pelta DA (1999) Protein structure prediction with evolutionary algorithms. Proceedings of the Genetic & Evolutionary Computing Conf (GECCO 1999).

83. Shmygelska A, Hoos HH (2003) An Improved Ant Colony Optimization Algorithm for the 2D HP Protein Folding Problem. Canadian Conference on AI 2003: 400–417.
84. Shmygelska A, Hoos HH (2005) An ant colony optimization algorithm for the 2D and 3D hydrophobic polar protein folding problem. BMC Bioinformatics 6:30.
85. Chu D, Till M and Zomaya A (2005) Parallel Ant Colony Optimization for 3D Protein Structure Prediction using the HP Lattice Model, Proc. of the 19th IEEE International Parallel and Distributed Processing Symposium (IPDPS'05), IEEE Computer Society Press.
86. Meksangsouy P and Chaiyaratana N (2003) DNA fragment assembly using an ant colony system algorithm, in Proc. of Congress on Evolutionary Computation (CEC 2006), IEEE Press, USA.
87. Ando S and Iba H (2001) Inference of gene regulatory model by genetic algorithms, Proc. Congress on Evolutionary Computation (CEC 2001), vol. 1, pp. 712–719.
88. Behera N and Nanjundiah V (1997) Trans-gene regulation in adaptive evolution: a genetic algor-ithm model, Journal of Theoretical Biology, vol. 188, pp. 153–162.
89. Ando S and Iba H (2000) Quantitative Modeling of Gene Regulatory Network - Identifying the Network by Means of Genetic Algorithms, The Eleventh Genome Informatics Workshop, 2000.
90. Ando S and Iba H (2001) The Matrix Modeling of Gene Regulatory Networks - Reverse Engineering by Genetic Algorithms, Proc. Atlantic Symposium on Computational Biology and Genome Information Systems and Technology.
91. Tominaga D, Okamoto M, Maki Y, Watanabe S and Eguchi Y (1999) Nonlinear Numerical optimization technique based on a genetic algorithm for inverse Problems: Towards the inference of genetic networks, Computer Science and Biology (Proc. German Conf. on Bioinformatics), pp. 127–140.
92. Branden CI and Tooze J (1999) Introduction to Protein Structure: 2nd edition. Garland Publishing, New York, 2nd edition.
93. Liu Y and Beveridge DL (2002) Exploratory studies of ab initio protein structure prediction: multiple copy simulated annealing, amber energy functions, and a generalized born/solvent accessibility solvation model. Proteins, 46.
94. Unger R and Moult J (1993) A genetic algorithm for 3d protein folding simulations. In 5th Proc. Intl. Conf. on Genetic Algorithms, pp. 581–588.
95. Pokarowski P, Kolinski A and Skolnick J (2003) A minimal physically realistic protein-like lattice model: Designing an energy landscape that ensures all-or-none folding to a unique native state. Biophysics Journal, 84: pp. 1518–26.
96. Kitagawa U and Iba H (2002) Identifying Metabolic Pathways and Gene Regulation Networks with Evolutionary Algorithms, in Evolutionary Computation in Bioinformatics, Fogel GB and Corne DW (Eds.) Morgan Kaufmann.
97. Shayne CG (2005), Drug Discovery Handbook, Wiley-Interscience.
98. Madsen U. (2002), Textbook of Drug Design and Discovery, CRC Press, USA.
99. Venkatasubramanian V, Chan K and Caruthers JM (1995). Evolutionary Design of Molecules with Desired Properties Using the Genetic Algorithm, J. Chem. Inf. Comp. Sci., 35, pp. 188–195.
100. Glen RC and Payne AWR (1995) A Genetic Algorithm for the Automated Generation of Molecule within Constraints. J. Computer-Aided Molecular Design, 9, pp. 181–202.

5

Time Course Gene Expression Classification with Time Lagged Recurrent Neural Network

Yulan Liang[1*] and Arpad Kelemen[2]

[1] Department of Biostatistics
University at Buffalo the State University of New York
Buffalo NY 14214, USA
yliang@buffalo.edu

[2] Department of Neurology, Buffalo Neuroimaging Analysis Center
The Jacobs Neurological Institute, University at Buffalo
the State University of New York, 100 High Street
Buffalo NY 14203, USA
akelemen@buffalo.edu

Summary. Heterogeneous types of gene expressions may provide a better insight into the biological role of gene interaction with the environment, disease development and drug effect at the molecular level. In this chapter for both exploring and prediction purposes a Time Lagged Recurrent Neural Network with trajectory learning is proposed for identifying and classifying the gene functional patterns from the heterogeneous nonlinear time series microarray experiments. The proposed procedures identify gene functional patterns from the dynamics of a state-trajectory learned in the heterogeneous time series and the gradient information over time. Also, the trajectory learning with Back-propagation through time algorithm can recognize gene expression patterns vary over time. This may reveal much more information about the regulatory network underlying gene expressions. The analyzed data were extracted from spotted DNA microarrays in the budding yeast expression measurements, produced by Eisen et al. The gene matrix contained 79 experiments over a variety of heterogeneous experiment conditions. The number of recognized gene patterns in our study ranged from two to ten and were divided into three cases. Optimal network architectures with different memory structures were selected based on Akaike and Bayesian information criteria using two-way factorial design. The optimal model performance was compared to other popular gene classification algorithms, such as Nearest Neighbor, Support Vector Machine, and Self-Organized Map. The reliability of the performance was verified with multiple iterated runs.

5.1 Introduction

Understanding the function of each gene in the human/animal genome is not a trivial task. Learning the gene interactions with the changing environment, with the development of a disease or under different treatment is an even greater challenge and critical to improve human life. DNA microarrays allow the measurement of expression levels for thousands of genes, perhaps all genes of a cell or an organism, within

Y. Liang and A. Kelemen: *Time Course Gene Expression Classification with Time Lagged Recurrent Neural Network*, Studies in Computational Intelligence (SCI) **94**, 149–163 (2008)
www.springerlink.com

a number of different experimental conditions [1]. As an important step, extracting the knowledge from heterogeneous types of gene expressions may provide a better insight into the biological role of gene interactions with disease development and drug effect at the molecular level. Heterogeneous types of gene expressions contain different experimental conditions. The experimental conditions may correspond to different time points under different dosages of a drug, measures from different individuals, different organs or different diseases. The dynamic patterns of genes expressed under different conditions can be useful indicators about gene state-trajectories and may reveal possible states and trajectories of disease and treatment effects [2–8]. Also, the analysis of the gene state patterns can help identifying important and reliable predictors of diseases, such as cancer, in order to develop therapies and new drugs [9]. Biologists, computer scientists and statisticians have had more than a decade of research on the use of microarrays to model gene expressions [2, 10–12]. However, most of the studies are interested in the genes that co-express in homogeneous conditions, but there are few works on heterogeneous types of gene expressions. Moreover, most of these studies focus on the mean profiles of the gene expression time course, which can make the clustering or classification of gene expressions largely simplified but ignores the important time updated (varied) information.

One feature of gene expression data in time course microarray experiment is that it includes a large number of attributes with high correlation and with high level noise. Because of its massive parallelism, potential for fault and noise tolerance, an Artificial Neural Network (ANN) based information processing is capable of taking the task to deal with this feature. ANNs can adapt their structure in response to the change of the gene expressions under different conditions in order to extract knowledge, which contributes to a deep understanding of gene interactions and identifies certain causal relationships among the genes with diseases and drugs [13–14].

The study of the heterogeneous gene expressions under different experimental conditions in a multivariate nonlinear time series may involve the study of dynamic changing of the statistical variations of non-stationary processes of gene expressions. There are several types of artificial neural networks for temporal processing, which can be used to model the natural characteristics of the gene changing under different conditions and update the information in the training data over time. Recurrent Neural Networks (RNNs) have the ability of dealing with time varying input and output and they can define neurons as states of the network [15]. The output of the hidden layer is fed back to the input layer via time delay. An internal state of the network encodes a representation of some characteristics or a biological mechanism of gene interactions, based on the transition function of the state from a recursive neural network, eventually to control the production of the internal information. State space model can be viewed as a special case of RNN, which combines a stochastic process with observation data model uniformly based on the recursive neural network. Hidden Markov processes can also be used to model the gene activity systems in which the gene states are unobservable, but can be represented by a state transition structure determined by the state parameters and the state transition matrix while processing the patterns over time. Time Lagged Recurrent Neural Networks

(TLRNNs) are extensions of conventional RNNs and outperform them in the terms of network size. A TLRNN use short memory structure instead of static topology networks to develop advanced classification systems and use a complex learning algorithm: Back-Propagation Through Time (BPTT) to learn the temporal pattern [16–17]. This dynamic learning process is well suited to the heterogeneous time series gene expression domain. TLRNNs have been used in nonlinear time series prediction, system identification and temporal pattern classification.

The goal of this chapter is to investigate the performance of heterogeneous types of multivariate time series data using time lagged recurrent neural networks with dynamic trajectory learning. The question we are interested in is whether the dynamic heterogeneous gene activity patterns can be well identified or classified through the trajectory learning with a time lagged recurrent neural network. Gene expression time series data not only exhibit very high noise level, but is also significantly non-stationary. The study of gene expressions under different experimental conditions in a multivariate nonlinear time series may involve the study of dynamic changing of the statistical variations of non-stationary processes of gene expressions. Time Lagged Recurrent Neural Networks are used to model the natural characteristics of gene changing under different conditions and update the information in the training data over time.

To deal with non-stationarity in the gene data, one approach is to build models based on a short time period or window only, such as Time Lagged Recurrent Neural Networks, which use the short memory structure to confine the input for temporal processing. Another way is to try to remove the non-stationarity using data transformation. Both approaches were performed in the application discussed in this chapter. With the presence of high level noise in the gene expression, training is difficult, and the random correlations with recent data can make the model to be based on the earlier data difficult, and it is likely to develop into an inferior model. So before building the appropriate model, data preprocessing have to be done in order to achieve desired classification and prediction performance.

In gene expression data under different experimental conditions with different time points there is a high dependence among the inputs and a high correlation among the samples, so the training is not statistically independent. One way to deal with the dependence of inputs is to include additional inputs, called lagged variables or a tapped delay in the network. Thus, one can train an ordinary network with these targets and lagged variables. Using only inputs as lagged target values are called "autoregressive models", which are widely studied statistical models. Using lagged variables means that the inputs include more than one time constant, which makes the network "dynamic" instead of using one time point (present data) with static structure. The dynamic part is called "memory structure". Such a neural network models the human brain's work in the aspect of short term memory, which essentially helps to remember the recent past events. To use lagged variables, we have to consider which lags and which input variables to include in the network, how many hidden units to use, etc. This corresponds to the design of the memory structure of the network.

The use of a recurrent neural network with time lag is important from the viewpoint of the "curse of dimensionality" and ill-conditioned problems. Trying to take into account a greater history with a Feed Forward Neural Network means increasing the number of delayed inputs, which results in an increase in the input dimension. This is called the "curse of dimensionality". If we have a small number of data points then increasing the dimensionality of the space rapidly leads to the point where the data is very sparse, in which case it provides a very poor representation of the mapping [13]. Comparing with classical time series model, TLRNNs implement Nonlinear Moving Average (NMA) models. With global feedback from the output to the hidden layer, they can be extended to Nonlinear AutoRegressive Moving Average (NARMA) models.

The rest of the chapter is divided as follows: in Section 5.2 we describe how the data was acquired and preprocessed. In Section 5.3 TLRNNs, statistical criteria for searching for the optimal model and related learning algorithms are presented. Experimental results are given in section 5.4. We survey related work in section 5.5 and finally we provide some concluding remarks in section 5.6.

5.2 Data Acquisition and Preprocessing

5.2.1 Data Extraction

The widely studied set of yeast expression measurements data, produced by Eisen et al. [18-20] contained 2465 genes. Each data point represented the ratio of expression levels of a particular gene under two different conditions: CY5 and CY3 with red and green fluorescence intensity, respectively. The gene matrix contained 79 time points over a variety of heterogeneous experimental conditions, which are important biological parameters. The data was generated from spotted arrays using samples collected at various time points during diauxic shift, mitotic cell division cycle, sporulation, temperature, reducing shocks, and so on. We extracted the data from the Stanford genome research web site (http://www-genome.stanford.edu). In our study we used two third of the data for training and the rest for testing.

5.2.2 Functional Class Extraction

If one claims to be able to predict some gene patterns or classes with certain accuracy, one should be questioned about the definition of gene patterns used, whether the patterns to be identified are biologically meaningful, and whether the biologists and pathologists actually care about them.

Classification of biological function of gene expression is essentially a classification of molecular roles for all genes and proteins. The features of gene expressions and the complexity of the genetic information make this task daunting, but it can be dealt with by ontology design, which attempts to classify and further process various aspects of molecule functions under highly qualitative and rich features of domains.

Table 5.1. Three gene functional classes and their sizes for Eisen's data

Class	Size
Ribosomal protein genes	121
Transcription protein genes	159
Secretion protein genes	96

Table 5.2. Ten gene functional classes and their sizes for Eisen's data

Class	Size
CELL CYCLE	168
CHROMATIN	48
CYTOSKELETON	72
DNA	103
mRNA	103
NUCLEAR	43
PROTEIN	477
SECRETION	116
TRANSCRIPTION	136
TRANSPORT	129

The training labels of Eisen's data were extracted from the Saccharomyces cerevisiae functional catalogue databases. There are over 145 classes of gene functional classes in the databases. Some types of the gene function classes such as cell cycle could be used to distinguish types of cancers. So once the construction of a reliable and effective classifier to learn gene functional patterns has been completed, we can predict unknown genes and identify different types of diseases.

We studied Eisen's experimental data at three levels:

- Identify two classes of gene functional patterns: 121 genes that code for ribosomal proteins and 2346 genes that code for non-ribosomal proteins.
- Identify three classes of gene functional patterns: the three classes and their sizes are listed in Table 5.1.
- Identify multiple classes of gene functional patterns: four to ten classes. Evaluate network performance when the number of gene functional patterns is increased. The selected classes and their corresponding sizes are given in Table 5.2.

5.2.3 Data Preprocessing

Figure 5.1 gives scatter plots of Eisen's data with three classes of gene expression patterns under different experimental conditions. The plots provide us some useful information of the data, e.g. there is no linear association between alpha 0 and other variables, there are some outliers and also there are possible potential clusterings, e.g. triangles are grouped together.

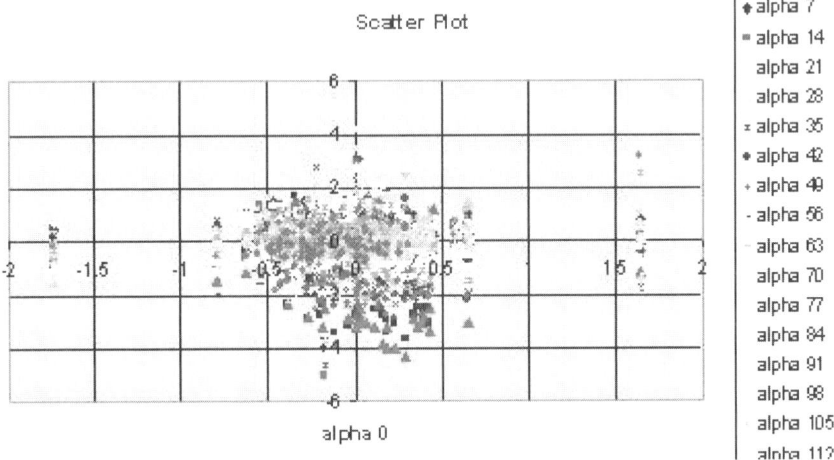

Fig. 5.1. Scatter plots of three classes of gene expression patterns (see Table 5.1) for Eisen's data (p87) under a variety of heterogeneous experimental conditions over 79 time points. Each experimental condition(time point) corresponds to one color with one shape; x-axis: experimental condition diauxic shift at alpha 0; y-axis: alpha 7 and so on.

Figure 5.2 provides time series plots for three classes of gene expression patterns under 79 experimental conditions (time points). The plot shows that the data is non-stationary, since the means and variances change with time.

5.2.4 Smoothing the Data

As it can be seen in the time series plot (Figure 5.2), the data oscillates with high frequency and high amplitude and is non-stationary, which makes direct modeling difficult. To remove these factors the raw time series data was transformed through first order difference and log compression. First order differencing reduces the non-stationarity of the time series. It can handle nondeterministic (stochastic) trends and remove the long-term trend. Log transformation can reduce the number of outliers and stabilize the variance. Figure 5.3 shows the time series plot after differencing and log transformation. As it can be seen in the figure, the transformations made the data more stationary than before transformations (Figure 5.2).

5.2.5 Input Selection

79 inputs may be too many for a Time Lagged Recurrent Neural Network, which is difficult to train, particularly if the data is noisy, and may result in overfitting problems, which do not provide good generalization. In order to select the neural network inputs, a statistical analysis has been carried out to determine the correlations between the inputs (time point) and the outputs (the class or pattern of genes).

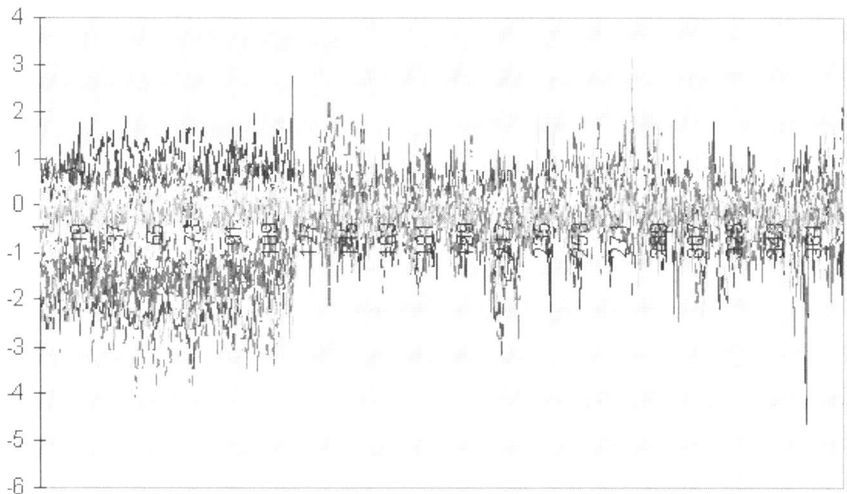

Fig. 5.2. Time series plot of three classes of functional gene expression patterns (see Table 5.1) for Eisen's data (p87) under a variety of heterogeneous experimental conditions over 79 time points. x axis: time points; y-axis: the ratio of expression levels of a particular gene under two different conditions: CY5 and CY3, respectively. Each gene time series was specified with given color, such as the gene "ORF YBL090W" was plotted with one color and so on. The means and variances of the time series are changing over time, which show that the series are nonstationary.

The Pearson correlation coefficients of inputs and outputs were computed first then the acceptance threshold was setup based on the p-values: if the p-value of the correlation coefficient was less than 0.0001, then correlation was considered and we accepted it as input, otherwise we dropped it. The selected inputs and computed correlation coefficients are given in Table 5.3. This way the number of inputs was reduced from 79 time points to 47. Several input permutation runs were also employed in order to find the combination, which produce the lowest error in the testing set. After filtering out the low correlation inputs, the data were fed into the Time Lagged Recurrent Neural Network.

When selecting input variables for a model, one must be careful not to include false predictors. A false predictor is a variable or input that is strongly correlated with an output class, but that is not available in a realistic prediction scenario. False predictors can easily sneak into a model because the process of extracting time lagged information from a database is difficult. The selection of the number of inputs is a delicate process. If the number of inputs is too small then noise makes it hard to find the true patterns in the data. On the other hand, if the number of inputs is too large then the non-stationarity of the data makes the data with statistics less relevant for the task when constructing the classifier.

One important advantage of using Time Lagged Recurrent Neural Networks is that they can use the memory function and the memory layer to confine the input,

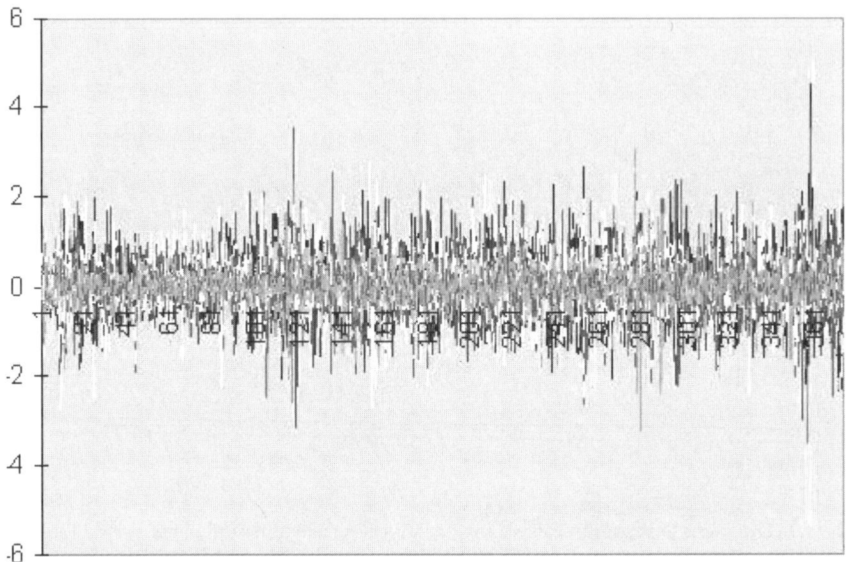

Fig. 5.3. Time series plot of three classes of functional gene expressions for Eisen's data (ribosomal, transcription, and secretion) under heterogeneous conditions after differencing and log transformation. x axis: time points; y-axis: the ratio of expression levels of a particular gene under two different conditions: CY5 and CY3, respectively. Each gene time series was specified with given color such as the gene named ORF YBL090W was plotted with one color and so on. The means and variances stay approximately constant, that indicates that the transformation made the time series closer to stationary.

which can be considered as further input preprocessors to select the inputs, and can reduce the redundant information and detect false predictors.

5.3 Design of Time Lagged Recurrent Neural Network

Time Lagged Recurrent Neural Networks are extensions of conventional Recurrent Neural Networks with short-term memory structures and local recurrent connections. We used the general network architecture with three layers and the feedback connection from the hidden layer back to the input layer. The input layer used the inputs delayed by L time points before presented to the network. Training of the TLRNN was done with Back-Propagation Through Time with trajectory learning and the parameters were learned via examples.

5.3.1 Memory Structures

There are several memory structures at the input layer to choose from. We have applied one time point delay and Gamma memory function to the data. In order to

Table 5.3. Input selection: Pearson correlation coefficients of inputs and outputs (class) for Eisen's data, Prob > |r| under H0: Rho=0

Inputs	Correlation coefficients	Inputs	Correlation coefficients
alpha0	−0.29655	cdc15210	−0.33316
alpha56	−0.26218	spo0	−0.22003
alpha63	−0.21622	spo2	0.63531
alpha70	−0.20749	spo5	0.61170
alpha84	−0.34190	spo7	0.58318
alpha91	−0.26009	spo9	0.41863
alpha98	−0.44897	spo511	−0.67660
alpha105	−0.2258	spoearly	0.67581
alpha112	−0.39606	spomid	0.67611
Elu0	0.39482	heat10	0.57206
Elu60	−0.43226	heat20	0.77191
Elu90	−0.57070	heat40	0.60434
Elu120	−0.63138	heat80	0.57094
Elu150	−0.59320	heat160	0.47246
Elu180	−0.52208	dtt15	−0.25992
Elu210	−0.48014	dtt60	0.51186
Elu240	−0.37021	dtt120	0.75011
Elu270	0.34820	cold20	0.32741
Elu300	−0.25142	cold40	0.38936
cdc1570	−0.26273	cold160	0.59074
cdc1590	−0.32061	diaua	0.33820
cdc15110	−0.40302	diauf	0.65262
cdc15130	−0.32735	diaug	0.67092
cdc15150	−0.23325		

search for the best network structure, the Akaike Information Criteria and Bayesian Information Criteria were applied. The Gamma memory function provided the lower value of AIC/BIC and the higher classification accuracy.

5.3.2 Learning Algorithms

BPTT can adapt the depth of the memory using different types of learning rules, instead of changing the number of inputs. Initial depth of the memory was setup to 10, which was later adapted by the network according to the Gamma memory function. The best learning rule for each layer for the studied data was back-propagation with gradient descent and momentum, where the momentum was setup to 0.7. As an activation function, tangent sigmoid worked best for the given data on the hidden layer and log sigmoid function on the output layer.

5.3.3 Statistical Criteria for Model Selection

The goal of model selection is to find the best network architecture that can achieve the balance between data fitting and model complexity in order to avoid overfitting and to maximize generalization performance. In a Time Lagged Recurrent Neural Network there are several dynamic parameters, such as the number of hidden neurons, the depth in the samples, and the number of trajectories in the search space that have to be optimized in order to achieve optimal model. The depth in the samples parameter can be adapted through BPTT. Two-way factorial arrays were designed to search for the best values of the trajectory and the number of hidden neurons. In this application the number of trajectories is ranged from 2 to 20 and the number of hidden nodes also ranged from 2 to 20. Statistical criteria, such as the Akaike Information Criteria and the Bayesian Information Criteria were computed in order to determine the optimal values for optimal network size and structure. We consider the best neural network to be the one with the highest classification accuracy and the lowest AIC/BIC. In case AIC and BIC don't agree we prefer BIC. The best model was chosen for the rest of the gene classification and future predictions.

5.4 Experimental Results

5.4.1 Two Classes: Ribosome and Non-Ribosome Protein Genes

After 1000 epochs of training the MSE dropped below 0.000059. The mean and the standard deviation of the correct classification rate for the testing data was 99.427% $\pm$ 0.366% with 10 independent runs. This result is even better than the reported result by the prediction algorithm "CLEAVER", with a correct classification accuracy of 99.229834% for the same data [12].

Nearest Neighbor with Mahalanobis Distance and Self-Organized Map methods were also employed for comparison study, which gave the correct classification rates of 97.39% and 98.53%, respectively. Hierarchical Bayesian Neural Network with regularization was also employed to the same data, which provided 99.3932% correct classification rate.

5.4.2 Three Classes: Ribosomal, Transcription and Secretion Gene Functional Classes

Table 5.4 provides the computed statistical criteria for model selection. The average values of AIC and BIC of five independent runs are shown. Table 5.5 reports generalization error rates for the same runs. As it can be seen in Table 5.4 the AIC and BIC values increase rapidly and approximately linearly with the number of hidden nodes, but their values only increase slowly with the number of trajectories. The best AIC and BIC values are highlighted and they are concentrated at 2 and 4 hidden neurons and at 5 and 12 trajectories. Regarding Table 5.5 most of the low error rates are reported around the upper left corner, which corresponds to low number of hidden

Table 5.4. Factorial array for model selection with Back-Propagation Through Time and dynamic trajectory learning for Time Lagged Recurrent Neural Network for Eisen's data. Average values of 5 runs of AIC and BIC. T: number of trajectories; H: number of hidden neurons

T/H	2	4	8	12	15	20
2	1861/2143	**3300/4289**	7661/8698	11398/12970	14273/16234	19021/21629
5	**1846/2127**	3828/4367	7638/8694	11404/12978	14316/16263	19092/21699
8	1953/2234	3788/4327	7636/8692	11444/13018	14293/16254	19091/21700
10	1904/2184	3861/4400	7631/8687	11410/12985	14154/16345	19046/21653
12	**1773/2250**	3722/4582	7688/8744	11460/13033	14195/17325	19128/21735
15	1945/2225	3815/4354	7679/8735	11505/13078	14288/16250	19111/21719
18	1907/2187	3807/4346	7624/8680	11420/12992	14282/16243	19059/21666
20	1936/2217	3834/4373	7665/8721	11445/13018	14378/16339	19139/21747

Table 5.5. Generalization error rate in percentages with Time Lagged Recurrent Neural Network for Eisen's data. T: number of trajectories; H: number of hidden neurons

T/H	2	4	8	12	15	20
2	**4.42**	**3.96**	5.72	4.70	5.30	5.01
5	**3.74**	4.97	6.46	5.61	5.09	7.97
8	5.73	4.36	5.29	7.99	6.16	7.72
10	4.53	7.43	5.61	4.63	5.90	8.66
12	5.43	6.53	9.15	9.31	7.36	11.60
15	5.96	4.50	9.71	10.41	5.51	9.41
18	4.39	**3.48**	5.75	6.23	5.03	5.90
20	6.03	5.26	8.39	6.36	10.58	12.66

nodes with low number of trajectories. This is a good indication, meaning that the two tables mostly agree with each other. The optimal value from both tables, which provide the lowest generalization error rate and lowest BIC can be found at 2 hidden nodes with 5 trajectories and its value is 3.74. However, there is an even lower error rate at 4 hidden nodes and 18 trajectories, but we don't prefer it since it has high AIC/BIC. Since the number of classes to be recognized for this study is only three, it is not surprising that small number of hidden nodes and small number of trajectories can provide good performance. Results show that if we increase the number of patterns (classes) to be recognized, the number of trajectories and the number of hidden nodes have to be increased in order to get optimal performance. Table 5.5 also shows that the learning capability (generalization performance) of the model varies with the number of trajectories and the number of hidden neurons and these two may be largely determined by the complexity of the patterns to be recognized.

Table 5.6 provides results of some other popular learning approaches for gene expression classification for comparison purposes: results of Nearest Neighbor with Mahalanobis Distance (NNMD), Self Organized Map (SOM) and Support Vector Machine (SVM) are shown. Table 5.6 shows the means and standard deviations

Table 5.6. Correct classification rates with standard deviations of five runs for different methods for Eisen's data

Methods	Correct classification rate (%) ± STD (%)
NNMD	73.28 ± 0.012
SVM	74.65 ± 0.002
SOM	80.44 ± 0.053
TLRNN	**95.61 ± 0.018**
JERNN	94.04 ± 0.015

Table 5.7. Correct classification rates of Time Lagged Recurrent Neural Network with Back-Propagation Through Time and dynamic trajectory learning corresponding to the number of classes for Eisen's data

Number of patterns (classes)	Correct classification rate (%)
3	96.52
4	87.14
5	85.06
6	76.15
10	62.14

of the correct classification rates for five independent runs. SVM in this case did not provide the highest performance as opposed to most gene expression studies. The reasons may come from the heterogeneous expression data and the existence of multiple classes; TLRNN particularly performs well for this kind of time series data. We have also applied another popular recurrent neural network, the Jordan/Elman Recurrent Neural Network (JERNN) for our data set. As it can be seen in Table 5.6 the TLRNN worked best for the heterogeneous time series gene expression data.

5.4.3 Multiple Functional Classes

The data distribution for more broad gene functional classes is given in Table 5.2. The correct classification rates with TLRNN are given in Table 5.7, which are based on the optimal structure given by the AIC/BIC. As it can be seen in the table the correct classification rate decreases with the number of classes, which is not surprising. Again, as we have discussed above, both the number of hidden nodes and the number of trajectories increased as the number of classes increased in order to achieve better performance.

5.5 Related Works

A large number of approaches have been proposed, implemented and tested by computer scientists and statisticians in order to discover or identify the gene functional patterns with microarray experiments [26–27]. For example, a genetic network approach was discussed and developed by Thieffry and Thomas [28] and D'haeseleer, et al. [10]. Time series was studied by Socci and Mitra [29] and so on. Self-organized hierarchical neural network was done by Herrero, et al. [11]. Unsupervised neural network and associated memory neural network was done by Azuaje [30] and Bicciato, et al. [31], classification and diagnostic prediction of cancers using gene expression profiling and artificial neural networks was investigated by Khan et al. [32]. Comparison of discrimination methods for the classification of tumors using gene expression data was done by Dudoit et al. [33]. We reported Bayesian neural network and regularised neural network approaches earlier [34–35]. Previous study showed that traditional statistical models can provide some insight into gene expressions and has precise results, but the weaknesses of statistical models are that they can not capture the dynamic changing of gene expressions from time to time well and are sensitive to noise and assumptions. Neural networks are more efficient and flexible for studying gene expressions. We, as an addition to our efforts reported in this chapter currently explore other kinds of neural network models for discovering correlation in gene patterns, and refine the Jordan/Elman neural network approach to study the heterogeneous time series gene expression patterns.

5.6 Conclusion

In this chapter, TLRNNs with BPTT and dynamic trajectory learning were proposed and explored in order to investigate multiple gene functional patterns with heterogeneous microarray experiments. Results show that the Time Lagged Recurrent Neural Network worked better than Nearest Neighbor with Mahalanobis Distance, Support Vector Machine and Self Organized Map. For the SVM this is a little surprise, since most well known results using SVM provided the highest performance and it has properties of dealing with high level noise and large number of attributes, which both exist in the gene expression data. The possible reasons may be found in the heterogeneous time series gene expression data and the existence of multiple classes. Another reason for the good performance of TLRNN is that it can iteratively construct the network for temporal patterns, train the weights, and update the time information. According to the results, the best generalization capability largely depends on the complexity of the patterns, which can be learned by TLRNN with BPTT and trajectory learning through monitoring the complexity of the trajectory with distinct types of states. With the increase in the number of gene functional patterns the generalization performance decreased. However, with changing the number of trajectories and the number of hidden nodes, the performance of the model can be improved based on the statistical criteria for model selection. In order to speed up the search for the best network architecture for dynamic parameters, such as the number of hidden neurons and the number of trajectories, two or three way factorial design with statistical criteria can be employed.

References

1. Gasch AP, Spellman PT, Kao CM, Carmel-Harel O, Eisen MB, Storz G, Botstein D, and Brown PO (2000) Genomic Expression Programs in the Response of Yeast Cells to Environmental Changes. Mol. Biol. Cell 11: 4241–4257.
2. Holter N, Maritan A, Cieplak M, Fedoroff N, and Banavar J (2001) Dynamic modeling of gene expression data. PNAS, USA 98: 1693–1698.
3. Ramoni M, Sebastiani P, and Kohane I (2002) Cluster analysis of gene expression dynamics. PNAS, 99, 9121– 9126.
4. Yeung KY and Ruzzo WL (2001) Principal component analysis for clustering gene expression data. Bioinformatics, 17, 763–774.
5. Yeung KY, Medvedovic M, and Bumgarner RE (2003) Clustering gene-expression data with repeated measurements. Genome Biology, 4, R34.1–R34.17.
6. Hastiel T, Tibshirani R, Eisen MB, Alizadeh A, Levy R, Staudt L, Chan WC, Botstein D, and Brown PO (2000) Gene shaving as a method for identifying distinct sets of genes with similar expression patterns. Genome Biology 1(2): research0003.1–0003.21.
7. Romualdi C, Campanaro S, Campagna D, Celegato B, Cannata N, Toppo S, Valle G, Lanfranchi G (2003) Pattern recognition in gene expression profiling using DNA array: a comparative study of different statistical methods applied to cancer classification. Human Molecular Genetics, Vol. 12, No. 8 pp. 823–836.
8. Neal SH, Madhusmita M, Holter NS, Mitra M, Maritan A, Cieplak M, Banavar JR, and Fedoroff, VF (2000) Fundamental patterns underlying gene expression profiles: Simplicity from complexicity. PNAS, 97:8409–8414.
9. Brown MP, Grundy WN, Lin D, Cristianini N, Sugnet CW, Furey TS, Junior MA, and Haussler D (2000) Knowledge-based analysis of microarray gene expression data by using supported vector machines. PNAS. USA vol. 97(1): 262–267.
10. D'haeseleer P, Liang S, and Somogyi R (1999) Gene expression analysis and genetic networks modeling. Pacific Symposium on Biocomputing.
11. Herrero J, Valencia A, and Dopazo J (2001) A hierarchical unsupervised growing neural network for clustering gene expression patterns. Bioinformatics, 17:126–136.
12. Raychaudhuri S, Sutphin PD, Stuart JM, and Altman RB (2000) CLEAVER: A publicly available web site for supervised analysis of microarray data.
13. Bishop CM (1995) Neural Networks for Pattern Recognition. Oxford University Press.
14. Haykin S (1999) Neural Networks. Prentice Hall Upper Saddle River, New Jersey.
15. Pearlmutter BA (1995) Gradient calculation for dynamic recurrent neural networks. IEEE Transactions on Neural Networks 6(5):1212–1228.
16. Stornetta WS, Hogg T, and Huberman B (1988) A dynamic approach to temporal pattern processing, 750–759. In: Anderson DZ (Ed.) Neural information processing systems. Springer Verlag.
17. Werbos PJ (1993) Backpropagation Through Time: What it does and how to do it. Proc. of the ICNN, San Francisco, CA.
18. Chu S, DeRisi J, Eisen M, Mulholland J, Botstein D, and Brown PO (1998) The Transcriptional Program of Sporulation in Budding Yeast. Science 282, 699–705.
19. Eisen MB, Spellman PT, Brown PO, and Botstein D (1998) Cluster analysis and display of genome-wide expression patterns. PNAS, Vol. 95, Issue 25, 14863–14868.
20. Spellman PT, Sherlock G, Zhang MW, Iyer VR, Anders K, Eisen MB, Brown PO, and Futcher B (1998) Comprehensive identification of cell cycle regulated genes of the yeast Saccharomyces cerevisiae by microarray hybridization. Mol. Biol. Cell 9: 3273–3297.

21. Jansen R and Gerstein M (2000) Analysis of the yeast transcriptome with structural and functional categories: characterizing highly expressed proteins. Nucleic Acids Research, Vol. 28, No. 6 1481–1488.
22. Miller DA and Zurada JM (1998) A dynamical system perspective of structural learning with forgetting. IEEE Transactions on Neural Networks, 9(3): 508–515.
23. Akaike H (1974) A new look at the statistical model identification. IEEE Transactions on Automatic Control 19:716–723.
24. Stone M (1974) Cross-validatory choice and assessment of statistical predictions (with discussion). Journal of the Royal Statistical Society, Series B, 36, 111–147.
25. Li W, Sherriff A, and Liu X (2000) Assessing risk factors of human complex diseases by Akaike and Bayesian information criteria (abstract). Am J Hum Genet 67(Suppl): S222.
26. Golub TR, Slonim, DK, Tamayo P, Huard C, Gaasenbeek M, Mesirov JP, Coller H, Loh, ML, Downing JR, Caligiuri MA, Bloomfield CD, and Lander ES (1999) Molecular classification of cancer: Class Discovery and class prediction by gene expression monitoring. Science, 286: 531–537.
27. Alter O, Brown PO, and Botstein D (2000) Singular value decomposition for genome-wide expression data processing and modeling. PNAS, 97, 10101–10106.
28. Thieffry D and Thomas R (1998) Qualitative analysis of gene networks. Pacific Symp. Biocomp, 3:77–88.
29. Socci ND and Mitra P (1999) Time series analysis of yeast S. cerevisiae cell cycle expression data. In Bronberg-Bauer E, De Beucklaer A, Kummer U, and Rost U (eds) Proceedings of Workshop on computation of Biochemical Pathways and Genetic Networks, Heidelberg.
30. Azuaje F (2001) Unsupervised neural network for discovery of gene expression patterns in B-cell lymphoma. Online Journal of Bioinformatics, Vol 1: 26–41.
31. Bicciato S, Pandin M, Didone G, and Bello C (2001) Analysis of an Associative Memory neural Network for Pattern Identification in Gene expression data. BIOKDD01, Workshop on Data mining in Bioinformatics.
32. Khan J, Wei JS, Ringner M, Saal LH, Ladanyi M, Westermann F, Berthold F, Schwab M, Antonescu CR, Peterson C, and Meltzer PS (2001) Classification and diagnostic prediction of cancers using gene expression profiling and artificial neural networks. Nature Med. Vol.7, No. 6. 673–679.
33. Dudoit S, Fridlyand J, and Speed T (2002) Comparison of discrimination methods for the classification of tumors using gene expression data. JASA, 97 (457):77–87.
34. Liang, Y and Kelemen A (2005) Temporal Gene Expression Classification with Regularised Neural Network. International Journal of Bioinformatics Research and Applications, 1(4), pp. 399–413.
35. Liang, Y and Kelemen A (2004) Hierarchical Bayesian Neural Network for Gene Expression Temporal Patterns. Journal of Statistical Applications in Genetics and Molecular Biology: Vol. 3: No. 1, Article 20.

6

Tree-Based Algorithms for Protein Classification

Róbert Busa-Fekete[1], András Kocsor[1,2,3], and Sándor Pongor[4,5]

[1] Research Group on Artificial Intelligence of the Hungarian Academy of Sciences and University of Szeged, Aradi vértanúk tere 1., H-6720 Szeged, Hungary
{busarobi,kocsor}@inf.u-szeged.hu

[2] Research Group on Artificial Intelligence NPC., Petőfi S. Sgt. 43., H-6725 Szeged, Hungary

[3] Applied Intelligence Laboratory Ltd., Petőfi S. Sgt. 43., H-6725 Szeged, Hungary

[4] Bioinformatics Group, International Centre for Genetic Engineering and Biotechnology, Padriciano 99, I-34012 Trieste, Italy
pongor@icgeb.org

[5] Bioinformatics Group, Biological Research Centre, Hungarian Academy of Sciences, Temesvári krt. 62, H-6701 Szeged, Hungary

Summary. The problem of protein sequence classification is one of the crucial tasks in the interpretation of genomic data. Many high-throughput systems were developed with the aim of categorizing the proteins based only on their sequences. However, modelling how the proteins have evolved can also help in the classification task of sequenced data. Hence the phylogenetic analysis has gained importance in the field of protein classification. This approach does not just rely on the similarities in sequences, but it also considers the phylogenetic information stored in a tree (e.g. in a phylogenetic tree). Eisen used firstly phylogenetic trees in protein classification, and his work has revived the discipline of phylogenomics. In this chapter we provide an overview about this area, and in addition we propose two algorithms that well suited to this scope. We present two algorithms that are based on a weighted binary tree representation of protein similarity data. *TreeInsert* assigns the class label to the query by determining a minimum cost necessary to insert the query in the (precomputed) trees representing the various classes. Then *TreNN* assigns the label to the query based on an analysis of the query's neighborhood within a binary tree containing members of the known classes. The algorithms were tested in combination with various sequence similarity scoring methods (BLAST, Smith-Waterman, Local Alignment Kernel as well as various compression-based distance scores) using a large number of classification tasks representing various degrees of difficulty. At the expense of a small computational overhead, both TreeNN and TreeInsert exceed the performance of simple similarity search (1NN) as determined by ROC analysis, at the expense of a modest computational overhead. Combined with a fast tree-building method, both algorithms are suitable for web-based server applications.

Key words: Proteomics, Protein classification, Phylogenetics, Phylogenomics

R. Busa-Fekete et al.: *Tree-Based Algorithms for Protein Classification*, Studies in Computational Intelligence (SCI) **94**, 165–182 (2008)
www.springerlink.com

6.1 Introduction

The categorization of biological objects is one of the fundamental and traditional tasks of the life sciences. For instance, the categorization of organisms into a hierarchical "Tree of life" leads to a complex model that summarizes not just the taxonomic relationships between the species, but also the putative time-course of evolution as we understand it today. With the advent of molecular biology in the 1970's, the categorization of genes and proteins itself became an important subject of research. Sequences of individual proteins can for instance be compared using string distance measures, and one can build trees that closely resemble the hypothetical "Tree of life". The categorization of protein structures, on the other hand, began from a different perspective: protein structures reveal a few fundamental molecular arrangements (like alpha-helices and beta-sheets) that can combine in a variety of ways and give rise to characteristic molecular shapes. Finally, the recent advent of genomics research - the wholesale analysis of the gene and protein repertoire of a species - led to yet another perspective with an emphasis on biological function. According to this approach, the known genes/proteins are categorized into *a priori* determined, empirical categories that reflect our current knowledge on the cellular and biochemical functions. As proteins carry many, perhaps most of the known biological functions, they play a prominent role in the functional analysis of genomes.

The methods of protein classification fall into three broad categories: i) Methods based on pairwise comparison, i.e. ones that work by comparing an unknown object (protein sequence or structure) with members of an a priori classified database of protein objects. The results are ranked according to the similarities and the strongest similarities are evaluated in terms of biological or statistical significance, after which a query is assigned to the class of the most similar object. ii) Methods based on consensus (or aggregate) descriptions, i.e. ones that are used to analyze distant sequence similarities that cannot readily be determined based on a simple similarity analysis. Here we first prepare a consensus description for all the classes of a protein sequence database, then we compare the unknown query with each of the consensus descriptions. As with the previous methods, the strongest similarities are evaluated and used to assign the protein to the given class. There are various methods for preparing consensus descriptions, including regular expressions, frequency matrices and Hidden Markov Models. The above methods are described in textbooks and are periodically reviewed. iii) A more recent type of protein classification methods attempts to use an external source of knowledge in order to increase the classification sensitivity. The external source of knowledge is the phylogenetic classification of an organism, i.e. the knowledge that is accumulated in the fields of taxonomy and molecular phylogeny. This approach is called phylogenomics (for a recent review see [1]) and is closely linked to the notions of orthologs (proteins that share both ancestry and function) and paralogs (proteins that share a common ancestry but carry different functions). The practical goal of phylogenomics is to reveal the true orthologous relationships and use them in the process of classification. Like ii), phylogenomic methods are used for distant similarities that cannot be decided by simple comparisons like those mentioned in i).

The aim of the present chapter is to describe two protein classification algorithms that make use of tree structures. The chief difficulties of protein classification arise from the fact that the databases are large, noisy, heterogeneous and redundant; that the classes themselves are very different in terms of most of their characteristics; that the assignments are often uncertain. For these reasons there is a constant need for better and faster algorithms that can cope with the growing datasets, and tree-based approaches are promising in this respect [2]. Even though trees are often used in molecular phylogenies and phylogenomics, the motivation of our work is quite different since we are not trying to reveal or to use the taxonomic relationships between species or proteins. We employ trees - especially weighted binary trees - as a simple and computationally inexpensive formalism to capture the hidden structure of the data and to use it for protein classification.

The rest of this chapter is structured as follows. Section 6.2 provides a brief overview of related work, with a strong focus on the protein comparison and tree-building methods used here. Afterwards Sections 6.3 and 6.4 respectively describe the two algorithms called *TreeNN* and *TreeInsert*. *TreeNN* (Section 6.2) is based on the concept of a distance that can be defined between leaves of a weighted binary tree. *TreeNN* is a pairwise comparison type algorithm (see i) above), where the distance function incorporates information encoded in a tree structure. Given a query protein and an *a priori* classified database, the algorithm first constructs a common tree that includes the members of the database and the query protein. In the subsequent step the algorithm attempts to assign labels to an unknown protein using the known class labels found in its neighborhood within the tree. A weighting scheme is applied, and the class label with the highest weight is assigned to the query. *TreeInsert* (Section 6.4) is based on the concept of tree insertion cost, this being a numerical value characterizing the insertion of a new leaf at a given point of a weighted binary tree. The algorithm finds the point with minimum insertion cost in a tree. *TreeInsert* uses the tree as a consensus representation so it is related to the algorithms described above in ii). Given an unknown protein and protein classes represented by precalculated weighted binary trees, the query is assigned to the tree into which it can be inserted at the smallest cost. In the description of both algorithms we first give a conceptual outline that summarizes the theory as well as its relation to the existing approaches i–iii. This is followed by the formal description of the algorithm, the principle of the implementation, and some possible heuristic improvements. Then we round off the chapter with a brief discussion and some conclusions in Section 6.5. As for the databases and the classification tasks used for benchmarking the algorithms, these are described in the Appendix.

6.2 Related Work

The algorithms described in this work belong to the broad area of protein classification, and have been summarized in several recent reviews and monographs [3]. In particular, we will employ the tools developed for protein sequence comparison that are now routinely used by researchers in various biological fields. The Smith-Waterman [4] and the Needleman-Wunsch algorithms [5] are exhaustive sequence comparison algorithms, while BLAST [6] is a fast heuristic algorithm. All of these programs calculate a similarity score that is high for similar or identical sequences and zero or below some threshold for very different sequences. Methods of molecular phylogeny build trees from the similarity scores obtained from the pairwise comparison of a set of protein sequences. The current methods of tree building are summarized in the splendid textbook by J. Felsenstein [7]. One class of tree/building methods the so-called distance based methods are particularly relevant to our work since we use one of the simplest method, namely Neighbour-Joining (NJ) [8], to generate trees from the data.

Protein classification supported by phylogenetic information is sometimes termed phylogenomics [1, 9]. The term covers an eclectic set of tools that combine phylogenetic trees and external data-sources in order to increase the sensitivity of protein classification [1]. Jonathan Eisen's review provides a conceptual framework for combining functional and phylogenetic information and describes a number of cases where functions cannot be predicted using sequence similarity alone. Most of the work summarized by Eisen is devoted to the human evaluation of small datasets by hand. The first automated annotation algorithm was introduced by Zmasek and Eddy [10] who used explicit phylogenetic inference in conjunction with real-life databases. Their method applies the gene tree and the species tree in a parallel fashion, and it can infer speciation and duplication events by comparing the two distinct trees. The worst case running time of this methods is $O\left(n^2\right)$, and the authors used the COG dataset [11] to show that their method is applicable for automated gene annotation.

Not long ago Lazareva-Ulitsky et al. employed an explicit measure to describe the compatibility of a phylogenetic tree and a functional classification [12]. Given a phylogenetic tree overlaid with labels of functional classes, the authors analyzed the subtrees that contain all members of a given class. A subtree is called perfect if its leaves all belong to the one functional class and an ideal phylogenetic tree is made up of just perfect subtrees. In the absence of such a perfect subdivision, one can establish an optimal division i.e. one can find subtrees that contain the least "false" labels. The authors defined a so-called tree measure that characterizes the fit between the phylogenetic tree and the functional classification, and then used it to develop a tree-building algorithm based on agglomerative clustering. For a comprehensive review on protein classification see [1].

6.3 *TreeNN*: Protein Classification via Neighborhood Evaluation within Weighted Binary Trees

6.3.1 Conceptual Outline

Given a database of a priori classified proteins that are compared to each other in terms of a similarity/dissimilarity measure[1], and a query protein that is compared to the same database in terms of the same similarity/dissimilarity measure, one can

[1] "Similarity measures" and "dissimilarity measures" are inversely related: a monotone decreasing transformation of a similarity measure leads to a dissimilarity measure.

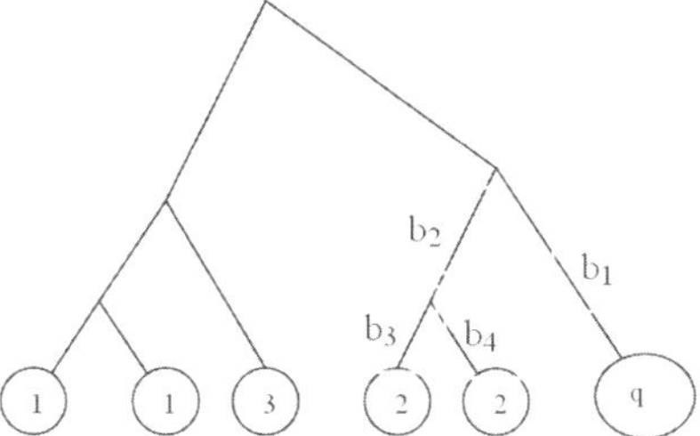

Fig. 6.1. A weighted tree of proteins overlayed with class labels

build a weighted, binary tree that will contain proteins in each leaf. If we now assign the known class labels to the proteins, all leaves except the unknown query will be labelled, as schematically shown in Figure 6.1.

First let us denote the length of the unique path between two leaves L_i and L_j by $p(L_i, L_j)$. Here $p(L_i, L_j)$ is an integer representing the number of edges along the path between L_i and L_j. We can define the closest neighbourhood $K(L)$ of a leaf L as the set of leaves for which there is no L_j leaf such that $p(L, L_j) < p(L, L_i)$. For instance the closest neighbours of q in Figure 6.1 are both members of class 2 and are three steps apart from q. These leaves are parts of the 3-neighbourhood of q (i.e. the set of leaves for which the path between q and them at the most 3). If the tree is a weighted binary tree we can define the leaf distance $l(L_i, L_j)$ between two leaves L_i and L_j as the sum of the branch-weights along the unique path between L_i and L_j. For instance the leaf distance of q from one of its closest neighbors q in the tree is $b_1 + b_2 + b_3$. Finally let us suppose that we know the value of a pairwise similarity measure (such as a BLAST score) between any pair of leaves L_i and L_j, whose value will be denoted by $s(L_i, L_j)$, and that we build a weighted binary tree using the $s(L_i, L_j)$ values. Within this tree we can also calculate the value of the leaf distance $l(L_i, L_j)$.

The *TreeNN* algorithm is a weighted nearest neighbour method that applies as weights a similarity/dissimilarity measure between the proteins constituting the tree and is calculated within the closest neighbourhood of the query within the tree. More precisely, let us assume that we have leaves from m different classes and an indicator function $I : \{L_1, ..., L_n\} \rightarrow \{1, ..., m\}$ that assigns the class labels to the proteins represented by the leaves of the tree. The aggregate similarity measure $R(j, L_q)$ of each of the m classes ($j \in \{1, ..., m\}$) will be an aggregate of the similarity measures or leaf distances obtained between the query on one hand and the members of the class within the closest neighbourhood on the other, calculated via an aggregation operator Θ (such as the sum, product or maximum):

$$R(j, L_q) = \underset{L_i \in K(L) \wedge I(L_i) = j}{\Theta} s(L_i, L_q) \tag{6.1}$$

or

$$\widetilde{R}(j, L_q) = \underset{L_i \in K(L) \wedge I(L_i) = j}{\Theta} l(L_i, L_q) \tag{6.2}$$

The first aggregated value (Eq. (6.1)) for classes is calculated using the original similarity values. This implementation just utilized the topology of the phylogenetic tree, while the second implementation (Eq. (6.2)) also takes into account the edge lengths.

We calculate aggregate measures for each of the classes and the class with the highest weight will be assigned to the query L_q. This analysis is similar to that for the widely used kNN principle, the difference being that we restrict the analysis to the tree neighbourhood of the query (and not simply to the k most similar proteins) and we can use a leaf distance, as shown in Eq. (6.1). As for as the aggregation operator, we could for instance use summation, but we can also use the average operator or the maximum value operator.

In order to increase the influence of the tree structure on the ranking, we introduce a further variant of *TreeNN*, in which the similarity measures $s(L_i, L_j)$ are divided by the path lengths between L_i and L_j. In this manner the weighted aggregate similarity measure becomes

$$W(j, L_q) = \underset{L_i \in K(L) \wedge I(L_i)=j}{\Theta} \left(\frac{s(L_i, L_q)}{p(L_i, L_q)} \right) \tag{6.3}$$

or

$$\widetilde{W}(j, L_q) = \underset{L_i \in K(L) \wedge I(L_i)=j}{\Theta} \left(\frac{l(L_i, L_q)}{p(L_i, L_q)} \right) \tag{6.4}$$

This formula ensures that the leaves farther away within the tree from L_q will have a smaller influence on the classification than the nearer neighbours.

6.3.2 Description of the Algorithm

Input:

- A distance matrix containing the all vs. all comparison of a dataset consisting of a query protein and of an a priori classified set of data.

Output:

- A class label assigned to the query protein

First a weighted binary tree is built from the data. The leaves of this tree are proteins and we select the set of closest tree-neighbours (minimum number of edges from the query). Then we apply a classification rule that might be one of the following:

TreeNN: Assigns to the query the class label with the highest aggregate similarity calculated according to Eq. (6.1) or (6.2).

Weighted TreeNN: Assigns to the query the class label with the highest aggregate similarity calculated according to Eq. (6.3) or (6.4).

The time complexity of the algorithm mainly depends on the tree-building part. For example the Neighbour-Joining method has an $O(n^3)$ time complexity. We have to build up a tree with $n+1$ leaves as each protein will be classified, hence this algorithm has an $O\left(tn^3\right)$ time complexity overall where t denotes the cardinality of the test set. Finding the closest tree-neighbours for a leaf can be carried out in linear time, hence it does not cause any extra computational burden.

Use in classification. The algorithm can be directly used both in two-class and multi-class classification. The size of the database influences both the time requirement of the similarity search and that of the tree-building. The latter is especially important since the time-complexity of tree building is $O\left(n^3\right)$. We can substantially speed up the computation if we include into the tree just the first r similarity/dissimilarity neighbours of the query (e.g. the first 50 BLAST neighbours). On the other hand class imbalance can cause an additional problem since an irrelevant class that has many members can easily outweigh smaller classes. An apparently efficient balance heuristic is to build the tree from the members of the first t ($t \leq 10$) classes nearest to the query, where each class is represented by a maximum of r ($r \leq 4$) members.

6.3.3 Implementation

A computer program was implemented in MATLAB that uses the NJ algorithm as encoded in the Bioinformatics Toolbox package [13]. The detailed calculation has four distinct steps:

1. An all vs. all distance matrix is calculated from the members of an *a priori* classified database using a given similarity/dissimilarity score (BLAST, Smith-Waterman etc) and the results are stored in CVS (Comma Separated Values).
2. The query protein is compared with the same database and the same similarity/dissimilarity score, and the first r sequences are selected for tree-building, choosing one of the heuristics mentioned above.
3. A small $([r+1] \times [r+1])$ distance matrix is built using the precomputed data of the database on the one hand and the query vs. database comparison on the other, and a NJ tree is built.
4. The query's label is assigned using the *TreeNN* algorithm in the way described in Section 6.3.2.

This implementation guarantees that the all vs all comparison of the database is carried out only once.

6.3.4 Performance Evaluation

The performance of *TreeNN* was evaluated by ROC analysis and error rate calculations in the way described in the Appendix (Section 6.6.4). For comparison we also include the results obtained by simple nearest neighbour analysis (1NN). In each table below the best scores of each set are given in bold, and the columns with the heading *Full* concern the performances of *TreeNN* without a heuristic (i.e. we considered all elements of the training set). Here we apply the *TreeNN* methods for a two class problem, thus the parameter t is always equal to 2. For aggregation operators we tried the sum, the average and the maximum operators (but the results are not shown here), and the latter (more precisely, maximum for similarity measures and minimum for distance measures) had a slightly but consistently superior performance, so we used this operator to generate the data shown below. For the calculation of the class weights we applied the scoring scheme that is based on the similarity measures given in Eq. (6.1) for *TreeNN* and Eq. (6.3) for *Weighted TreeNN*. Tables 6.1 and Table 6.2 show the *TreeNN* results for ROC analysis and error rate calculations, respectively. In the next two tables (Table 6.3 and 6.4) we show the performance of the *Weighted TreeNN* using the same settings as that used for *TreeNN* in Tables 6.1 and 6.2. The results, along with the time requirements (wall clock times) are summarized in Tables 6.5. The time requirements of *Weighted TreeNN* is quite

Table 6.1. ROC values of *TreeNN* with and without a heuristic on the COG and 3PGK datasets

	1NN	*TreeNN*		
		Full	$r = 3$	$r = 10$
COG				
BLAST	0.8251	**0.8381**	0.8200	0.7226
Smith-Waterman	0.8285	0.8369	**0.8438**	0.7820
LAK	0.8249	0.8316	**0.8498**	0.7813
LZW	**0.8155**	0.7807	0.7750	0.7498
PPMZ	0.7757	0.8162	**0.8162**	0.7709
3PGK				
BLAST	0.8978	0.9580	**0.9699**	0.9574
Smith-Waterman	0.8974	0.9582	**0.9716**	0.9587
LAK	0.8951	0.9418	**0.9688**	0.9641
LZW	0.8195	0.8186	**0.9040**	0.8875
PPMZ	0.8551	0.9481	**0.9556**	0.7244

Table 6.2. ROC values of Weighted *TreeNN* with and without a heuristic on the COG and 3PGK datasets

	1NN	*TreeNN*		
		Full	$r = 3$	$r = 10$
COG				
BLAST	0.8251	**0.8454**	0.8206	0.8016
Smith-Waterman	0.8285	0.8474	**0.8492**	0.8098
LAK	0.8249	0.8417	**0.8540**	0.8133
LZW	0.8195	**0.9356**	0.9040	0.9228
PPMZ	0.8551	**0.9797**	0.9673	0.8367
3PGK				
BLAST	0.8978	**0.9760**	0.9589	0.9579
Smith-Waterman	0.8974	**0.9761**	0.9547	0.9510
LAK	0.8951	0.9612	**0.9719**	0.9354
LZW	**0.8195**	0.7365	0.7412	0.8183
PPMZ	**0.8551**	0.8140	0.8018	0.7767

Table 6.3. Error rates of *TreeNN* with and without a heuristic on the COG and 3PGK datasets

	1NN	*TreeNN*		
		Full	$r = 3$	$r = 10$
COG				
BLAST	14.7516	**10.3746**	11.6270	15.1469
Smith-Waterman	13.4940	10.5381	**9.9996**	13.0470
LAK	13.3817	10.8644	**9.7976**	12.3784
LZW	16.7301	**13.2106**	13.9285	14.4467
PPMZ	15.0174	**11.6331**	11.9598	13.2246
3PGK				
BLAST	42.1046	35.4026	**32.2360**	35.8291
Smith-Waterman	42.1046	35.6582	**32.2360**	35.5694
LAK	42.0856	33.4081	**32.1928**	34.0542
LZW	36.5293	35.1731	33.8335	**30.4403**
PPMZ	34.6671	37.2146	**32.1706**	37.4445

Table 6.4. Error rates of the Weighted *TreeNN* with and without a heuristic on the COG and 3PGK datasets

	1NN	*TreeNN*		
		Full	$r = 3$	$r = 10$
COG				
BLAST	14.7516	**10.3746**	11.6270	15.1469
Smith-Waterman	13.4940	10.5381	**9.9996**	13.0470
LAK	13.3817	10.8644	**9.7976**	12.3784
LZW	16.7301	**13.2106**	13.9285	14.4467
PPMZ	15.0174	**11.6331**	11.9598	13.2246
3PGK				
BLAST	42.1046	35.4026	**32.2360**	35.8291
Smith-Waterman	42.1046	35.6582	**32.2360**	35.5694
LAK	42.0856	33.4081	**32.1928**	34.0542
LZW	36.5293	35.1731	33.8335	**30.4403**
PPMZ	34.6671	37.2146	**32.1706**	37.4445

Table 6.5. Time requirements for the *TreeNN* method on the COG dataset in seconds

Elapsed time(in second)/Method		1NN	*TreeNN* ($r = 100$)
Preprocessing	BLAST	-	-
	Other	-	15.531
Evaluation	BLAST	0.223	0.223
Evaluation	Other	-	0.109

similar to that of *TreeNN* because the two differ only in the calculation of class aggregate values (cf. Eqs. (6.3) and (6.4)).

The above results reveal a few general trends. First, *TreeNN* and its weighted version outperforms the 1NN classification in terms of the error rate. We should mention here that this improvement in error rate is apparent even when we use a heuristic. As for AUC, the results on the COG database are comparable with those of 1NN, both

Table 6.6. The performance of the *TreeNN* using leaf distances and the original similarity measures

Comparison of the *TreeNN*	*TreeNN*			
	Using similarities		Using leaf distances	
	ROC	Error Rate	ROC	Error Rate
BLAST	0.9699	35.4026	0,9509	36,978
Smith-Waterman	0.9582	35.6582	0,9529	36,979

with and without a heuristic. Moreover they are noticeably better than 1NN on the 3PGK dataset. The fact that the precision improves while the time requirements are comparable with that of the very fast 1NN algorithm is a good sign and confirmation that our approach is a promising one (Table 6.5).

We calculated the time requirements for the methods we employed in a real life scenario. We first assumed that we had an *a prior* classified dataset containing 10000 elements. Applying the 1NN we simply needed to find the most similar protein in this dataset to the query protein, and the class label of this was assigned to the query protein. This approach did not need additional preprocessing step, so these process time requirements were just equal to the calculation of the similarity measure between the query and the *a prior* classified dataset. The *TreeNN* method however required some additional running time. This was because after the *TreeNN* had chosen the *r* most similar element from the known dataset (according to the heuristic in Section 6.3.2) it built up a phylogenetic tree for these elements. This step suggested some additional preprocessing time requirements. In our experiments the parameter was set to 100. As the experiments showed, applying *TreeNN* did not bring about any significant growth in time requirements.

Table 6.6 lists a comparison of the performance of the *TreeNN* algorithm when we used the original similarities/distances of proteins and when we used the leaf distances just according to Eqn. (6.2). The results of these tests clearly show that the performance of the classifiers was only marginally influenced by the measure (sequence similarity measure vs. leaf distances) we chose in the implementation of the algorithm.

6.4 *TreeInsert*: Protein Classification via Insertion into Weighted Binary Trees

6.4.1 Conceptual Outline

Given a database of a priori classified proteins, we can build separate weighted binary trees from the members of each of the classes. A new query protein is then assigned to the class to which it is nearest in terms of *insertion cost (IC)*. A query protein will then be assigned to the class whose *IC* is the smallest. First we note that insertion of a new leaf into a weighted binary tree is the "amount of fitting" into the original tree. In this algorithm we consider an insertion optimal if the query protein is the best suited

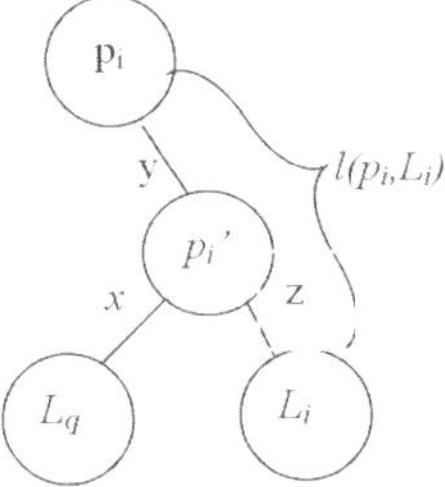

Fig. 6.2. The insertion of the new leaf next to L_i

compared to every other possible insertion. Second, note that IC can be defined in various ways using the terminology introduced in Section 6.3. The insertion of a new leaf L_q next to leaf L_i is depicted in Figure 6.2.

In this example we insert the new element L_q next to the i-th leaf so we need to divide the edge between L_i and its parent into two parts with a novel inner point p_i'. According to Figure 6.2 we can express the relationship of the new leaf L_q to the other leaves of the tree in the following way: $l(L_j, L_q) = l(p_i, L_j) + y + z$ if $i \neq j$. The $l(L_i, L_q)$ leaf distance between the ith leaf and L_q is just equal to $x + z$. This extension step of the leaf distances means that all relations in the tree remain the same, and we have only to determine the new edge lengths x, y and z. The place of p_i' on the divided edge and the weights of the edge that are between L_q and its parent (denoted by x in Figure 6.2) have to be determined so that the similarities and the tree-based distances will be as close as possible. With this line of thinking we can formulate the insertion task as the solution of the following system of equations:

$$\min_{0 \leq x,y} \left(\sum_{j=1}^{n} \left(s(L_j, L_q) - l(L_i, L_q) \right) \right)^2 \tag{6.5}$$
$$\text{s.t.} \quad x + y = l(p_i, L_i)$$

This optimization task determines the value of the three unknown edge lengths x, y and z, and the constraints ensures that the leaf-distance between L_i and its parent remains unchanged. With this in mind we can define the insertion cost for a fixed leaf.

Definition 1. *Let T be a phylogenetic tree, and let its leaves be $L_1, L_2, \ldots, L_n$. The leaf insertion cost $IC(L_q, L_i)$ of a new leaf L_q next to L_i is defined as the branch length of x found by solving the optimisation task in Eq. (6.5).*

Our goal here is to find the position of the new leaf in T with the lowest leaf insertion cost. This is why we define the insertion cost of a new leaf for the whole tree using the Definition 1 in the following way:

Definition 2. *Let T be a phylogenetic tree, and its leaves be $L_1, L_2, \ldots, L_n$. The insertion cost $IC(L_q)$ of a new leaf L_q into T is the minimal leaf insertion cost for T:*

$$IC(L_q) = \min\{IC(L_q, L_1), \ldots, IC(L_q, L_n)\} \tag{6.6}$$

In preliminary experiments we tried several possible alternative definitions for the insertion cost *IC* (data not shown), and finally we chose the branch length x (Figure 6.2) as the definition. This value provides a measure of dissimilarity: it is zero when the insertion point is vicinal to a leaf that is identical with the query. The *IC* for a given tree is the smallest value of x found within the tree.

6.4.2 Description of the Algorithm

Input:

- A weighted binary tree built using the similarity/dissimilarity values (such as a BLAST score) taken between the elements of a protein class.
- A set of comparison values taken between a query protein on the one hand and the members of the protein class on the other, using the same similarity/dissimilarity value as we used to construct the tree. So for instance, when the tree was built using BLAST scores, the set of comparison values were a set of BLAST comparison values.

Output:

- The value of the insertion cost calculated according to Definition 2.

The algorithm will evaluate all insertions that are possible in the tree. An insertion of a new leaf next to an old one requires the solution of an equation system that consists of n equations, where n is the number of leaves. This has a time complexity of $O(n)$. The number of possible insertions for a tree having n leaves (i.e. we insert next to all leaves) is n. Thus calculating the insertion for a new element has a time complexity of $O(n^2)$. One can reduce the time complexity using a simple empirical consideration: we just assume that the optimum insertion will occur in the vicinity of the r leaves that are most similar to the query in terms of the similarity/dissimilarity measure used for the evaluation. If we use BLAST, we can limit the insertions to the r nearest BLAST neighbours of the query. This will reduce the time complexity of the search to $O(rn)$.

Use in classification. If we have a two-class classification problem, we will have to build a tree both for the positive class and the negative class, and we can classify the query to the class whose *IC* is smaller. In practical applications we often have to classify a query into one of several thousand protein classes, such as the classes of protein domains or functions. In this case the class with the smallest *IC* can be chosen. This is a simple nearest neighbour classification which can be further refined by adding an *IC* threshold above which the similarities shall not be considered. In order to decrease the time complexity, we can also exclude from the evaluation those classes whose members did not occur among the r proteins most similar to the query. Protein databases are known to consist of classes very different in size. As the tree size does not influence the insertion cost, class imbalance will not represent a problem to *TreeInsert* when calculations are performed.

6.4.3 Implementation

We used the Neighbour-Joining algorithm for tree-building as given in the MATLAB Bioinformatics Toolbox [13]. In conjunction with the sequence comparison methods listed in Section 6.2, the programs were implemented in MATLAB. The execution of the method consists of 2 distinct steps, namely:

1. The preprocessing of the database into weighted binary trees and storage of the data in Newick file format [14]. For this step, the members of each class were compared with each other in an all vs. all fashion, and the trees were built using the NJ algorithm. For a large database like COG (51 groups 5332 sequences) the entire procedure takes 5.95 Seconds on a Pentium IV Computer (3.0 GHz processor).
2. First, the query is compared with the database using a selected similarity/dissimilarity measure and the data are stored in CSV file format. Next, the query is inserted into a set of class-representation trees, and the class with the optimal (smallest) *IC* value is chosen.

6.4.4 Performance Evaluation

The performance of *TreeInsert* was evaluated via ROC analysis and via the error rate as described in Section 6.6.4. For comparison we also include here the results obtained by simple nearest neighbour analysis (1NN). The results, along with the time requirements (wall clock times) are summarized in Table 6.9. Our classification tasks were the same as those in Section 6.3.4, thus the parameter t (number of considered class) was always equal to 2. The dependence of the performance on the other tuneable parameter r (the number of elements per class) is shown in Tables 6.7 and 6.8.

Table 6.7. ROC analysis results (AUC values) for the *TreeInsert* algorithm on the COG and 3PGK datasets. Here several different implementations were used

	1NN	*TreeNN*		
		Full	$r = 3$	$r = 10$
COG				
BLAST	0.8251	**0.8741**	0.8441	0.8708
Smith-Waterman	0.8285	**0.8732**	0.8474	0.8640
LAK	0.8249	0.8154	0.8276	**0.8734**
LZW	0.8155	0.7639	0.8243	**0.8316**
PPMZ	0.7757	0.8171	0.8535	**0.8682**
3PGK				
BLAST	0.8978	**0.9473**	0.8984	0.9090
Smith-Waterman	0.8974	**0.9472**	0.8977	0.9046
LAK	0.8951	**0.9414**	0.8851	0.9068
LZW	0.8195	**0.8801**	0.8009	0.8421
PPMZ	0.8551	0.8948	0.8646	**0.9123**

Table 6.8. Error rate values for the *TreeInsert* algorithm on the COG and 3PGK datasets. As before, several different implementations were used

	1NN	*TreeNN*		
		Full	$r = 3$	$r = 10$
COG				
BLAST	14.7516	**10.6127**	17.3419	17.3419
Smith-Waterman	**13.4940**	13.8218	17.9189	17.9189
LAK	13.3817	**11.3340**	15.9436	15.9436
LZW	16.7301	**13.8962**	20.0073	20.0073
PPMZ	15.0174	11.3386	**8.3167**	8.3167
3PGK				
BLAST	42.1046	**20.2009**	25.4544	35.7754
Smith-Waterman	42.1046	**20.3730**	24.7976	36.0115
LAK	42.0856	**20.2009**	25.8242	39.5036
LZW	36.5293	**15.7901**	37.0648	26.4240
PPMZ	34.6671	**14.4753**	32.3829	28.9935

Table 6.9. Time requirements of the *TreeInsert* methods on the COG dataset. Here n means the number of classes in question

Elapsed time (in second)/Method		1NN	*TreeInsert (r = 100)*
Preprocessing	BLAST	-	2232.54
	Other	-	1100
Evaluation	BLAST	0.223	0.223
Evaluation	Other	-	$0.029 * n$

In most of the test cases *TreeInsert* visibly outperforms 1NN in terms of ROC AUC and error rate. What is more, the *TreeInsert* method achieves the best results when we consider all the possible insertions, not just those of the adjacent leaves. This probably means that the insertion cost is not necessarily correlated with the similarity measure between the proteins.

When we examined the classification process using TreeInsert we found that we needed to carry out a preprocessing step before we evaluated the method. This preprocessing step consisted of the building of the phylogenetic trees for each class in the training dataset. Following the testing scheme we applied in Section 6.3.4, we assumed that the training dataset contained 1000 classes, and the classes contained 100 elements on average. Thus this step caused a significant growth in the running time. But when we investigated this method from a time perspective we found that this extra computation cost belonged to the offline time requirements. The evaluation of this method hardly depended on the number of classes in question because we had to insert an unknown protein into the phylogenetic trees of the known protein family. Table 6.9 describes this dependency, where n here denotes the number of classes.

6.5 Discussion and Conclusions

The problem of protein sequence classification is one of the crucial tasks in the interpretation of genomic data. Simple nearest neighbour (kNN) classification based on fast sequence comparison algorithms such as BLAST is efficient in the majority of the cases, i.e. up to $70-80\%$ of the sequences in a newly sequenced genome can be classified in a reassuring way, based on their high sequence similarity. A whole arsenal of sophisticated methods has been developed in order to evaluate the remaining $20-30\%$ of sequences that are often known as "distant similarities". The most popular current methods are "consensus" descriptions (see ii) in the Introduction that are based about the multiple alignment of the known sequence classes. A multiple alignment can be transformed either into a Hidden Markov model or a sequence profile; both are detailed, structured descriptions that contain sequence position-specific information on the multiple alignments. A new sequence is then compared with a library of such descriptions. These methods use some preprocessing that requires CPU time as well as human intervention. Also the time of the analysis (evaluation of queries) can be quite substantial, especially when these are compared to BLAST runs. The golden mean of sequence comparison is to develop classification methods that are as fast as BLAST but are able to handle the distant similarities as well.

The rationale behind applying tree-based algorithms is to provide a structured description that is simple and computationally inexpensive, but still may allow one to exceed the performance of simple kNN searches. *TreeNN* is a kNN type method that first builds a (small) tree from the results of the similarity search and then performs the classification in the context of this tree. *TreeInsert* is a consensus type method that has a preprocessing time as well as an evaluation time. Both *TreeNN* and *TreeInsert* exceed the performance of simple similarity searches and this is quite promising for future practical applications. We should remark here however that the above comparisons were made on very difficult datasets. On the other hand we used two-class scenarios, whereas the tasks in genome annotation are multiclass problems. Nevertheless, both *TreeNN* and *TreeInsert* can be applied in multiclass scenarios without extensive modifications so we are confident that they will be useful in these contexts. According to preliminary results obtained on the Protein Classification Benchmark collection [15] it also appears that, in addition to sequence comparison data, both algorithms can be efficiently used to analyse protein structure classification problems, which suggests that they might be useful in other fields of classification as well.

6.6 Appendix: Datasets and Methods

6.6.1 Datasets and Classification Tasks

In order to characterize of the tree-based classifier algorithms described in this work we designed classification tasks. A classification task is a subdivision of a dataset into +train, +test, -train and -test groups. Here we used two datasets.

Dataset A was constructed from evolutionarily related sequences of an ubiquitous glycolytic enzyme, 3-phosphoglycerate kinase (3PGK, 358 to 505 residues in length). 131 3PGK sequences were selected which represent various species of the Archaean, Bacterial and Eukaryotic superkingdoms [16]. 10 classification tasks were then defined on this dataset in the following way. The positive examples were taken from a given superkingdom. One of the phyla (with at least 5 sequences) was the test set while the remaining phyla of the kingdom were used as the training set. The negative set contained members of the other two superkingdoms and were subdivided in such a way that members of one phylum could be either test or train.

Dataset B is a subset of the COG database of functionally annotated orthologous sequence clusters [11]. In the COG database, each COG cluster contains functionally related orthologous sequences belonging to unicellular organisms, including Archaea, Bacteria and unicellular Eukaryota. Of the over 5665 COGs we selected 117 that contained at least 8 eukaryotic sequences and 16 additional prokaryotic sequences (a total of 17973 sequences). A separate classification task was defined for each of the 117 selected COG groups. The positive group contained the Archaean proteins randomly subdivided into +train and +test groups, while the rest of the COG was randomly subdivided into -train and -test groups. In a typical classification task the positive group consisted of 17 to 41 Archaean sequences while the negative group contained 12 to 247 members, both groups being subdivided into equal test and train groups.

6.6.2 Sequence Comparison Algorithms

Version 2.2.4 of the BLAST program [6] had a cutoff score of 25. The Smith-Waterman algorithm [4] we used was implemented in MATLAB [13], while the program implementing the local alignment kernel algorithm [23] was obtained from the authors of the method. Moreover, the BLOSUM 62 matrix [24] was used in each case.

Compression based distance measures (CBMs) were used in the way defined by Vitányi et al. [17]. That is,

$$CBM(X,Y) = \frac{C(XY) - \min\{C(X),C(Y)\}}{\max\{C(X),C(Y)\}} \tag{6.7}$$

where X and Y are sequences to be compared and $C(.)$ denotes the length of a compressed string, compressed by a particular compressor C, like the LZW algorithm or the PPMZ algorithm [18]. In this study the LZW algorithm was implemented in MATLAB while the PPMZ2 algorithm was downloaded from Charles Bloom's homepage (`http://www.cbloom.com/src/ppmz.html`).

6.6.3 Distance-Based Tree Building Methods

Distance-based or the distance matrix methods of tree-building are fast and quite suitable for protein function prediction. The general idea behind each is to calculate a measure of the similarity between each pair of taxons, and then to find a tree that predicts the observed set of similarities as closely as possible. In our study we used two popular algorithms, the Unweighted Pair-Group Method using Arithmetic Averages (UPGMA) [19], and the Neighbour-Joining (NJ) algorithm [8]. Both algorithms here are based on hierarchical clustering. UPGMA employs an agglomerative algorithm which assumes that the evolutionary process can be represented by an ultrametric tree: or, in other words, that it satisfies the "molecular clock" assumption. On the other hand, NJ is based on divisive clustering and produces additive trees. The time complexity of both methods is $O\left(n^3\right)$. In our experiments we used the UPGMA and the NJ algorithms as implemented in the Phylip package [20].

6.6.4 Evaluation of Classification Performance (ROC, AUC and Error rate)

The classification was based on nearest neighbour analysis. For simple nearest neighbour classification (1NN) a query sequence is assigned to the *a priori* known class of the database entry that was found most similar to it in terms of a distance/similarity measure (e.g. BLAST, Smith-Waterman etc.). For a tree-based classifier (Section 6.4), the query was assigned to the class that was nearest in terms of insertion costs (*TreeInsert*) or with the highest weight (*TreeNN*).

The evaluation was carried out via standard receiver operator characteristic (ROC) analysis, which characterizes the performance of learning algorithms under varying conditions like misclassification costs or class distributions [21]. This method is especially useful for protein classification as it includes both sensitivity and specificity, based on a ranking of the objects to be classified [22]. In our case the ranking variable was the nearest similarity or distance value obtained between a sequence and the positive training set. Stated briefly, the analysis was then carried out by plotting sensitivity vs 1-specificity at various threshold values, then the resulting curve was integrated to give an "area under curve" or AUC value. Note here that AUC $= 1.0$ for a perfect ranking, while for random ranking AUC $= 0.5$ [21]. If the evaluation procedure contains several ROC experiments (10 for Dataset A and 117 for Dataset B), one can draw a cumulative distribution curve of the AUC values. The integral of this cumulative curve, divided by the number of classification experiments, lies in a $[0, 1]$ interval with the higher values representing better performances.

Next, we calculated the classification error rate - which is the fraction of errors (false positives and false negatives) within all the predictions. Thus $ER = (fp+fn)/(tp+fp+tn+fn)$.

6.7 Acknowledgements

A. Kocsor was supported by the János Bolyai fellowship of the Hungarian Academy of Sciences. Work at ICGEB was supported in part by grants from the Ministero dell. Universita'e della Ricerca (D.D. 2187, FIRB 2003 (art. 8), "Laboratorio Internazionale di Bioinformatica").

References

1. Sjölander, K. (2004) Phylogenomic inference of protein molecular function: advances and challenges. Bioinformatics, Vol. 20, pp. 170–179.
2. Marco Cuturi, Jean-Philippe Vert. (2004) The Context Tree Kernel for Strings. Neural Networks, Volume 18, Issue 8, special Issue on NN and Kernel Methods for Structured Domains.
3. Mount, D.W. (2001) Bioinformatics. 1 ed. Cold Spring Harbor Laboratory Press, Cold Spring Harbor.
4. Smith, T.F. and Waterman, M.S. (1981) Identification of common molecular subsequences, J. Mol. Biol., 147, 195–197.
5. Needleman, S. B., Wunsch, C. D. (1970): A general method applicable to the search for similarities in the amino acid sequence of two proteins. J. Mol. Biol. 48:443–453.
6. Altschul, S.F., Gish, W., Miller, W., Myers, E.W. and Lipman, D.J. (1990) Basic local alignment search tool, J Mol Biol, 215, 403–410.
7. Felsenstein, J. (2004) Inferring phylogenies, Sinauer Associates, Sunderland, Massachusetts.
8. Saitou N., Nei M. (1987) The neighbor-joining method: a new method for reconstructing phylogenetic trees. Mol Biol Evol. Jul; 4(4):406–25.
9. Eisen, J.A. (1998) Phylogenomics: improving functional predictions for uncharacterized genes by evolutionary analysis. Genome Res. Mar; 8(3):163–7.
10. Zmasek, C.M. and Eddy, S.R. (2001) A simple algorithm to infer gene duplication and speciation events on a gene tree. Bioinformatics, 17, 821–828.
11. Tatusov, R.L., Fedorova, N.D., Jackson, J.D., Jacobs, A.R., Kiryutin, B., Koonin, E.V., Krylov, D.M., Mazumder, R., Mekhedov, S.L., Nikolskaya, A.N., Rao, B.S., Smirnov, S., Sverdlov, A.V., Vasudevan, S., Wolf, Y.I., Yin, J.J. and Natale, D.A. (2003) The COG database: an updated version includes eukaryotes, BMC Bioinformatics, 4, 41
12. Lazareva-Ulitsky B., Diemer K., Thomas PD.: On the quality of tree-based protein classification. Bioinformatics. 2005 May 1; 21(9):1876–90.
13. MathWorks, T. (2004) MATLAB. The MathWorks, Natick, MA.
14. Newick file format: http://evolution.genetics.washington.edu/phylip/newicktree.html
15. Sonego P., Pacurar M., Dhir S., Kertész-Farkas A., Kocsor A., Gáspári, Z., Leunissen, J.A.M. and Pongor S. (2007) A Protein Classification Benchmark collection for machine learning. Nucleic Acids. Res., *in press*.
16. Pollack, J.D., Li, Q. and Pearl, D.K. (2005) Taxonomic utility of a phylogenetic analysis of phosphoglycerate kinase proteins of Archaea, Bacteria, and Eukaryota: insights by Bayesian analyses, Mol Phylogenet Evol, 35, 420–430
17. Cilibrasi, R. and Vitányi, P.M.B. (2005) Clustering by compression, IEEE Transactions on Information Theory, 51, 1523–1542
18. Kocsor, A., Kertész-Farkas, A., Kaján, L. and Pongor, S. (2005) Application of compression-based distance measures to protein sequence classification: a methodological study, Bioinformatics (22), pp 407–412.
19. Rohlf, FJ. (1963) Classification of Aedes by numerical taxonomic methods (Diptera: Culicidae). Ann Entomol Soc Am 56:798–804.
20. Phylip package, http://evolution.genetics.washington.edu/phylip.html
21. Egan, J.P. (1975) Signal Detection theory and ROC Analysis. New York.
22. Gribskov, M. and Robinson, N. L. (1996) Use of receiver operating characteristic (ROC) analysis to evaluate sequence matching. Computers and Chemistry, 20(1):25–33, 18.
23. Saigo, H., Vert, J.P., Ueda, N. and Akutsu, T. (2004) Protein homology detection using string alignment kernels, Bioinformatics, 20, 1682–1689.
24. Henikoff, S., Henikoff, J.G. (1992) Amino acid substitution matrices from protein blocks. Proc Natl Acad Sci U S A. 1992 Nov 15; 89(22):10915–9.

7

Covariance-Model-Based RNA Gene Finding: Using Dynamic Programming versus Evolutionary Computing

Scott F. Smith

Dept. of Electrical and Computer Engineering, Boise State University, Boise, Idaho, 83725-2075, USA
sfsmith@boisestate.edu

Summary. This chapter compares the traditional dynamic programming RNA gene finding methodolgy with an alternative evolutionary computation approach. Both methods take a set of estimated covariance model parameters for a non-coding RNA family as given. The difference lies in how the score of a database position with respect to the covariance model is computed. Dynamic programming returns an exact score at the cost of very large computational resource usage. Presently, databases are prefiltered using non-structural algorithms such as BLAST in order to make dynamic programming search feasible. The evolutionary computing approach allows for faster approximate search, but uses the RNA secondary structure information in the covariance model from the start.

7.1 Introduction

The initial focus of interpreting the output of sequencing projects such as the Human Genome Project [1] has been on annotating those portions of the genome sequences that code for proteins. More recently, it has been recognized that many significant regulatory and catalytic functions can be attributed to RNA transcripts that are never translated into protein products [2]. These functional RNA (fRNA) or non-coding RNA (ncRNA) molecules have genes which require an entirely different approach to gene search than protein-coding genes.

Protein-coding genes are usually detected by gene finding algorithms that generically search for putative gene locations and then later classify these genes into families. As an example, putative protein-coding genes could be identified using the GENESCAN program [3]. Classification of these putative protein-coding genes could then be done using profile hidden Markov models (HMMs) [4] to yield families of proteins (or protein domains) such as that in Pfam [5]. It is not necessary to scan entire genomes with an HMM since a small subset of the genome has already been identified by the gene finding algorithm as possible protein-coding gene locations. Unlike protein-coding genes, RNA genes are not associated with promoter regions and open reading frames. As a result, direct search for RNA genes using only

S.F. Smith: *Covariance-Model-Based RNA Gene Finding: Using Dynamic Programming versus Evolutionary Computing*, Studies in Computational Intelligence (SCI) **94**, 183–208 (2008)
www.springerlink.com

generic characteristics has not been successful [6]. Instead, a combined RNA gene finding and gene family classification is undertaken using models of a gene family for database search over entire genomes. This has the disadvantage that RNA genes belonging to entirely novel families will not be found, but it is the only currently available method that works. It also means that the amount of genetic information that needs to be processed by the combined gene finder and classifier is much larger than for protein classifiers.

Functional RNA is made of single-stranded RNA with intramolecular base pairing. Whereas protein-coding RNA transcripts (mRNA) are primarily information carriers, functional RNA often depends on its three dimensional shape for the performance of its task. This results in conservation of three dimensional structure, but not necessarily primary sequence. The three dimensional shape of an RNA molecule is almost entirely determined by the intramolecular base pairing pattern of the molecule's nucleotides. There are many examples of RNA families with very little primary sequence homology, but very well conserved secondary structure (see pp. 264–265 in [7]). It is very difficult to find RNA genes without taking conservation of secondary structure into account.

Most homology search algorithms such as BLAST [8], Fasta [9], Smith-Waterman [10], and profile HMMs only model primary sequence and are therefore not well suited for RNA gene search. These algorithms are in the class of regular grammars in the Chomsky hierarchy of transformational grammars [11]. In order to capture the long-range interactions embodied in RNA secondary structure, one needs to move up one level in the Chomsky hierarchy to a context-free grammar. The extension of the regular-grammar-based HMM to a context-free grammar is a covariance model (CM) [12].

The structure of covariance models and model parameter estimation from a secondary-structure-annotated multiple alignment of a RNA gene family is the subject of the next section. The use covariance models for gene search by a specific non-coding RNA database (Rfam) will be examined in Section 2. It will be seen that the traditional dynamic-programming method of scoring database locations with respect to a covariance model is so computationally intensive that filters are normally first used to reduce the amount of searched database by orders of magnitude. The advantages and drawbacks of these filters are discussed in Section 3. An alternative to filtering is introduced in Section 4, where an evolutionary-computation CM-based search method is shown. Finally, conclusions are drawn and discussion of work that remains to be done is undertaken.

7.2 Review of Covariance Models for RNA Gene Finding

Covariance models can be viewed as an extension of profile hidden Markov models such that covariation in nucleotides at model positions that are widely separated in sequence, but physically connected as base pairs is captured statistically. Profile hidden Markov models are a specific form of hidden Markov model in which state transitions have a unidirectional flow from the start (5' in RNA/DNA or N-terminal in proteins) to the end (3' in RNA/DNA or C-terminal in proteins) of the model's consensus sequence. Similarly, a CM has unidirectional flow state transitions, but a more complicated connection topology. Profile hidden Markov models have five different types of states (start, match, insert, delete, and end). A CM has seven distinct state types (start, match pair, match/insert left, match/insert right, delete, bifurcate, and end). Finally, both types of models associate a group of states with each sequence position in the consensus sequence of the model. For the profile HMM, one match, one insert, and one delete state is associated with each consensus position (with possible exception of the first and/or last position). For the CM, a group of states (called a node) is associated with each consensus based-pair of positions and consensus unpaired position.

Both profile HMM and CM parameters are estimated from a group of nucleotide or protein sequences know as a family. In the case of the CM, it is also necessary to have a consensus secondary structure. This secondary structure may either be observed experimentally, or predicted from the sequence. In the case of non-coding RNA genes, prediction could be done with the Mfold [13] or RNAPredict [14] programs for example. The sequences may be either in the form of a multiple alignment or unaligned. For clarity of exposition, it is assumed here that the sequences are available in aligned form. In this case, the structure of the HMM or CM is determined by selecting alignment columns as either conserved or as insertion columns. Conserved columns are associated with some form of match state in the model. The most abundant symbol (nucleotide or amino acid) in each conserved column is taken as the consensus symbol for that position in the consensus sequence. The consensus sequence has length equal to the number of conserved multiple alignment columns.

7.2.1 CM Model Structure Determination

A multiple alignment that could be used to form a CM for the U12 family of non-coding RNA genes is shown in Figure 7.1. Included in the alignment are the seven sequences used by the Rfam database (described later) to form its CM of the family. These sequences are called seed sequences in Rfam and the resulting model has been used to find seven additional U12 family members. The four rows following the seven sequences contain consensus information. These four rows show consensus secondary structure, consensus sequence, CM node type assigned to the consensus column, and CM model branch letter code (used for reference to Figure 7.3 below) respectively. In the consensus structure rows, the symbol "-" indicates an unpaired conserved column, the symbols < and > represent the left (5'-side) and right (3'-side) halves of two base-paired conserved columns, and "." represents a non-conserved column. Note that there is no indication of which < column base pairs with which > column. This is because the structure is assumed to not have any pseudoknots. If the actual structure does have pseudoknots, then some of the base-paired columns have to be treated as if they were not base-paired. This results in some loss of power in the model, but the CM is not capable of representing pseudoknots and the increase in computational complexity of a model that can handle general pseudoknots is too high. If pseudoknots are disallowed, then the base-pairing notation is unambiguous.

Non-pseudoknotted structures have all base pairs such that all other base pairs are either completely inside or completely outside of them.

Of the six types of CM nodes, three do not emit consensus symbols (S = start, B = bifurcation, and E = end), two emit a single consensus symbol (L = left and R = right), and one emits two consensus symbols (P = pair). Therefore, only L, R, and P node types appear in the node type rows of Figure 7.1. L and R nodes are associated with a single multiple alignment column. Each P node is associated with two columns, one with secondary structure notation < and one with notation >. The multiple alignment shown has 156 columns, 149 of which represent consensus sequence positions. Four of the columns are assigned to R nodes, 63 to L nodes, and 41 pairs of columns to P nodes for a total of 4 + 63 + 2*41 = 149 consensus columns. Any time that a column could be assigned to either an L or an R node, the L node is preferred by convention, so there are generally many more L than R nodes in covariance models.

First, let us consider models that do not allow for insertions or deletions with respect to the consensus sequence. Such a profile HMM in shown in Figure 7.2 and a CM drawn at a level of detail that does not show insertion and deletion possibilities is shown in Figure 7.3. The HMM has only match (M) states. The arrow at the top of the figure shows that the model is written to process nucleotide sequences in the 5' to 3' direction. Each match state is associated with four emission probabilities, one for each of the four possible nucleotides. The nucleotide with highest probability is shown in parentheses inside the box for the M state and is equivalent to the consensus symbol for the conserved column of the multiple alignment represented by the M state. All transition probabilities is this model are equal to 1 since there is never any transition choice.

The node structure of the CM for the U12 RNA family is shown in Figure 7.3. The arrows pointing to the right and/or left of the R, L, and P nodes represent emission of a consensus symbol. This is a condensed version of the node structure where the numbers next to the emission arrows indicate that there is a vertical chain of that many nodes of identical type. The circled lower-case letters correspond to the branch codes in Figure 7.1. The first two consensus columns in Figure 7.1 (on the 5' end) correspond to two L nodes near the top of Figure 7.3 and the last three consensus columns in Figure 7.1 (on the 3' end) correspond to three R nodes just below in Figure 7.3. These two L nodes and three R nodes are all in branch "a". CM node diagrams always take the form of a binary tree. The top node in the tree is always an S node (called the root start node) and the bottoms of all terminal branches are always E nodes. Branches split into two at B nodes and the two children of any B node are always S nodes (called a left child start node and right child start node respectively).

Two of the branches in Figure 7.3 are surrounded by dashed boxes (branches a and b). All of branch a and the top portion of branch b are shown in expanded form in Figure 7.4. As shown by the arrows, the left side of a CM branch should be read from top to bottom and the right side of a branch from bottom to top in order to read consensus symbols in left-to-right multiple-alignment order (5' to 3' order). The a branch shows the first two consensus symbols are UG and the last three consensus symbols are CCG. The nodes actually represent a set of emission

```
1 .UGCCUUAAACUUAUGAGUAAGGAAAAUAACAACU......CGGGGUGAC
2 .UGCCUUAAACUUAUGAGUAAGGAAAAUAACGAUU......CGGGGUGAC
3 .UGCCUUAAACUUAUGAGUAAGGAAAAUAACGAUU......CGGGGUGAC
4 AUGUCUUAAACUUAUGAGUAAGGAAAAUAACGAUUGUUAUUCGGGGUGAU
5 .UGCCUUAAACUUAUGAGUAAGGAAAAUAACGAUU......CGGGGUGAC
6 AUGCCUUAAACUUAUGAGUAAGGAAAAUAACGAUU......CGGGGUGAC
7 AUGUCUUAAACUUAUGAGUAAGGAAAAUAACGAUUGUUAUUCGGGGUGAU
  .--<<<<<---------->>>>>--------<<<<......<<<<-----
  .UGCCUUAAACUUAUGAGUAAGGAAAAUAACGAUU......CGGGGUGAC
  .LLPPPPPLLLLLLLLLLPPPPPLLLLLLLLPPPP......PPPPLLLLL
  .aabbbbbbbbbbbbbbbbbbbbccccccccdddd......ddddddddd

1 GCCCGAGUCCUCACUACUGAUGUGAGAGGAAUUUUUGUGCGGGUACAGGU
2 GCCCGAGUCCUCACUGCUUAUGUGAGAAGAAUUUUUGAGCGGGUAUAGGU
3 GCCCGAAUCCUCACUGCUAAUGUGAGACGAAUUUUUGAGCGGGUAAAGGU
4 GCCCGAAUCCUCACUGCUAAUGUGAGACGAAUUUUUGAGCUGGUAAAGGU
5 GCCCGAAUCCUCACUGCUAAUGUGAGACGAAUUUUUGAGCGGGUAAAGGU
6 GCCCGAAUCCUCACUGCUAAUGUGAGACGAAUUUUUGAGCGGGUAAAGGU
7 GCCCGAAUCCUCACUGCUAAUGUGAGACGAAUUUUUGAGCUGGUAAAGGU
  ->>>>>>>><<<<<------->>>>>-----------<<<<<<<---<<<
  GCCCGAAUCCUCACUGCUAAUGUGAGACGAAUUUUUGAGCGGGUAAAGGU
  LPPPPPPPPPPPPPLLLLLLLPPPPPLLLLLLLLLLLPPPPPPPLLLPPP
  dddddddddeeeeeeeeeeeeeeeeefffffffffffggggggggggggg

1 CGUCCCC.GGGUGACCCGCUUACUUCGCGGGAUGCCCAGGUGCAAUGAUCUGCCCG
2 UGCAAUCUGAGCGACCCGCCUACUUUGCGGGAUGCCUGGGUGACGCGAUCUGCCCG
3 CGCCCUCAAGGUGACCCGCCUACUUUGCGGGAUGCC....................
4 CGCCCCUAAGGUGACCAGCCUACUUUGCGGGAUGCCUAGGAGUCGCGAUCUGCCUG
5 CGCCCUCAAGGUGACCCGCCUACUUUGCGGGAUGCC....................
6 CGCCCUCAAGGUGACCCGCCUACUUUGCGGGAUGCCUGGGAGUUGCGAUCUGCCCG
7 CGCCCCUAAGGUGACCAGCCUACUUUGCGGGAUGCCUAGGAGUCGCGAUCUGCCUG
  <<<<----->>>>>>>->>>>>>>--<<<<<<<<<--------->>>>>>>>>---
  CGCCCUCAAGGUGACCCGCCUACUUUGCGGGAUGCCUAGGAGUCGCGAUCUGCCCG
  PPPPLLLLLPPPPPPPRPPPPPPPLLPPPPPPPPPLLLLLLLLLPPPPPPPPPRRR
  gggggggggggggggggggggggghhhhhhhhhhhhhhhhhhhhhhhhhhhhhaaa
```

Fig. 7.1. U12 multiple alignment of seven seed family members

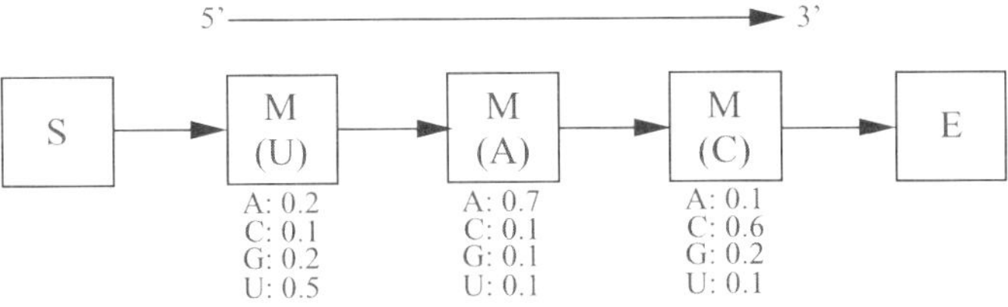

Fig. 7.2. A profile HMM with no insert or delete states

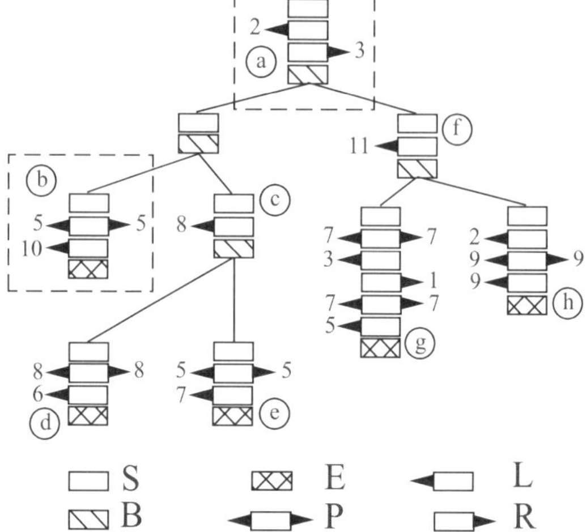

Fig. 7.3. Condensed U12 CM node structure

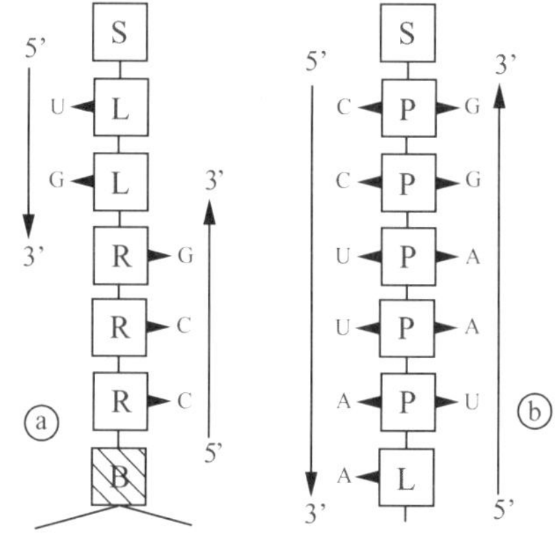

Fig. 7.4. Expanded portions of U12 CM node structure

probabilities. The L node directly below the root start node emits U with highest probability, by the probabilities of A, C, and G may be nonzero. L and R nodes have four such match probabilities (one for each possible nucleotide) and P nodes have sixteen probabilities (one for each possible pair of nucleotides). At the node structure level, the CM is similar to the HMM without insert or delete states in that transitions from a child node to a parent node happen with probability 1. At the bifurcations, the two submodels represented by the two subtrees are simply joined as one contiguous sequence with the left child subsequence on the left and the right child subsequence on the right.

Profile HMMs are normally augmented with insert and delete states as shown in Figure 7.5. The delete states (D states) are silent states that do not emit any symbols. These states simply allow one or more consensus positions in the model to be

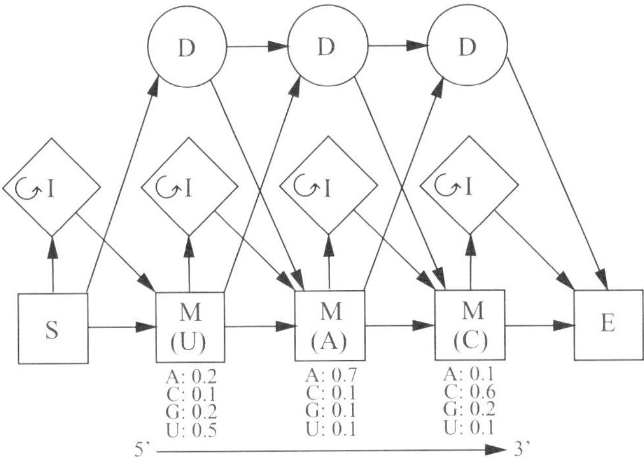

Fig. 7.5. A profile HMM with insert and delete states

skipped. The deletion penalty is imposed by the normally lower transition probabilities associated with the arrows to, from, and between D states as compared to the higher transition probabilities between M states. Affine delete gap penalties are available by having higher transition probabilities on D-to-D transitions than on M-to-D or D-to-M transitions. This allows multiple sequential deletions to be penalized less heavily than the same number of scattered deletions. This is consistent with observed gaps in nature and with gap penalties used in algorithms such as Smith-Waterman. The insert states (I states) have loop arrows inside the state diamond symbols to remind us that insert states always have self-loop transitions (both in HMMs and in CMs). This allows more than one possible insertion between consensus symbols. Affine insertion gap penalties are possible with differing self-loop and I-to-M/M-to-I transition probabilities. Unlike constant gap initiation and gap continuation penalties commonly used in algorithms such as Smith-Waterman, the gap penalties in an HMM or CM are position specific and can be different for insertions versus deletions. This leads to more flexibility, but also a large number of free parameters.

While it is possible to also include I-to-D and D-to-I transitions, Figure 7.5 omits these. Direct insertion to deletion transitions are rarely observed in real data and inclusion of these transitions just adds to the number of free parameters. The lack of these transitions is referred to as "plan seven" in the HMMER literature [15] (a program which estimates and scores profile HMMs). Seven refers to the number transitions leaving the D-I-M state triple associated with a consensus model position (including the I-state self loop). The alternative "plan nine" HMM architecture is not as commonly used. The standard CM is equivalent to a plan nine HMM in the sense that direct deletion to insertion transitions (and vice versa) are allowed. Investigation of the effect of removing these transitions in the CM case do not appear to have been undertaken to date.

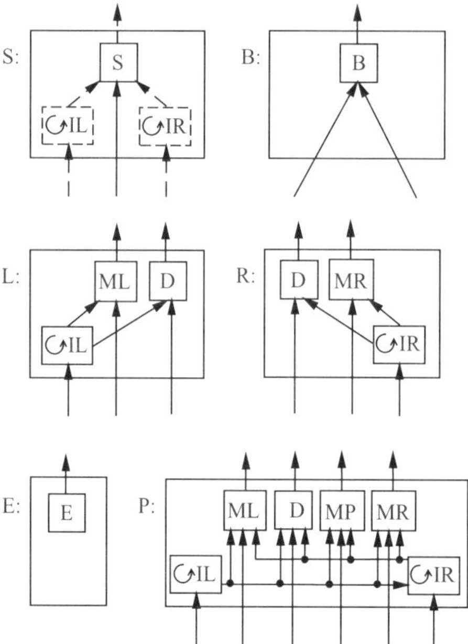

Fig. 7.6. Internal state structures of CM nodes

The equivalent to adding I and D states to the HMM is to allow non-consensus states within the nodes of the CM. Figure 7.6 shows the internal state structure of the six types of CM nodes. Each type of node contains a consensus state plus zero to five non-consensus states. The S, B, L, R, E, and P nodes have consensus states S, B, ML, MR, E, and MP respectively. Emitting nodes (L, R, and P) have D (delete) states that allow the consensus emitting state to be bypassed. The P node also contains two states that allow only the right or left half of the consensus pair to be missing in the database sequence. The ML state in the P node allows the right half of the consensus pair to be absent and the MR state allows the left half to be absent. IL and IR states allow additional database symbols to be inserted between consensus positions. These insert states have self-loop transitions, as indicated by the circular arrows inside the IL and IR state boxes such that any number of symbols may be inserted. The choice of which insert states to place in which node types is done such that there is no ambiguity as to which insert state in which node is responsible for insertions between a given couple of sequentially adjacent consensus locations. There are three sub-types of S nodes. The root S node has both IL and IR states. The right child S node has only an IL state. The left child S node has no insert states.

There are two levels of states within each node. The consensus state, D state (if present), and delete-related states (ML and MR in P nodes) are in the top level. The bottom level contains any insert states (IL or IR) that may be present. This implies that any insertions are added to the database sequence before any consensus matching is done (since the model is evaluated from the leaves toward the root). All top level states in a given node have transitions to all states in the parent node. Bottom level states (insert states) only have transitions to top level states in the same node and to themselves. As a result, the arrows entering or leaving a node in Figure 7.6 represent a bundle of transitions whose number depends on the type of parent or child node. The IL- or IR- to-D transitions are clearly seen in Figure 7.6, but the D-to -IL or -IR transitions are only implicitly shown. These transitions make the standard CM architecture equivalent to a plan nine profile HMM.

7.2.2 Mapping a Database Sequence to a Covariance Model

In order to fit a database sequence to a covariance model one starts at the E states and works up the CM tree toward the root S state. Each E state models a null sequence and with no database symbols mapped to it. Transitioning from a child state to a parent state maps zero, one, or two database symbols to the model. Non-emitting parent states map no symbols, single-emitting states (IL, IR, ML, and MR) map a single symbol to the model, and the pair-emitting state (MP) maps two symbols. The transition adds a log-likelihood ratio transition score to the overall model score. If the parent state is an emitting state, a log-likelihood ratio emission score is also added.

Figure 7.7 shows the effect of moving from a top-level state T in a child node (of any type) to the top-level MR state in a parent R node. This has the effect of matching a database symbol to a model consensus symbol and inserting zero or more database symbols between the existing mapped database symbols and the consensus-matched symbol. If at least one database symbol is to be inserted, the first transition is from the child-node top-level state to the IR state of the parent R node. The length of the mapped sequence increases by one and the overall score increases by the T-to-IR transition score plus the IR state emission score for the inserted symbol (the emitted symbol is G in Figure 7.7). If an additional database symbol is inserted, the mapped sequence again increases in length by one and the overall score increases by the IR-to-IR transition score plus the IR state emission score of the new symbol. This continues until all inserted symbols are finished. Finally, the consensus-matched symbol (U in the figure) increases the mapped sequence length by one and increases the score by the IR-to-MR transition score plus the MR state emission score for the

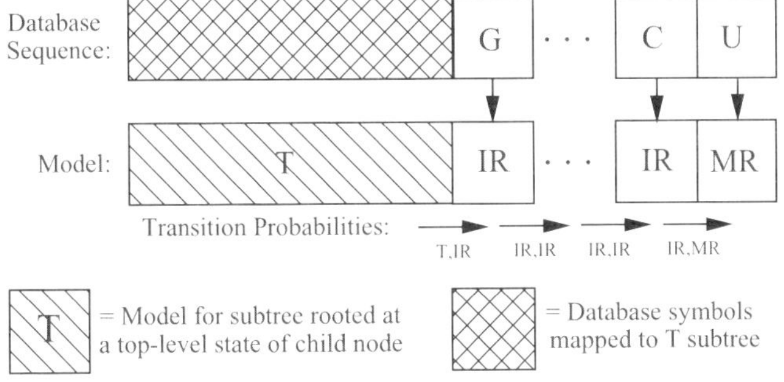

Fig. 7.7. Building to the right with an R node

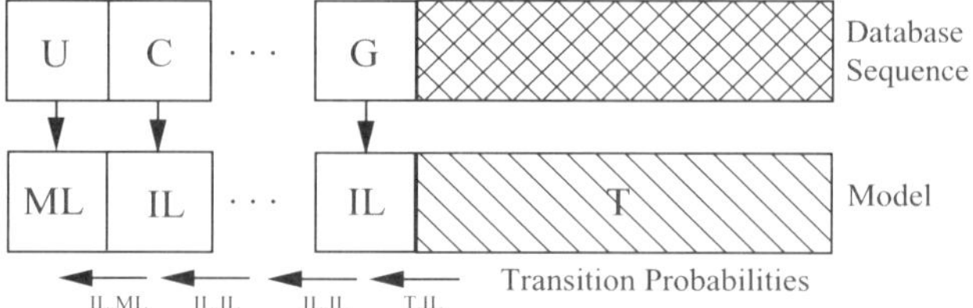

Fig. 7.8. Building to the Left with an L node

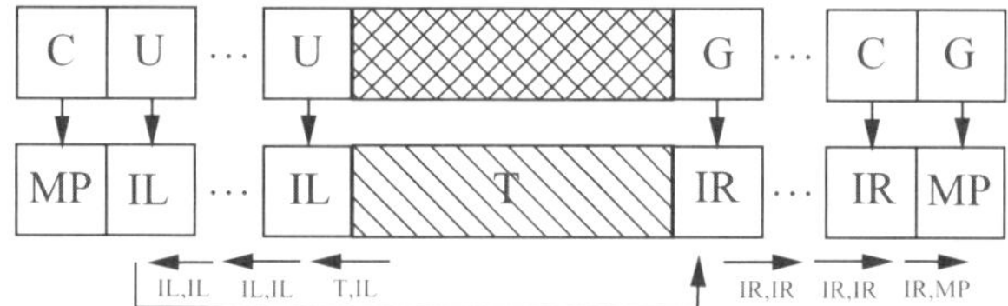

Fig. 7.9. Building both ways simultaneously with a P node

matched symbol. Not shown in the figure is the alternative that the model position is deleted. In this case, a single transition is made from the T state to the D state. The mapped sequence length remains unchanged and the score increases by the T-to-D transition score only.

The effect of moving from a top-level state T in a child node to the top-level ML state in a parent L node is shown in Figure 7.8 and is an exact mirror of the situation in Figure 7.7 for an R parent node. It can be noted that structure generated by a model with only a single branch of R nodes or L nodes is very similar to the profile HMM. Each node contains one match, one insert, and one delete state. In fact, the single-emission nodes of the CM are simply modeling the primary sequence homology of the non-base-paired portions of consensus alignment. Also note, there is no need to have IR states in L nodes or to have IL states in R nodes. Insertions to the outside of the sequence represented by a top-level state are generated by the next node up the tree (possibly by the root start node, which is why the root start node has both IR and IL states).

All of the advantage of using a CM over an HMM is embodied in the MP states of the P nodes. It is the sixteen distinct emission probabilities for each of the possible nucleotide pairs that allows covariation to be detected. Typically, there will be one pair with very high probability (indicating both sequence homology and secondary structure homology) and other canonical or wobble base pairs with lesser, but still high, probabilities. Figure 7.9 shows the effect of a transition from a child-node top-level state T to the MP state of a parent P node. The IL state can be visited zero or more times and the IR state can be visited zero or more times in between. The number of IL visits does not need to equal the number of IR state visits. Even though MP appears twice in Figure 7.9, this is a single visit of the MP state which emits two match symbols. The score increases by the sum of the transition score and the emission score for each of these transitions.

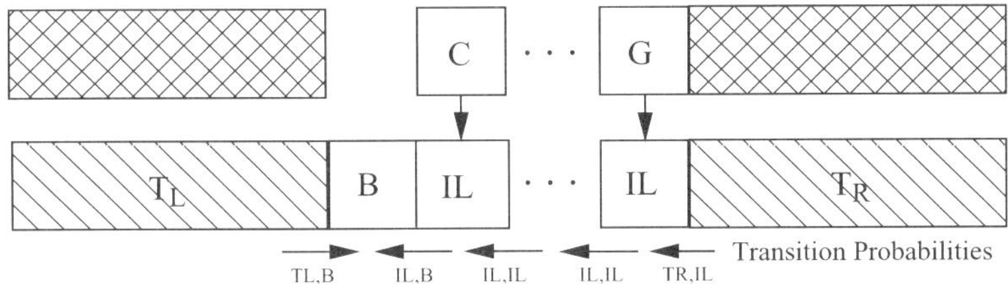

Fig. 7.10. Joining two submodels with a B node

Using the S, L, R, P, and E nodes it is possible to construct a single branch of a CM tree that can describe any secondary structure as long as all base pairs are fully nested. For instance, the structures «-«---»>-> and «<--»-> can be described, but «-»-««--»» and «-»«<--» >-«-» can not. The later two structures require one and two bifurcations respectively. Roughly speaking, the branches of the CM model tree correspond to stem-loops in the RNA secondary structure. Figure 7.10 shows how a B state joins two submodels represented by its left and right child states (the TL and TR states). These states are always start states and are part of a left-child S node and a right-child S node respectively. It is necessary that either the left-child S node contain an IR state or the right-child S node contain an IL state in order to allow inserted database symbols to the right of the rightmost consensus position of the left-child submodel. Since left emissions are always preferred to right emissions by tradition whenever either would be possible, right-child S nodes are chosen to contain an IL state and left-child S nodes do not contain any insert states.

Depending on the number of insert and delete state visits the length of the sequence mapped by a transit of the model can be longer or shorter than that of the consensus sequence. In order to find the optimal mapping of a database sequence to the model, one should consider all possible insertion and deletion patterns. If the database sequence is an entire chromosome, the database sequence could be several hundred million bases in length. It would be possible in theory to search insertion patterns that include total inserted symbol counts of many millions of database symbols. In practice, true RNA genes are unlikely to have so many insertions and even if they did, they would be rejected by a scoring scheme that would add up large numbers of gap extension penalties. In order to make database search possible with deterministic algorithms such as dynamic programming, dynamic programming a cutoff on the maximum database sequence length that can be represented at any state in the model is made. Often this cutoff is less than one and one half times the consensus sequence length of the RNA family. The length of the database sequence mapped to a CM state is equal to the consensus sequence length represented by the model subtree rooted at that state plus the net number of insertions less deletions made up to that state.

The traditional method of using a CM for database search is to use dynamic programming. A score is calculated for the database subsequences ending at every possible position in the database and for every possible subsequence length up to the length cutoff discussed above. At each possible database position, the maximum over all subsequence lengths explored for that position is taken as the database position score. Database position scores exceeding a selected threshold cause the database position to be declared as the ending position of a putative gene for the RNA family represented by the CM.

The dynamic programming algorithm starts at the E states. These states represent null sequences and are assigned a score of zero for all possible database subsequence end positions and the mapped subsequence length is taken to be zero. Score evaluation then progresses from the E states up the tree towards the root S state. The scores at the root S state are the scores used to determine the putative gene locations. At each model state, the best possible score for a database subsequence ending at each possible database position and for every possible database subsequence length between zero and the length cutoff are evaluated. These best scores are used by parent states to calculate the best possible scores for the submodel rooted at the parent node.

A more formal description of the dynamic programming algorithm used to generate scores for a database sequence with respect to a covariance model of an RNA family is given in Figure 7.11. The algorithm uses a triple-nested loop over database end position, subsequence length, and CM state. All database end positions $\mathbf{j}$ in the range 0 to the length of the database sequence $\mathbf{L}$ are examined. All database subsequence lengths in the range 0 to the length cutoff $\mathbf{D}$ are examined (with subsequence lengths that would extend past the start of the database sequence ignored). Finally scores are generated for each state number $\mathbf{v}$, where there are $\mathbf{M+1}$ states and the root start state is numbered 0. States are indexed such that the index of a child state is always higher than that of a parent state. Evaluating state scores in reverse state index order ensures that the score for a submodel at a child state is always available when needed by its parent state. The scores calculated at each position, subsequence length, and state are given by $\mathbf{s(j,d,v)}$.

For E states, the score is zero for null subsequences and minus infinity for any other subsequence. This ensures that any subsequence other than the null subsequence is discarded by the maximum operation when finding the best score further up the tree. Delete and start states are computationally identical. Neither adds any database symbols to the mapping. Only the transition score $\mathbf{trans(c,v)}$ from the child

```
for j = 0 to L
 for d = 0 to min(D,j)
  for v = M to 0
   case type(v) is
    E: if d == 0 then s(j,0,v) = 0; else s(j,d,v) = -Infinity
    D or S: max(over children c)[s(j,d,c) + trans(c,v)]
    L: if d == 0 then s(j,0,v) = -Infinity; else s(j,d,v) =
      emit(l,v) + max(over children c)[s(j,d-1,c)+trans(c,v)]
    R: if d == 0 then s(j,0,v) = -Infinity; else s(j,d,v) =
      emit(r,v) + max(over children c)[s(j-1,d-1,c)+trans(c,v)]
    P: if d < 2 then s(j,d,v) = -Infinity; else s(j,d,v) =
      emit(l,r,v) + max(over children c)[s(j-1,d-2,c)+trans(c,v)]
    B: max(over k in 0 to d)[s(j-k,d-k,lc)+s(j,k,rc)]
```

Fig. 7.11. Algorithm for dynamic-programming CM scoring

state **c** to parent state **v** is needed here. The L, R, and P state types add database symbols to the mapping and therefore change the database position and/or subsequence length mapped when compared to the mapping passed from the child state. L-type states include both IL and ML states, R-type states include both IR and MR states, and P-type states only appear as MP states. L- and R-type states add one symbol to the mapping, so the resulting mapping can not have length less than 1. Therefore the score is set to minus infinity for length 0. P-type states add two symbols to the mapping and therefore a score of minus infinity (meaning impossible) is set for lengths of 0 or 1. For L-type states, the resulting database end position remains unchanged, but mapping a symbol on the left increases the subsequence length by one, so the score for length **d** depends on a child score for length **d** – **1**. For R-type states, adding a symbol on the right moves the end position one place right, so the score for length **d** and position **j** depends on child length **d** – **1** and position **j** – **1**. For P-type states, the score for length **d** and position **j** depends on child length **d** – **2** and position **j** – **1**. For L-, R-, and P-type states, the change in score is both an emission score (**emit**) and a transition score (**trans**). The emission score depends on the database symbols found on the right (**r**) and left (**l**) respectively. Finally, bifurcation finds the best two submodels whose lengths add up to the subsequence length score to be evaluated and which are contiguous along the database.

It should be clear that the amount of computation involved in the dynamic programming algorithm is very large. In fact, this is the central drawback to application of the algorithm. The remainder of this chapter explores current use of covariance models for RNA gene finding and approaches to reducing the computational cost of CM-based database search.

7.3 Application of CM-Based RNA Gene Finding: The Rfam Database

Over five hundred families of non-coding RNA sequences have been identified and modeled in the publically-available Rfam database [16]. Groups of sequences are formed into families using reference to the literature, aligned, and annotated with secondary structure using either experimentally-derived or computer-predicted structures. These carefully hand curated multiple alignments are referred to as "seed" alignments in Rfam. The structure of the CM model tree is then generated from the consensus secondary structure annotation of the alignment. Transition and emission scores for the CM are estimated from observed frequencies of nucleotides and gaps at the various multiple alignment columns. A prior distribution is then combined with the observed frequencies in an attempt to correct for limited sample size. This has the effect of eliminating log-likelihood ratio scores that are minus infinity due to observed counts of zero. Such scores are undesirable because they rule out certain patterns merely because they did not happen to be observed in the training data even though there is no theoretical reason to believe that they are impossible.

A package of programs used by Rfam to estimate covariance models and for database search is called Infernal [17]. This package is publicly available, including source code. The Rfam site includes both alignments of the original seed sequences for families and combinations of seed sequences and new sequences found through database search. The database also includes the parameter files of the estimated covariance models. These parameter files are particularly useful when exploring methods other than dynamic programming for CM-based database search since the parameters can be transformed into other formats suitable for alternative searches.

The amount of computational power needed for direct search of the available genomic data using dynamic programming and covariance models is excessive. In order the trim the amount of data searched by orders of magnitude, a filtering operation is first applied to the database. In the case of Rfam, the filtering method is to use BLAST [8] to score the database with respect to the consensus sequence of the model. It is hoped that the new RNA genes will have enough primary sequence homology with the existing family members that their score will be raised enough above the background noise to be retained in the portion of data passed to the full CM search. The extent to which this hope is true in practice is not very well studied. In the following section of this chapter, another proposed filtering method from the literature will be discussed. In section 4, we will discuss an evolutionary computation alternative to filtering the database (and to using preset length cutoffs).

In examining the possibility of using evolutionary computation, data extracted from the Rfam database is used. In particular, fourteen sequences belonging to the U12 ncRNA family (accession number RF00007) are used for testing. The parameter file for this Rfam family is also used. Of the fourteen sequences, seven are from the seed family used to estimate the model parameters and seven were found through database search using the model. The U12 family [18] [19] [20] are small nuclear RNA (snRNA) which form a complex with specific proteins to function as part of the minor spliceosome. The function of U12 is to remove introns from pre-mRNA. The U12 ncRNA acts in a way similar to that of the U2 ncRNA in the major spliceosome.

7.4 Filters to Reduce Database Search Cost

A major problem with using a filter to reduce the amount of genomic data to be searched with a CM is that there may not be enough primary sequence homology to keep the true gene in the retained data set. With the BLAST method of database reduction, there is no known way to set the score threshold to guarantee retention. However, Weinberg and Ruzzo [21] have recently come up with a way to guarantee that a profile HMM filter will not discard any portion of the database that contains a subsequence that the dynamic- programming CM search would score as a putative RNA gene. The procedure involves extracting from the CM parameter files equivalent profile HMM parameters that ignore the joint probability information inherent in the P state emission probabilities. The maximum additional score that could come from the secondary structure information in the CM with a perfect database match can be calculated and subtracted from the score threshold to be used with the CM search. The result is the minimum primary sequence score contribution that must come from the database sequence in order for the overall CM score to exceed the

CM threshold. Portions of the database which do not meet this minimum score contribution are found when the HMM score does not exceed this minimum primary sequence contribution.

Disadvantages of the HMM method are that the HMM is much slower than BLAST (although significantly faster than full CM search) and that the reduction in database size varies greatly from one RNA family model to another. No comprehensive study of the speedup of this method has been undertaken. The Weinberg and Ruzzo paper looks at only 34 of the over 500 families. Extrapolating from this paper, it is still predicted to take tens of CPU years with a modern desktop computer to search all Rfam families on the 8-gigabase database of the study. Since both the number of known RNA families and the amount of genomic data are rapidly expanding, this amount of computation is still too much.

Weinberg and Ruzzo [22] have also recently come up with an heuristic filter that is an alternative to BLAST and appears to perform better than BLAST. This heuristic filter is based on a profile HMM and as such does not use secondary structure information at the filtering stage.

7.5 An Alternative to Filters Using Evolutionary Computation

In this section, the use of evolutionary computation (EC) as an alternative to filtering followed by dynamic programming search is examined. Secondary structure information will be used from the start on the entire database. Also, no sequence length cutoff is employed. The results of the non-exhaustive EC-based search will also be compared to those of a simple hill-climbing algorithm which is also non-exhaustive, but does not have the ability to escape local minima. The ability to jump out of a local minimum is shown to be crucial to the algorithm.

7.5.1 Dynamic Programming Efficiency

To motivate why the traditional dynamic-programming exhaustive search might not be the most efficient way to find RNA genes using a CM, the observed usage of search space regions is first examined. Dynamic programming finds the best score at each model state for each database end position and each database subsequence length ending at that position (up to a predefined cutoff length). The first observation made is that only a small range of the subsequence lengths evaluated at a given state are normally observed in real data. These subsequence lengths cluster about the consensus sequence length for the submodel represented by the subtree rooted at the state. In what follows, *length deviation* will be defined as the length of the database subsequence generating a score at a given state minus the length of the consensus sequence represented by the state. Length deviation is therefore equivalent to the number of inserted symbols minus the number of deleted model positions in the submodel mapping of a given state.

The actual usage of subsequence lengths at various states of the Rfam U12 CM model for the fourteen known U12 family members (seven seed and seven discovered members) is shown in Table 7.1. The "top" and "bottom" designations and the

Table 7.1. Subsequence length use in observed U12 data

State	Branch	Consensus Length	Obs Max	Obs Min	DP Max	DP Min
root S	a	149	+6	−6	+11	−149
bottom R	a	145	+6	−6	+16	−145
top P	b	20	+4	−3	+140	−20
bottom L	b	1	+1	0	+159	−1
top L	c	47	+6	−4	+113	−47
bottom L	c	40	+6	−4	+120	−40
top P	d	22	+6	−1	+138	−22
bottom L	d	1	0	0	+159	−1
top P	e	17	0	−3	+143	−17
bottom L	e	1	0	0	+159	−1
top L	f	77	+1	−3	+83	−77
bottom L	f	67	+1	−2	+93	−67
top P	g	37	+1	−1	+123	−37
bottom L	g	1	0	0	+159	−1
top L	h	29	0	−2	+131	−29
bottom L	h	1	0	0	+159	−1

branch letter refer to Figure 7.3. For example, "bottom R" and branch "a" is the consensus MR state in the R node and the bottom of the "a" branch of the CM tree. The consensus sequence length is 145 for the model subtree rooted at this MR state since two more R nodes and two more L nodes are in the CM tree above it and the overall consensus sequence length is 149. Using dynamic programming, all subsequence lengths in the range 0 to 160 are investigated at every state (since the cutoff length for this model is chosen to be 160 in Rfam). The last four columns of the table show length deviations from the consensus length at each model state. The dynamic programming (DP) maximum and minimum length deviations are always 160 minus the consensus length and the negative of the consensus length respectively. The observed maximum and minimum length deviations are shown in the fourth and fifth table columns respectively. These are seen to cluster near zero and to be generally much smaller than the dynamic programming limits.

One possible way to make the dynamic programming algorithm more efficient by about one to two orders of magnitude is to specify state-dependent minimum and maximum length deviations (or equivalently, minimum and maximum subsequence lengths). This requires extra complexity in the search code. The model input file needs to be augmented with the state-dependent search limits. These limits would need to be determined by a program that automatically extracts the observed length deviations at each state from the seed sequence multiple alignment. A buffer region about the observed deviations needs to be added to allow for deviations of true family members that are outside the range observed in the seeds. The statistical analysis needed for a good choice of buffer region size is nontrivial.

7.5.2 CM-Based Search Without Dynamic Programming

Another way to improve efficiency is to expand the search about the zero length deviation solution [23]. This could be done with either deterministic (such as hill-climbing) or randomized (such as genetic algorithm) search methods. In either case, the initial step is to determine the scores of an ungapped mapping of a database subsequence to the covariance model at every database position. This is equivalent to evaluating every consensus state for zero length deviation only and assigning a score of minus infinity if the length deviation is not zero or the state is a non-consensus state. Unlike filtering with BLAST, Fasta, or an HMM, the ungapped scoring method employs base-pairing information from the start on all portions of the database. It is also several orders of magnitude faster than the full dynamic programming CM search due to a number of factors. First, only consensus states are evaluated for about a factor of three reduction in evaluated states. Second, there is no need to add state transition scores in this initial sweep of the database since they only contribute an additive constant to the score at every database position, for a computational reduction of about a factor of two. Third, only one subsequence length is evaluated for a reduction by a factor of the cutoff length (often two to three orders of magnitude). Forth, bifurcation states are very expensive relative to other states since they need to check every possible allotment of subsequence length between the two branches. This results in bifurcations having a computational complexity that is higher than other states by a factor about equal to the cutoff length. Since the function of the bifurcation is not needed without gaps, this saves another two to three orders of magnitude. Overall, ungapped scoring of a database is somewhere in the range of three to eight orders of magnitude faster than full dynamic programming scoring. The proportionate speedup is greater for models with very long consensus sequences (and long length cutoff), which are exactly the models that take longest with the conventional scoring method.

The clustering of the true subsequence lengths about zero length deviation as shown in Table 7.1 for U12 is a general phenomenon. What is less clear is whether the scores of the best solutions improve monotonically as the length deviation is changed from zero to its true value. If all possible insertions of two contiguous symbols are attempted and all possible deletions of three contiguous symbols are attempted, then a large number of scores are generated for length deviations of $+2$ and -3. If all possible combinations of simultaneous double insertions and triple deletions are tried, then a much larger number of scores for length deviations of -1 are created. The alignment patterns with the double insertion and triple deletion may be much different than that of a single deletion (which also has a length deviation of -1). Thus it is not clear that an algorithm that searches by trying every possible single insertion and every possible single deletion at each model position will necessarily move closer to the true solution. In fact, it has been found for the U12 family that such a simplistic hill-climbing approach does not work as well as a randomizing algorithm that is capable of escaping from local minima.

In addition to the possibility of focusing the search on model mappings with relatively few insertions and deletions, there is also the possibility of focusing the search around database locations that have high scores with suboptimal alignments. The initial ungapped sweep of the database should give generally higher scores near true RNA gene family members than on unrelated portions of the database. This is a result of matching at least some part of the database sequence that does not happen to have gaps relative to the consensus sequence. The search can start with an expansion about the ungapped alignment for a relatively large number of high-scoring database positions. As some database positions start to show score improvements and others not, the search can move to focus only on those database positions showing either very high initial scores or somewhat lower, but improving, scores. Finally, once the number of database positions in narrowed sufficiently, it is possible to resort to full dynamic programming search of the neighborhoods around the very highest scoring positions.

It is helpful to have a fixed-length representation of the alignment of a database subsequence to the consensus sequence of the CM. The representation used here is taken from the literature on protein threading using evolutionary computation [24]. A vector of non-negative integers of length equal to the length of the consensus sequence is used. If a vector element is 0, then the corresponding consensus model symbol is deleted. If the vector element is 1, then the model symbol is matched and there are no inserted database symbols to the right of this consensus position. If the vector element is a value **n** greater than 1, then **n** − **1** database symbols are inserted. Figure 7.12 shows the correct alignment vectors for the seven seed sequences of the U12 family (see Figure 7.1 for comparison). Each alignment is a vector of 149 integers and the break of six spaces is only in the figure to show correspondence between the values and the original multiple alignment (the actual representation contains no such spaces). The goal of the search algorithm is to expand the search around the initial solution vector (149 ones) toward the true alignment vectors as in Figure 7.12. Notice that there is nothing more to do in the case of database sequences 2 and 6 since the optimal alignment has no gaps with respect to the consensus sequence. There is no way to represent insertions to the left of the first position, so the representation is the same whether the sequence has a symbol in the first column or not. This is not a problem, since the alignment is local. The putative gene start position can be off by several bases due to initial insertions.

There are two components to a candidate search solution, the alignment vector and the location in the database sequence of the first alignment position. An alignment vector change is assumed to take a form that results in either adding or removing one or more contiguous insertions or deletions. Adding contiguous insertions involves increasing a single vector element by one or more. Removing contiguous insertions is done by decreasing a single vector element by one for more such that the resulting element value is greater than 0. Adding contiguous deletions is accomplished by changing a range of consecutive vector elements to 0. Removing contiguous deletions occurs when a range of consecutive zeros are all changed to 1.

When considering a change to a given candidate solution to potentially improve the score of the solution, two classes of variations are possible which will be called compensating and non-compensating. Non-compensating changes are those for which the alignment of the model to the database remains unchanged to the left

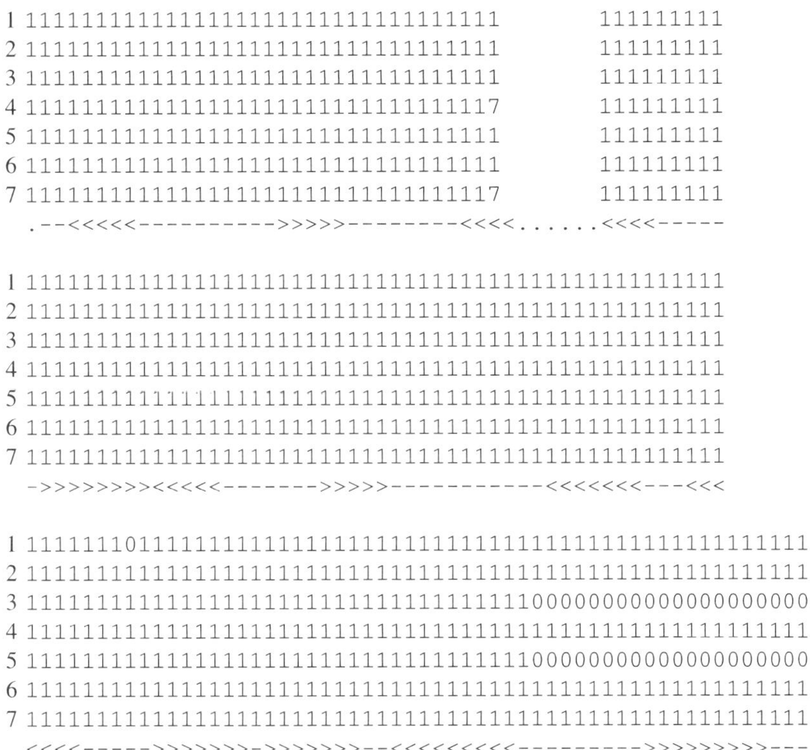

Fig. 7.12. U12 seed sequence alignment vectors

of the alignment vector alteration and compensating changes are those for which the model/database alignment is unchanged to the right. Non-compensating changes are the result of alignment vector changes with the database location of the first position unchanged. Compensating changes occur when the database start position is changed by an amount with equal magnitude but opposite sign to that of the total change in alignment vector values. Figure 7.13 shows how a compensating change to a candidate solution might improve a score whereas a non-compensating change does not. Initially the alignment vector is **11111111** and the database start position is at the first A in the portion of the database sequence shown (**GGAAUCACUG**) as shown at the top of the figure. The correct alignment vector is **11131111** and the correct database start position is two places further to the left. The initial alignment causes the last four database symbols shown to correctly align with the last four consensus symbols of the model. If a close (but not exactly correct) change to the alignment vector is tried such that the candidate vector is **11311111** without a compensating change to the database start position, the alignment gets worse. This non-compensating change is shown in the middle of the figure. If the same close change to the vector is tried, but with database start position compensation, then the situation is as in the bottom portion of the figure. Since the sum of the vector element values increased by 2

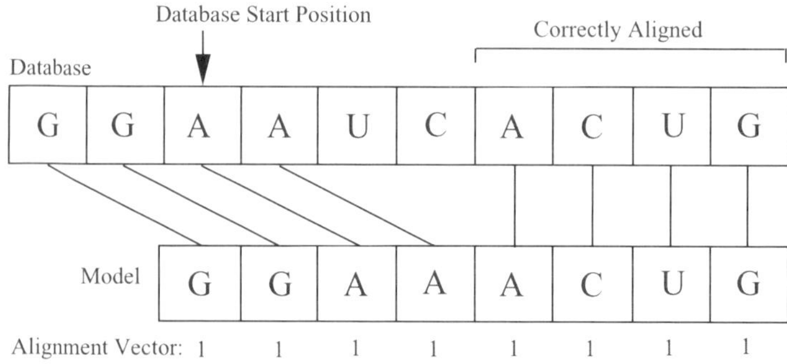

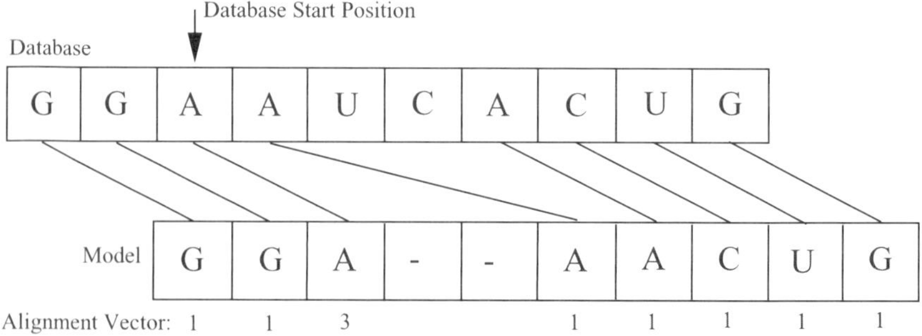

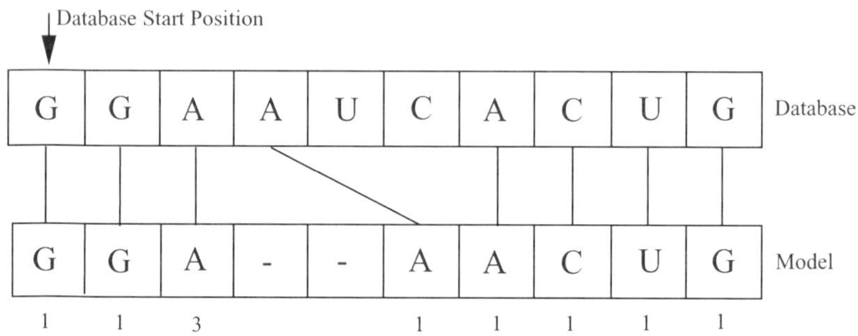

Fig. 7.13. Compensating and non-compensating candidate solution changes

when going from **11111111** to **11311111**, the compensating change is to decrease the starting position by 2. Now all but one of the positions aligns and the score should increase. Sometimes, the portion of the alignment that is contributing to a high score for the initial solution is due to a good alignment to the left of the vector change and uncompensated change may improve a score. Other times, the high-scoring alignment portion may be to the right of the vector change and compensated change may improve the score. In general, both should be tried.

The fitness function of an individual is the score of the database subsequence associated with the individual with respect to the covariance model. The database subsequence associated with an individual starts at a database location specified by the individual and continues for a length equal to the sum of element values of the alignment vector. The alignment vector specifies a unique path through the covariance model states and the score is found as the sum of log-likelihood ratio scores for each of the state transitions and symbol emissions for this unique tree parse. The search space is the set of all possible starting locations within the database as well as all possible combinations of non-negative alignment vector values such that the sum of the vector values plus the starting location does not exceed the end of the database. Single-point crossover is employed and as well as single mutations. The single mutations take either the form of changing a single alignment vector element to some other non-negative value (changing the number of insertions at a point) or taking a range of values and changing them to 0 (creating a contiguous deletion region). Single mutations can either be compensating or non-compensating as described above with a fifty percent probability of each likely a good choice. The probabilities of single mutations to small element values should be higher than those of large values (an exponentially decreasing probability would be a reasonable choice). Similarly, the probabilities of small deletion regions should be greater than large regions.

7.5.3 Experimental Results

In order to try out the idea of using a GA to search for good CM alignments in a database, an artificial dataset has been created which contains a mixture of U12 RNA genes and other ncRNA genes. This was done to keep the test database small and at the same time provide tempting incorrect targets for the algorithm searching for U12 genes. The other ncRNA genes contain stem-loop structures that are likely to be more similar to U12 genes than randomly chosen segments of genome. In future research, the GA search method should be applied to a real search, but this research has not yet progressed to that point. This is partly due to the fact that the current version of the software is written in MATLAB and needs to be rewritten in C and optimized for large-scale use.

The test dataset contains 15880 bases, such that about ten percent of the sequence is composed of true U12 genes and the remainder of randomly selected other ncRNA genes taken from Rfam. Since the true U12 genes are in the Rfam database, they are all able to pass the BLAST filter (true genes that might exist and can not get past the filter can not be in Rfam by definition). The initial ungapped scoring of each database position with the CM for U12 is shown in Figure 7.14. The true U12 genes start at database locations 446, 1039, 2475, 3858, 6096, 7406, 8196, 8880, 9705, 10774, 11624, 12428, 13493, and 14615. Twelve of these locations have peaks in the initial database sweep at or near the true position with scores that are the twelve highest scores out of the 15732 scores (148 positions at the end of the database are not scored with respect to the 149 position model). Two of the true U12 genes have peaks that are harder to discern from the background. The peak at 8196 has a score of about 5

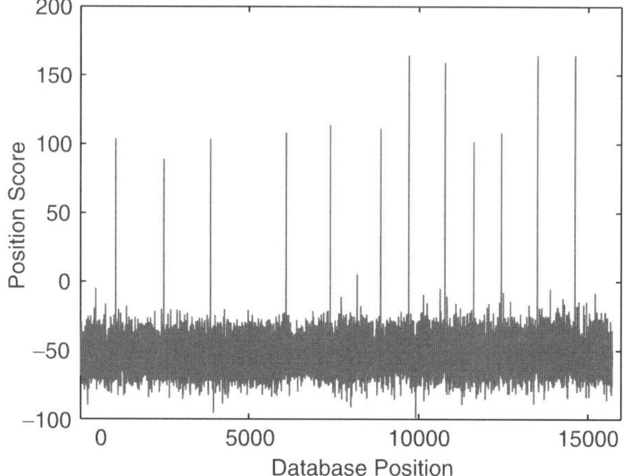

Fig. 7.14. Ungapped scores at each database position

and is the 17th highest peak and the peak at 446 has a score of just under 0 and is the 22nd highest peak. These scores are base-2 log values and therefore have units of bits. Neither a score of 0 nor 5 would be considered statistically significant in any reasonable database search. A search algorithm that can generate a better alignment for these two marginal cases is of primary interest, although improvements in the alignment of the other twelve is also indicative of a generally successful algorithm. The U12 gene at 446 was a seed sequence used to estimate the CM, but the gene at 8196 was not. Four of the true U12 genes do not have any gaps with respect to the consensus sequence and therefore the score from the initial ungapped database sweep is already optimal.

The 100 highest-scoring locations from the upgapped database scores are used as starting points for two search algorithms. The first is a simple hill-climbing algorithm and the other is a genetic algorithm. Each algorithm is permitted 700 candidate solution evaluations per starting point. The GA is run for 20 generations with 35 individuals per generation and the hill-climbing algorithm is run for six rounds of 116 evaluations each. It turns out that additional rounds for the hill-climbing algorithm would not be helpful since the algorithm has converged on a solution in all 14 true U12 gene cases by the sixth round. The choice of 700 evaluations per location is based on observed convergence of the GA.

With a step size of 1 for alignment vector changes in the hill-climbing algorithm, there are 4*149 = 596 possible changes per round. These changes are to increase or decrease a single vector element by 1 without start location compensation and increase or decrease a single element with compensation. Since this would use up almost all of the allocated evaluations in a single round, an alternative strategy is employed. Every fifth alignment vector element is allowed to change rather than every element. This can result in slightly suboptimal alignments if a true insertion or deletion is not at the allowed change location, but the suboptimality is small (see

Figure 7.13 for the effect of making a change in a position slightly different than the correct position). Using every fifth position results in 4*29 = 116 evaluations per round.

The GA uses both mutation and single-point crossover to create new individuals. The fittest individual is retained in the new generation (elitism). Four new individuals per generation are produced by single-point crossover without mutation of two individuals randomly selected from the twenty fittest with crossover points uniformly distributed along the alignment vector. All remaining individuals are produced using mutation without crossover. Twenty five individuals per generation are generated with a single mutation of an individual chosen randomly from the five fittest, where the mutation takes the form of increasing or decreasing an alignment vector position by 1. Half of these mutations are randomly selected to be compensating and the other half non-compensating. Finally, five individuals are produced each generation with a single mutation uniformly chosen in the range $+7$ to -1. Half of these mutations are compensating and half not.

Table 7.2 shows the scores of the best solutions in the final round or generation for the hill-climbing and GA algorithms. The hill-climbing algorithm is deterministic, so only one run is made because the result is always the same. The GA result is the mean over ten runs of the best solution in the final generation. The table also shows the source of the U12 gene sequence in terms of EMBL accession code and the nucleotide positions of the gene within the EMBL sequence. The seven seed sequences are identified with a cross-reference number to the sequences shown in Figure 7.2 of this chapter.

Overall, the performance of the two algorithms is rather similar. However, the two genes of greatest interest 442 and 8196 show that the hill-climbing algorithm did not improve the scores at all, whereas the GA made significant gains. The scores

Table 7.2. Experimental results on U12 dataset

Dataset Position	Hillclimb Score	GA Score	Accession Code (EMBL)	Nucleotide Positions	Fig 7.2 Seed No.
446	−0.74	43.68	L43844.1	2–149	1
1039	146.37	145.66	AC087420.4	142608–142466	
2475	124.95	123.88	AC112938.11	234142–234291	
3858	146.37	143.34	AL591952.9	131760–131611	
6096	110.92	133.57	AL669944.8	2483–2625	
7406	159.12	158.12	AC133939.4	22042–22191	
8196	5.24	40.85	AC132590.3	81080–80927	
8880	147.13	147.13	AL772347.6	146375–146226	
9705	164.47	164.47	L43843.1	2–150	2
10774	159.12	159.12	L43846.1	332–480	3
11624	160.80	160.30	J04119.1	2–150	7
12428	110.92	125.88	L43845.1	358–512	4
13493	164.47	164.47	Z93241.11	76642–7679	6
14615	164.47	164.47	AL513366.11	57717–57871	5

of the optimal alignments for these two sequences are 78.48 and 81.79 respectively, so the GA was only able to get about half of the score increase possible when measured in bits. Even so, the scores changed from statistically insignificant to very statistically significant. Two other cases, sequences 6096 and 12428 also show more improvement with the GA than with the hill-climbing algorithm. These results seem to indicate that there is some advantage to an algorithm with randomization that can jump out of local optima when doing CM-based RNA gene search.

7.6 Conclusions and Future Direction

We have seen that traditional dynamic programming scoring of database sequences with respect to ncRNA gene family covariance models can be rather inefficient due to consideration of may candidate alignment solutions that are far different that those observed in real genomic data. Dynamic programming scoring also requires the use of an arbitrary cutoff on the maximum allowed length of putative genes in the database and a primary-sequence-only filtering of the database in order to reduce required computational resources to a feasible level. Both the cutoff and filtering can cause loss of sensitivity. Dynamic programming further applies equal computational effort to all portions of the database retained by the initial filtering operation.

An alternative scoring method using genetic algorithms does not impose a length cutoff and uses the secondary structure information in the covariance model parameters right from the start. This method also has the potential to allow regions of the database which are not showing score improvement to be abandoned before excessive computational resources are applied in those regions. An exploratory investigation of the alternative scoring method has been applied to a set of known U12 genes and the results are encouraging. This experiment also gives some evidence that deterministic search algorithms which can not escape local optima may not be successful.

Much remains to be done to turn this alternative scoring idea into a standard functional RNA gene search methodology. Investigations on more ncRNA gene families need to be undertaken to determine how to best choose which database locations should be passed to the search algorithm based on the initial ungapped scan of the database. The details of when to abandon a search at a given database location need to be worked out. The experimental investigation above did not even attempt this as a fixed number (700) of evaluations was undertaken at each position. Other stochastic search methods such as simulated annealing need to be investigated to see if they might outperform the genetic algorithm. A parameter sweep needs to be undertaken for such things as the optimal number of individuals per generation and the ratio of crossed-over to mutated individuals.

Improvement to the search may also take the form of better direction for the mutation operator. There is information in the covariance model parameters as to the relative likelihood of an insertion or deletion at a particular point in the consensus sequence of the model. This information could be used to make mutations at these locations statistically more likely during mutation. Also, it may be possible to guess good mutation points by examining the contribution to the overall score of a candidate solution as a function of location in the consensus sequence normalized to the maximum score possible at the location. A drop off in this score-contribution measure at a particular location may be indicative of an insertion or deletion at that location.

In general the use of covariance models for RNA gene search is not nearly as well developed as the use of profile hidden Markov models for protein domain classification. With the increasing recognition of the importance of untranslated RNA to biological function, there should be significant interest in computational methods to study the function and structure of these molecules.

References

1. International Human Genome Sequencing Consortium (2004) Finishing the euchromatic sequence of the human genome. Nature 431:931–945
2. Gesteland R, Cech T, Atkins J (2006) The RNA world. Cold Spring Harbor Laboratory Press, New York
3. Burge C, Karlin S (1997) Prediction of complete gene structures in human genomic DNA. J Mol Biol 268:78–94
4. Eddy S (1998) Profile hidden Markov models. Bioinformatics 14:755–763
5. Finn R, Mistry J, Schuster-Bñockler B, Griffiths-Jones S, Hollich V, Lassmann T, Moxon S, Marshall M, Khanna A, Durbin R, Eddy S, Sonnhammer E, Bateman A (2006) Pfam: clans, web tools and services. Nucleic Acids Research 64:D247–D251
6. Rivas E, Eddy S (2000) Secondary structure alone is generally not statistically significant for detection of noncoding RNAs. Bioinformatics 16:583–605
7. Durbin R, Eddy S, Krogh A, Mitchison G (1998) Biological sequence analysis. Cambridge University Press, Cambridge UK
8. Altschul S, Gish W, Miller W, Myers E, Lipman D (1990) Basic local alignment search tool. J Mol Biol 215:403–410
9. Pearson W, Lipman D (1988) Improved tools for biological sequence comparison. Proc Natl Acad Sci 85:2444–2448
10. Smith T, Waterman M (1981) Identification of common molecular subsequences. J Mol Biol 147:195–197
11. Chomsky N (1959) On certain formal properties of grammars. Information and Control 2:137–167
12. Eddy S, Durbin R (1994) RNA sequence analysis using covariance models. Nucleic Acids Research 22:2079–2088
13. Zucker M (1989) Computer prediction of RNA structure. Methods in Enzymology 180:262–288
14. Wiese K, Hendricks A, Deschênes A, Youssef B (2005) Significance of randomness in P-RnaPredict - a parallel algorithm for RNA folding. IEEE Congress on Evolutionary Computation
15. Eddy S (2003) HMMER user's guide. http://hmmer.janelia.org
16. Griffiths-Jones S, Moxon S, Marshall M, Khanna A, Eddy S, Bateman A (2005) Rfam: annotating non-coding RNAs in complete genomes. Nucleic Acids Research 33:D121–D124

17. Eddy S (2005) Infernal user's guide. ftp://selab.janelia.org/pub/software/infernal/Userguide.pdf
18. Shukla G, Padgett R (1999) Conservation of functional features of U6atac and U12 snRNAs between vertebrates and higher plants. RNA 5:525–538
19. Tarn W, Steitz J (1997) Pre-mRNA splicing: the discovery of a new spliceosome doubles the challenge. Trends Biochem Sci 22:132–137
20. Otake L, Scamborova P, Hashimoto C, Steitz J (2002) The divergent U12-type spliceosome is required for pre-mRNA splicing and is essential for development in Drosophila. Mol Cell 9:439–446
21. Weinberg Z, Ruzzo W (2004) Faster genome annotation of non-coding RNA families without loss of accuracy. Int Conf Res Computational Molecular Biology (RECOMB) 243–251
22. Weinberg Z, Ruzzo W (2006) Sequence-based heuristics for faster annotation of non-coding RNA families. Bioinformatics 22:35–39
23. Smith S (2006) Covariance searches for ncRNA gene finding. IEEE Sym Computational Intelligence in Bioinformatics and Computational Biology (CIBCB) 320–326
24. Yadgari J, Amir A, Unger R (2001) Genetic threading. Constraints 6:271–292

8

Fuzzy Classification for Gene Expression Data Analysis

Gerald Schaefer[1], Tomoharu Nakashima[2] and Yasuyuki Yokota[2]

[1] School of Engineering and Applied Science, Aston University, U.K.
[2] Department of Computer Science and Intelligent Systems, Osaka Prefecture University, Japan

Summary. Microarray expression studies measure, through a hybridisation process, the levels of genes expressed in biological samples. Knowledge gained from these studies is deemed increasingly important due to its potential of contributing to the understanding of fundamental questions in biology and clinical medicine. One important aspect of microarray expression analysis is the classification of the recorded samples which poses many challenges due to the vast number of recorded expression levels compared to the relatively small numbers of analysed samples. In this chapter we show how fuzzy rule-based classification can be applied successfully to analyse gene expression data. The generated classifier consists of an ensemble of fuzzy if-then rules which together provide a reliable and accurate classification of the underlying data. Experimental results on several standard microarray datasets confirm the efficacy of the approach.

8.1 Introduction

Microarray expression studies measure, through a hybridisation process, the levels of genes expressed in biological samples. Knowledge gained from these studies is deemed increasingly important due to its potential of contributing to the understanding of fundamental questions in biology and clinical medicine. Microarray experiments can either monitor each gene several times under varying conditions or analyse the genes in a single environment but in different types of tissue. In this chapter we focus on the latter where one important aspect is the classification of the recorded samples. This can be used to either categorise different types of cancerous tissues as in [8] where different types of leukemia are identified, or to distinguish cancerous tissue from normal tissue as done in [2] where tumor and normal colon tissues are analysed.

One of the main challenges in classifying gene expression data is that the number of genes is typically much higher than the number of analysed samples. Also is it not clear which genes are important and which can be omitted without reducing the classification performance. Many pattern classification techniques have been employed to analyse microarray data. For example, Golub *et al.* [8] used a weighted voting scheme, Fort and Lambert-Lacroix [6] employed partial least squares and logistic

G. Schaefer et al.: *Fuzzy Classification for Gene Expression Data Analysis*, Studies in Computational Intelligence (SCI) **94**, 209–218 (2008)
www.springerlink.com

regression techniques, whereas Furey *et al.* [7] applied support vector machines. Dudoit *et al.* [5] investigated nearest neighbour classifiers, discriminant analysis, classification trees and boosting, while Statnikov *et al.* [16] explored several support vector machine techniques, nearest neighbour classifiers, neural networks and probabilistic neural networks. In several of these studies it has been found that no one classification algorithm is performing best on all datasets (although for several datasets SVMs seem to perform best) and that hence the exploration of several classifiers is useful. Similarly, no universally ideal gene selection method has yet been found as several studies [14, 16] have shown.

In this chapter we apply fuzzy rule based classification concepts to the classification of microarray expression data and show, based on a series of experiments, that it affords good classification performance for this type of problem. Several authors have used fuzzy logic to analyse gene expression data before. Woolf and Wang [19] used fuzzy rules to explore the relationships between several genes of a profile while Vinterbo *et al.* [18] used fuzzy rule bases to classify gene expression data. However, Vinterbo's method has the disadvantage that it allows only linear discrimination. Furthermore, they describe each gene by only 2 fuzzy partitions ('up' and 'down') while we also explore division into more intervals and show that by doing so increased classification performance is possible.

8.2 Methods

While in the past fuzzy rule-based systems have been mainly applied to control problems [17], more recently they have also been applied to pattern classification problems. Various methods have been proposed for the automatic generation of fuzzy if-then rules from numerical data for pattern classification [9–11] and have been shown to work well on a variety of problem domains.

Pattern classification typically is a supervised process where, based on set of training samples with known classifications, a classifier is derived that performs automatic assignment to classes based on unseen data. Let us assume that our pattern classification problem is an n-dimensional problem with C classes (in microarray analysis C is often 2) and m given training patterns $\mathbf{x}_p = (x_{p1}, x_{p2}, \ldots, x_{pn})$, $p = 1, 2, \ldots, m$. Without loss of generality, we assume each attribute of the given training patterns to be normalised into the unit interval $[0, 1]$; that is, the pattern space is an n-dimensional unit hypercube $[0, 1]^n$. In this study we use fuzzy if-then rules of the following type as a base of our fuzzy rule-based classification systems:

$$\begin{array}{l}\text{Rule } R_j\text{: If } x_1 \text{ is } A_{j1} \text{ and } \ldots \text{ and } x_n \text{ is } A_{jn} \\ \text{then Class } C_j \text{ with } CF_j, \quad j = 1, 2, \ldots, N,\end{array} \tag{8.1}$$

where R_j is the label of the j-th fuzzy if-then rule, $A_{j1}, \ldots, A_{jn}$ are antecedent fuzzy sets on the unit interval $[0, 1]$, C_j is the consequent class (i.e. one of the C given classes), and CF_j is the grade of certainty of the fuzzy if-then rule R_j. As antecedent fuzzy sets we use triangular fuzzy sets as in Figure 8.1 where we show a partition of the unit interval into a number of fuzzy sets.

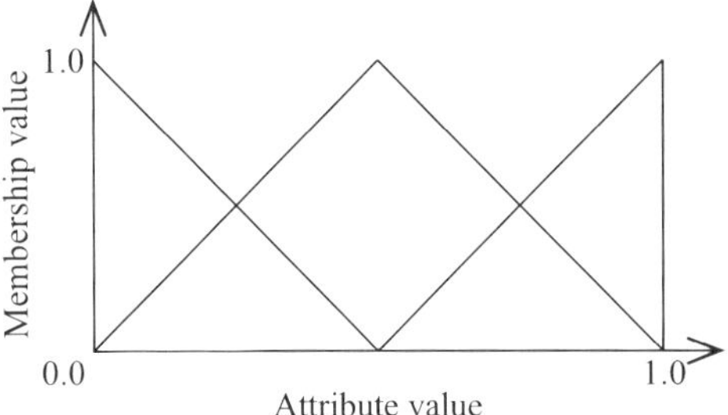

Fig. 8.1. Example triangular membership function ($L = 3$)

Our fuzzy rule-based classification system consists of N fuzzy if-then rules each of which has a form as in Equation (8.1). There are two steps in the generation of fuzzy if-then rules: specification of antecedent part and determination of consequent class C_j and the grade of certainty CF_j. The antecedent part of fuzzy if-then rules is specified manually. Then the consequent part (i.e. consequent class and the grade of certainty) is determined from the given training patterns [13]. In [12] it is shown that the use of the grade of certainty in fuzzy if-then rules allows us to generate comprehensible fuzzy rule-based classification systems with high classification performance.

8.2.1 Fuzzy Rule Generation

Let us assume that m training patterns $\mathbf{x}_p = (x_{p1}, \ldots, x_{pn})$, $p = 1, \ldots, m$, are given for an n-dimensional C-class pattern classification problem. The consequent class C_j and the grade of certainty CF_j of the if-then rule are determined in the following two steps:

1. Calculate $\beta_{\text{Class } h}(j)$ for Class h as

$$\beta_{\text{Class } h}(j) = \sum_{\mathbf{x}_p \in \text{Class } h} \mu_j(\mathbf{x}_p), \tag{8.2}$$

   where

$$\mu_j(\mathbf{x}_p) = \mu_{j1}(x_{p1}) \cdot \ldots \cdot \mu_{jn}(x_{pn}), \tag{8.3}$$

   and $\mu_{jn}(\cdot)$ is the membership function of the fuzzy set A_{jn}. In this chapter we use triangular fuzzy sets as in Figure 8.1.
2. Find Class $\hat{h}$ that has the maximum value of $\beta_{\text{Class } h}(j)$:

$$\beta_{\text{Class } \hat{h}}(j) = \max_{1 \le k \le C} \{\beta_{\text{Class } k}(j)\}. \tag{8.4}$$

If two or more classes take the maximum value, the consequent class C_j of the rule R_j can not be determined uniquely. In this case, specify C_j as $C_j = \phi$. If a single class $\hat{h}$ takes the maximum value, let C_j be Class $\hat{h}$. The grade of certainty CF_j is determined as

$$CF_j = \frac{\beta_{\text{Class } \hat{h}}(j) - \bar{\beta}}{\sum_h \beta_{\text{Class } h}(j)} \tag{8.5}$$

with

$$\bar{\beta} = \frac{\sum_{h \neq \hat{h}} \beta_{\text{Class } h}(j)}{C-1}. \tag{8.6}$$

8.2.2 Fuzzy Reasoning

Using the rule generation procedure outlined above we can generate N fuzzy if-then rules as in Equation (8.1). After both the consequent class C_j and the grade of certainty CF_j are determined for all N rules, a new pattern $\mathbf{x} = (x_1, \ldots, x_n)$ can be classified by the following procedure:

1. Calculate $\alpha_{\text{Class } h}(\mathbf{x})$ for Class h, $j = 1, \ldots, C$, as

$$\alpha_{\text{Class } h}(\mathbf{x}) = \max\{\mu_j(\mathbf{x}) \cdot CF_j | C_j = h\}, \tag{8.7}$$

2. Find Class h' that has the maximum value of $\alpha_{\text{Class } h}(\mathbf{x})$:

$$\alpha_{\text{Class } h'}(\mathbf{x}) = \max_{1 \leq k \leq C} \{\alpha_{\text{Class } k}(\mathbf{x})\}. \tag{8.8}$$

If two or more classes take the maximum value, then the classification of $\mathbf{x}$ is rejected (i.e. $\mathbf{x}$ is left as an unclassifiable pattern), otherwise we assign $\mathbf{x}$ to Class h'.

8.2.3 Rule splitting

It is generally known that any type of rule-based system suffers from the curse of dimensionality. That is, the number of generated rules increases exponentially with the number of attributes involved. Our fuzzy rule-based classifier is no exception, in particular considering that for successful classification of microarray data typically at least a few dozens genes are selected. For example, based on the selection of 50 genes, the classifier would generate $2^{50} = 1.1259 * 10^{15}$ rules even if we only partition each axis into two which is clearly prohibitive both in terms of storage requirements and computational complexity. We therefore apply a rule splitting step and limit the number of attributes in a fuzzy if-then rule to 2. As the number of combinations of attribute pairs is $\binom{50}{2} = 1225$ for 50 genes and as for two fuzzy sets for each attribute $2^2 = 4$ rules are necessary in total we need only $4 * 1225 = 4900$ rules, a significantly lower number than 2^{50}. Of course, techniques can be employed to further decrease this number; although we refrained from it in our experiments a rule pruning step similar to the one outlined in [18] can be applied to arrive at a smaller and more compact classifier rule base.

8.3 Results and Discussion

Before we report on the experimental results we obtained from our classification method we wish to point out a few important differences of our work compared to the fuzzy classifier employed by Vinterbo *et al.* in [18]. The algorithm in [18] represents a fairly simple fuzzy classification approach and provides only linear separation of classes. That is, separate classes can be divided by a hyperplane in feature space. In contrast, with our classifier it is also possible to perform non-linear separation. While at the moment this might be of little effect (due to the limited size of data samples) as has been shown in [3] with increasing sizes of datasets this could prove useful in the near future. Furthermore, our classifier employs the concept of grade of certainty which not only provides improved classification performance but can also provide an additional feedback and/or a means for pattern rejection (due to too low classification confidence). Finally, the classifier in [18] only employed 2 fuzzy partitions per gene to model up and down regulation. While this might seem intuitive it does not necessarily afford best classification performance. In our work we experimented with up to five partitions per attribute.

To demonstrate the usefulness and efficacy of our proposed approach we evaluated our proposed method on several gene expression data sets that are commonly used in the literature. In the following we characterise each dataset briefly:-

- Colon dataset [2]: This dataset is derived from colon biopsy samples. Expression levels for 40 tumor and 22 normal colon tissues were measured for 6500 genes using Affymetrix oligonucleotide arrays. The 2000 genes with the highest minimal intensity across the tissues were selected. We pre-process the data following [5], i.e. perform a thresholding [floor of 100 and ceil of 16000] followed by filtering [exclusion of genes with max/min < 5 and (max-min) < 500] and $\log_{10}$ transformation.
- Leukemia dataset [8]: Bone marrow or peripheral blood samples were taken from 47 patients with acute lymphoblastic leukemia (ALL) and 25 patients with acute myeloid leukemia (AML). The ALL cases can be further divided into 38 B-cell ALL and 9 T-cell ALL samples and it is this 3-class division that we are basing our experiments on rather than the simpler 2-class version which is more commonly referred to in the literature. Each sample is characterised by 7129 genes whose expression levels where measured using Affymetrix oligonucleotide arrays. The same preprocessing steps as for the Colon dataset are applied.
- Lymphoma dataset [1]: This dataset contains gene expression data of diffuse large B-cell lymphoma (DLBCL) which is the most common subtype of non-Hodgink's lymphome. In total there are 47 samples of which 24 are of germinal centre B-like and the remaining 23 of activated B-like subtype. Each sample is described by 4026 genes, however there are many missing values. For simplicity we removed genes with missing values from all samples.
- Ovarian dataset [15]: This data stems from experiments designed to identify proteomic patterns in serum that distingiush ovarian cancer from non-cancer. The proteomic patterns were obtained through mass spectroscopy and there are 91

non-cancer and 162 ovarian cancer samples. While this is not a gene expression dataset it shares many commonalities with such which is the reason why we have included it in our study. The relative amplitude of the intensity at each of the 15154 molecular mass/charge (M/Z) identities was normalised against the most and least intense values according to: $NV = (V - Min)/(Max - Min)$ where NV is the normalised and V is the original value while Min and Max are the minimum and maximum intensities in the data stream [14].

Although all datasets except for the Leukemia set represent 2-class problems due to the large number of genes involved any rule based classification system would consist of a very large number of rules and hence represent a fairly complex process. Also, not all genes are equally important for the classification task at hand. We therefore sort the significance of genes according to the BSS/WSS (the ratio of between group to within group sum of squares) criterion used in [5] and consider only the top 50 respectively 100 genes as input for our classification problem.

In a first step we train our classifiers on all samples available and perform the resulting classification performance. This of course provides only a partial indication as the training data and test data are identical. We therefore perform standard leave-one-out cross-validation where classifier training is performed on all available data except for the sample to be classified and this process is performed for all samples [1]. Fuzzy rule based classifiers with partition sizes L between 2 and 5 partitions for each gene were constructed following the process described in Section 8.2. To evaluate the achieved results we also implemented nearest neighbour and CART classifiers. The nearest neighbour classifier we employ searches through the complete training data to identify the sample which is closest to a given test input and assigns the identified sample's class. CART [4] is a classical rule based classifier which builds a recursive binary decision tree based on misclassification error of subtrees.

The results on the four datasets are given in Tables 8.1 to 8.4 where detailed performance on training and unseen (leave-one-out) test data is shown. Given are the number of correctly classified samples (CR), the number of incorrectly classified or unclassified samples (FR), and the classification accuracy (Acc.), i.e. the percentage of correctly classified samples.

Looking at the results for the Colon dataset which are given in Table 8.1, on training data the fuzzy classifier with $L = 5$ and the nearest neighbour classifier both achieve 100% classification accuracy based on 50 genes while for the case of 100 genes also the fuzzy classifier with $L = 4$ achieves perfect classification. More interesting of course is the performance on test data, i.e. the results of the leave-one-out cross validation we performed. Here for the case of 50 selected features the fuzzy classifier with 3 partitions performs best with a classification accuracy of 85.48% which corresponds to 9 incorrectly classified cases while nearest neighbour classification and CART produce 13 and 14 errors respectively. However when selecting the 100 top genes the nearest neighbour classifier performs slightly better than the fuzzy system. It is interesting to compare the performance of the fuzzy rule-based classifier

[1] It should be noted that the top 50 respectively 100 genes were selected solely based on the training set

Table 8.1. Classification performance on Colon dataset given in terms of number of correctly classified samples (CR), falsely classified or unclassified samples (FR), and classification accuracy (Acc.). Results are given both for training data and for leave-one-out cross validation. Experiments were performed with 50 and 100 selected genes respectively and with a varying number L of partitions per gene. For comparison results obtained using a nearest neighbour classifier and a rule-based CART classifier are also listed

n	classifier	training data CR	training data FR	training data Acc.	test data CR	test data FR	test data Acc.
50	fuzzy $L=2$	55	7	88.71	50	12	80.65
	fuzzy $L=3$	56	6	90.32	**53**	**9**	**85.48**
	fuzzy $L=4$	59	3	95.16	52	10	83.87
	fuzzy $L=5$	**62**	**0**	**100**	48	14	77.42
	nearest neighbour	**62**	**0**	**100**	49	13	79.03
	CART	59	3	95.16	48	14	77.42
100	fuzzy $L=2$	53	9	85.48	44	18	70.97
	fuzzy $L=3$	59	3	95.16	51	11	82.26
	fuzzy $L=4$	**62**	**0**	**100**	50	12	80.65
	fuzzy $L=5$	**62**	**0**	**100**	46	16	74.19
	nearest neighbour	**62**	**0**	**100**	**52**	**10**	**83.87**
	CART	60	2	96.77	45	17	72.58

Table 8.2. Classification performance on Leukemia dataset, laid out in the same fashion as Table 8.1

n	classifier	training data CR	training data FR	training data Acc.	test data CR	test data FR	test data Acc.
50	fuzzy $L=2$	68	4	94.44	66	6	91.67
	fuzzy $L=3$	71	1	98.61	68	4	94.44
	fuzzy $L=4$	**72**	**0**	**100**	67	5	93.06
	fuzzy $L=5$	71	1	98.61	66	6	91.67
	nearest neighbour	**72**	**0**	**100**	**70**	**2**	**97.22**
	CART	72	0	100	47	25	65.28
100	fuzzy $L=2$	67	5	93.06	63	8	87.50
	fuzzy $L=3$	71	1	98.61	**71**	**1**	**98.61**
	fuzzy $L=4$	**72**	**0**	**100**	69	3	95.83
	fuzzy $L=5$	**72**	**0**	**100**	67	5	93.06
	nearest neighbour	**72**	**0**	**100**	70	2	97.22
	CART	**72**	**0**	**100**	45	27	62.50

when using different numbers of partitions for each attribute. It can be seen that on this dataset the best performance is achieved when using 3 partitions (although on training data alone more partitions afford better performance). In particular it can be observed that the case with $L=2$ as used in the work of Vinterbo *et al.* [18] produces the worst results and hence confirms that increasing the number of fuzzy intervals as we suggest leads to improved classification performance. However, it can also be seen that applying too many partitions can decrease classification performance as is apparent in the case of $L=5$ on test data.

Table 8.3. Classification performance on Lymphoma dataset, laid out in the same fashion as Table 8.1

n	classifier	training data			test data		
		CR	FR	Acc.	CR	FR	Acc.
50	fuzzy $L = 2$	**47**	**0**	**100**	45	2	95.74
	fuzzy $L = 3$	**47**	**0**	**100**	46	1	97.87
	fuzzy $L = 4$	**47**	**0**	**100**	**47**	**0**	**100**
	fuzzy $L = 5$	**47**	**0**	**100**	44	3	93.62
	nearest neighbour	**47**	**0**	**100**	45	2	95.74
	CART	45	2	95.74	36	11	76.60
100	fuzzy $L = 2$	**47**	**0**	**100**	44	3	93.62
	fuzzy $L = 3$	**47**	**0**	**100**	44	3	93.62
	fuzzy $L = 4$	**47**	**0**	**100**	44	3	93.62
	fuzzy $L = 5$	**47**	**0**	**100**	39	8	82.98
	nearest neighbour	**47**	**0**	**100**	**47**	**0**	**100**
	CART	43	4	91.49	38	9	80.85

Table 8.4. Classification performance on Ovarian cancer dataset, laid out in the same fashion as Table 8.1

n	classifier	training data			test data		
		CR	FR	Acc.	CR	FR	Acc.
50	fuzzy $L = 2$	224	29	88.54	224	29	88.54
	fuzzy $L = 3$	249	4	98.42	249	4	98.42
	fuzzy $L = 4$	251	2	99.21	249	4	98.42
	fuzzy $L = 5$	248	5	98.02	247	6	97.63
	nearest neighbour	253	0	99.60	252	1	99.60
	CART	243	10	96.05	228	25	90.12
100	fuzzy $L = 2$	223	30	88.14	221	32	87.35
	fuzzy $L = 3$	248	5	98.02	248	5	98.02
	fuzzy $L = 4$	250	3	98.81	249	4	98.42
	fuzzy $L = 5$	250	3	98.81	249	4	98.42
	nearest neighbour	253	0	99.60	252	1	99.60
	CART	251	2	99.21	239	14	94.47

Turning our attention to the results on the Leukemia dataset which are given in Table 8.2 we see a similar picture. Again the worst performing fuzzy classifier is that which uses only two partitions per gene while the best performing one as assessed by leave-one-out cross validation is the case of $L = 3$. CART performs fairly poorly on this dataset with classification accuracies on the test data reaching only about 65% (despite perfect classification on training data) while nearest neighbour classification performs well again confirming previous observations that despite its simplicity nearest neighbour classifiers are well suited for gene expression classification [5]. The best classification results are achieved by the fuzzy classifier with $L = 3$ for the case of 100 selected genes with a classification accuracy of 98.61% and the nearest neighbour classifier with 97.22% for 50 selected genes.

Table 8.3 lists the results obtained from the Lymphoma dataset. Here all classifiers except CART achieve perfect classification on the training data. Perfect classification on test data is provided by the fuzzy classifier with $L = 4$ for 50 selected genes and by nearest neighbour classification based on 100 genes.

Finally, we examine the results obtained from the Ovarian dataset which are given in Table 8.4. Here we can see that once again CART provides the poorest classification while nearest neighbour classification achieves the best performance, misclassifying only 1 sample for both 50 and 100 selected genes. In contrast for the best fuzzy classifier 4 samples are misclassified or rejected which confirms previous observations that different classifiers are better suited for different datasets. Again, the case with $L = 2$ achieves significantly worse results for the fuzzy classifier compared to other partitions.

In summary we see that our fuzzy rule-based classifier provides good classification performance on all four datasets clearly outperforming classical rule-based classification and performing fairly similar to a nearest neighbour classifier. However, it should be noted that in our experiments the nearest neighbour classifier always provided a prediction while for our fuzzy classifier we rejected samples which could not uniquely classified (the false rate FR comprises both incorrectly classified and rejected cases). By randomly classifying rejected patterns we could have achieved improved classification accuracy, however this is not in our interest as a random classification hardly provides any insight in the actual expression level data. We also wish to again point out that restriction to 'up' and 'down' regulated partitions for fuzzy classification as in [18] has a negative impact on the classification performance. Our experiments suggest that selecting 3 or 4 fuzzy partitions for each gene can provide much improved classification accuracy. On the other hand using too many partitions as in the cae of $L = 5$ can also have negative effects on the classification performance.

8.4 Conclusions

In this chapter we proposed the application of fuzzy rule based classification for the analysis of gene expression data. The generated classifier consists of an ensemble of fuzzy if-then rules which together provide a reliable and accurate classification of the underlying data. In addition the structure of our classifier has the potential to contribute to the understanding of the underlying data as it is based on a combination of simple, human-understandable rules. Furthermore, for each classification the grade of certainty is provided, which represents the level of confidence the system has in the prediction of a specific sample, and which could hence be utilised in further stages of analysis.

References

1. A.A. Alizadeh, M.B. Eisen, E.E. Davis, C. Ma, I.S. Lossos, A. Rosenwald, J.C. Boldrick, H. Sabet, T. Tran, X. Yu, J.I. Powell, L. Yang, G.E. Marti, T. Moore, J. Hudson, L. Lu, D.B. Lewis, R. Tibshirani, G. Sherlock, W.C. Chan, T.C. Greiner, D.D. Weisenburger,

J.O. Armitage, R. Warnke, R. Levy, W. Wilson, M.R. Grever, J.C. Byrd, D. Botstein, P.O. Brown, and L.M. Staudt. Different types of diffuse large B-cell lymphoma identified by gene expression profiles. *Nature*, 403:503–511, 2000.

2. U. Alon, N. Barkai, D.A. Notterman, K. Gish, S. Ybarra, D. Mack, and A.J. Levine. Broad patterns of gene expression revealed by clustering analysis of tumor and normal colon tissues probed by oligonucleotide arrays. In *Proc. Natnl. Acad. Sci. USA*, volume 96, pages 6745–6750, 1999.
3. A. Ben-Dor, L. Bruhn, N. Friedman, I. Nachman, M. Schummer, and Z. Yakhini. Tissue classification with gene expression profiles. *Journal of Computational Biology*, 7:559–583, 2000.
4. L. Breiman, J.H. Friedman, R. Olshen, and R. Stone. *Classification and Regression Trees*. Wadsworth, 1984.
5. S. Dudoit, J. Fridlyand, and T.P. Speed. Comparison of discrimination methods for the classification of tumors using gene expression data. *Journal of the American Statistical Association*, 97(457):77–87, 2002.
6. G. Fort and S. Lambert-Lacroix. Classification using partial least squares with penalized logistic regression. *Bioinformatics*, 21(7):1104–1111, 2005.
7. T.S. Furey, N. Cristianini, N. Duffy, D.W. Bednarski, M. Schummer, and D. Haussler. Support vector machine classification and validation of cancer tissue samples using microarray expression data. *Bioinformatics*, 16(10):906–914, 2000.
8. T.R. Golub, D.K. Slonim, P. Tamayo, C. Huard, M. Gaasenbeek, J.P. Mesirov, H. Coller, M.L. Loh, J.R. Downing, M.A. Caligiuri, C.D. Bloomfield, and E.S. Lander. Molecular classification of cancer: class discovery and class prediction by gene expression monitoring. *Science*, 286:531–537, 1999.
9. M. Grabisch and F. Dispot. A comparison of some methods of fuzzy classification on real data. In *2nd Int. Conference on Fuzzy Logic and Neural Networks*, pages 659–662, 1992.
10. H. Ishibuchi and T. Nakashima. Improving the performance of fuzzy classifier systems for pattern classification problems with continuous attributes. *IEEE Trans. on Industrial Electronics*, 46(6):1057–1068, 1999.
11. H. Ishibuchi and T. Nakashima. Performance evaluation of fuzzy classifier systems for multi-dimensional pattern classification problems. *IEEE Trans. Systems, Man and Cybernetics - Part B: Cybernetics*, 29:601–618, 1999.
12. H. Ishibuchi and T. Nakashima. Effect of rule weights in fuzzy rule-based classification systems. *IEEE Trans. Fuzzy Systems*, 9(4):506–515, 2001.
13. H. Ishibuchi, K. Nozaki, and H. Tanaka. Distributed representation of fuzzy rules and its application to pattern classification. *Fuzzy Sets and Systems*, 52(1):21–32, 1992.
14. H. Liu, J. Li, and L. Wong. A comparative study on feature selection and classification methods using gene expression profiles and proteomic patterns. *Gene Informatics*, 13:51–60, 2002.
15. E.F. Petricon et al. Use of proteomic patterns in serum to identify ovarian cancer. *The Lancet*, 359:572–577, 2002.
16. A. Statnikov, C. Aliferis, I. Tsamardinos, D. Hardin, and S. Levy. A comprehensive evaluation of multicategory classification methods for microarray expression cancer diagnosis. *Bioinformatics*, 21(5):631–643, 2005.
17. M. Sugeno. An introductory survey of fuzzy control. *Information Science*, 30(1/2):59–83, 1985.
18. S.A. Vinterbo, E-Y. Kim, and L. Ohno-Machado. Small, fuzzy and interpretable gene expression based classifiers. *Bioinformatics*, 21(9):1964–1970, 2005.
19. P.J. Woolf and Y. Wang. A fuzzy logic approach to analyzing gene expression data. *Physiological Genomics*, 3:9–15, 2000.

9

Towards the Enhancement of Gene Selection Performance

D. Huang[1] and Tommy W.S. Chow[2]

[1] Department of Electric Engineering, City University of Hong Kong
dihuang@ee.cityu.edu.hk
[2] Department of Electric Engineering, City University of Hong Kong
eetchow@cityu.edu.hk

Summary. In a microarray dataset, the expression profiles of a large amount of genes are recorded. Identifying the influential genes from these genes is one of main research topics of bioinformatics and has drawn many attentions. In this chapter, we briefly overview the existing gene selection approaches and summarize the main challenges of gene selection. After that, we detail the strategies to address these challenges. Also, using a typical gene selection model as example, we show the implementation of these strategies and evaluate their contributions.

9.1 Introduction

Microarray techniques, such as cDNA chip and high-density oligonucleotide chip, are powerful biotechnological means because they are able to record the expression levels of thousands of genes simultaneously [1]. Systematic and computational analysis on microarray data enables us to understand phenological and pathologic issues in a genomic level [1, 2]. Recently, microarray analysis has been widely exploited to study gene expression tumor tissue, such as breast cancer and lung cancer, on a genome-wide scale. Early research results have confirmed that different biological subtypes of breast cancer are accompanied by differences in their transcriptional programs [27]. With the microarray technology, the relative expression levels of genes within a specific tissue sample can be measured simultaneously. Take breast cancer, one the most studied diseases, as an example, its breast tissue is found to be heterogeneous, with the cell types of epithelial, mesenchymal, endothelial, and lymphopoietic derivation [28]. In spite of these differences, it is confirmed that we can analytically estimate the influences of these cell types on the tumor's total pattern of gene expression. As a result, microarray technology enables researchers to conduct a tissue analysis computationally. More importantly, this technology is able to provide a very reliable and stable molecular model of tumors. Microarray data, however, always contains a huge gene set of up to ten thousands and a small sample set that may be as little as tens. Moreover, only a very small fraction of genes are informative for a certain task [3, 4]. For example, Singh et al. [5] used

D. Huang and T.W.S. Chow: *Towards the Enhancement of Gene Selection Performance*, Studies in Computational Intelligence (SCI) **94**, 219–236 (2008)
www.springerlink.com

the HG U95A array of Affymetrix as platform to record the expression profiles of about 10000 genes for studying prostate cancer. The biological studies summarized by SuperArray Bioscience Corporation (http://www.superarray.com/gene-array-product/HTML/OHS-403.html) suggested that only hundreds of genes were (putatively) biomarkers of prostate cancer. Different diseases are related to different gene sets. In a microarray dataset, the expression profiles of ten thousands of genes are recorded. Facing the staggering volume of gene sets, computational analysis of them has become essential in order to offer a comprehensive understanding of the mechanisms underlying biological phenomena and complex human diseases. To study human diseases on a genomic-wide scale, identifying the disease-causing genes is the first step. Successful identification of disease-causing genes has many merits. From clinical perspective, the use of microarray will enable physicians to distinguish normal and tumor tissue and subclassify cancer. Thus, physician will be able to make more accurate prediction of clinical outcome and response to systemic cancer therapy.

Using a small gene set, computational data analysis is conducted in a relatively low-dimensional data domain. This is very useful to deliver precise, reliable and interpretable results. Also, with the gene selection results, biology researchers can focus only on the marker genes, and confidently ignore the irrelevant genes. The cost of biological experiment and decision can thus be greatly reduced. In determining the disease-causing genes, there are two ways of handling the gene selection. First, genes are considered as a feature from the machine learning perspective. Second, genes are considered as functional groups. A group of genes that are functionally associated control or regularize a cellular process or molecular activity. Researchers have also been attempting to understand cancer at the level of gene functional group. Despite the similarity in the nature of gene selection as individual or functional groups, the computational mechanism between the two approaches are, however, very different. In this book chapter, we will mainly focus on gene selection. In the next section, gene selection models will be briefly reviewed. Your text goes here.

9.2 Brief Reviews on Gene Selection

To date, various machine learning models and statistical concepts have been directly applied or adapted to gene selection. Referring to [6, 7], a general model of gene selection frameworks is shown in Fig. 9.1. Based on a given and the currently-selected genes, a search engine generates a pool of gene subsets. From the pool, the best gene subset is detected according to an evaluation criterion. This generating-subsets-detecting-best process repeats until certain stopping criterion is met.

Obviously, there are two important components in a gene selection model. They are the search engine and the evaluation criterion. There are several search engines including ranking, optimal search, heuristic search and stochastic search. Generally, gene selection models are categorized as a filter model, a wrapper model and an embedded model according to the type of evaluation criterion. Filter model work independent of classification learning process, using computational or statistic concepts

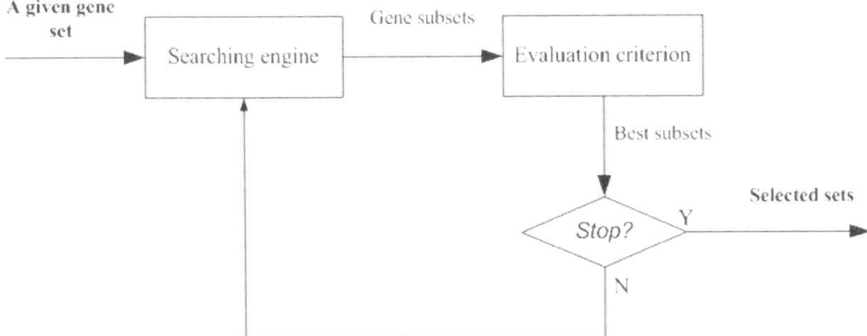

Fig. 9.1. A general gene selection model

to evaluate the classification capability of gene sets. On the other hand, wrapper and embedded models interact closely with classification learning. Wrapper models directly employ the recognition accuracy of a certain recognition procedure to evaluate the quality of feature subsets, while embedded models explore the parameters of certain training classification models to determine the importance of genes.

The gene-ranking is the simplest and earliest attempt for gene selection. In this scheme, the search engine includes only one action by which all genes are ranked in a descending order of their biological importance and then the top-ranked genes are finally marked as selected genes. Various statistical or computational concepts have been used to evaluate the importance of gene. For example, in [3], the distinguishing capabilities of genes are firstly evaluated in terms of the metric of linear separatability. Since this gene-ranking scheme does not take the interactions of genes into account, it is not comprehensive enough to tackle complicated situations and has been replaced by more sophisticated gene selection models. Until now, two types of gene selection model are employed in most studies. They are the filter models employing heuristic/stochastic search engines [8–11], and the embedded models [12–14].

In embedded gene selection models, many popular machine learning schemes have been explored, including Bayesian network, support vector machine, minimax probability machine, fuzzy-logic-based classifier, and penalized COX, etc. In a typical filter gene selection model [8], the search engine is a sequential forward search process (SFS), and a gene evaluation criterion is designed based on the "normalized" mutual information. The SFS, a typical heuristic search engine, begins with an empty gene subset and detect certain important genes at a time. Also, as for gene evaluation, the study holds the assumption that a good subset should have high relevance to classification task and have low redundancy among the involved genes. In another example of filter gene selection model [11], information gain is first used to filter out unimportant genes. Then among the remaining genes, the concept of Markov Blanket is explored to eliminate the less important genes in a sequential backward search process (SBS). The SBS, a typical heuristic search engine, selects all genes at the beginning and eliminate less important genes at a time. Also, evolutionary algorithm is also used to select the important genes and promising results was reported [10].

In this chapter, we place more emphasis on the filter model. Especially, our focuses lie in two topics that are very important but have been overlooked in most contexts of gene selection. The first one is about enhancing the effectiveness of gene selection. This objective can be obtained through modify search engines. The second topic is about the issue of effectively addressing the problem of overfitting. The problem of overfitting can be alleviated from the perspectives of both search engine and evaluation criterion.

To elaborate the technical details of these topics, we use a sequential forward gene selection method as an example, in which Bayesian discrimiant based criterion is used to evaluate the quality of genes. Below are the details of this conventional method.

Assuming that we have a classification dataset $D = \{X, C\} = \{(x_1, c_1),(x_2, c_2),......,(x_N, c_N)\}$. (x_i, c_i) represents a data sample, in which x_i is the input vector, and c_i records the class label of x_i. x_i is a M-dimensional vector, that is, a sample is described with the expression levels of M genes. These genes are represented as $F = \{f_1, f_2, ..., f_M\}$. Moreover, the samples in D are grouped into L classes denoted as $\{w_1, ..., w_L\}$. For a data sample (say, x_i), it has $c_i = w_k$, where $1 \leq k \leq L$. A gene subset evaluation criterion is represented by $\Phi(S)$, where S is a gene subset. Furthermore, without loss of generality, it is supposed that a large value of $\Phi(S)$ means a good S. Thus, the goal of a gene selection process is to maximize $\Phi(S)$ through adjusting S. In order to fulfill this goal, the SFS process can be briefed as followings.

Step 1 Set the selected gene subset S with empty.

Step 2 Repeat the followings until certain stopping conditions are met. Identify the most useful gene (say, g_u) from the unselected genes, and place it into S. g_u satisfies

$$\Phi(S + g_u) = max \Phi(S + g). \tag{9.1}$$

$\phi(S)$ is the evaluation criterion of the above process. Many concepts can be used as $\phi(S)$. For more information, please read ref. 15. In this example, Bayesian discriminant based criterion (BD) [15] is used as $\phi(S)$. Thus, for a given dataset D, BD is defined as

$$BD = \frac{1}{N} \sum_{i=1}^{N} \log \frac{p_s(c_i|x_i)}{p_s(\overline{c_i}|x_i)}, \tag{9.2}$$

where $\overline{c_i}$ means all the classes but class c_i, and $p_S(\cdot)$ represents a probability density function estimated based on the gene set S. In order to estimate the posterior probabilities $p(c|x)$ in (9.2), the margin probability $p(x)$ and the joint probability $p(x,c)$ should be firstly obtained. Parzen window [26] is an excellent method to build $p(x)$ and $p(x,c)$. Given the aforementioned dataset $D = \{X, C\}$, Parzen window estimators are modeled as

$$p(x,c) = \sum_{x_i \in class\ c} P(x_i) \kappa(x - x_i, h_i), \tag{9.3}$$

$$p(x) = \sum_{c} p(x,c) = \sum_{all\ x_i} P(x_i) \kappa(x - x_i, , h_i), \tag{9.4}$$

where $P(x_i)$ is the likelihood of $x = x_i$, k is the kernel function and h_i is the width of κ. With a proper selection of $\kappa(\cdot)$ and h, a Parzen window estimator can converge to the real probability density. Gaussian function is a popular choice as κ, that is,

$$\kappa(x-x_i,h_i) = \mathrm{G}(x-x_i,h_i) = \frac{1}{(2\pi h_i^2)^{\frac{M}{2}}} exp[-\frac{1}{2h_i^2}(x-x_i)(x-x_i)^T], \qquad (9.5)$$

where M is the dimension of x. The width h_i is set with $h_i = 2d(x_i,x_j)$, where $d(x_i,x_j)$ is the Euclidean distance between x_i and x_j, and x_j is the 3rd nearest neighbor of x_i. Following the general rule, it is $P(x_i) = 1/\mathrm{N}$. Also, according to the Bayes formula $p(c|x)$ can be modeled as

$$p(c|x) = \frac{p(x|c)P(c)}{p(x)} = \frac{p(x,c)P(c)}{p(x)} = \frac{\sum_{x_i \in classc} P(x_i)\kappa(x-x_i,h_i)}{\sum_{all\ x_i} P(x_i)\kappa(x-x_i,h_i)}. \qquad (9.6)$$

With the above estimate of $p(c|x)$, the gene evaluation criterion BD (9.2) can be calculated.

This chapter focuses on enhancing the effectiveness of filter gene selection models. The discussions can be conducted from two perspectives. The first one is to modify search engines, while the other is to address the problem of overfitting. In the following sections, these issues will be discussed one by one.

9.3 Modifying Search Engines

In a filter model, the evaluation criterion and search engine play equally important roles. As described above, evaluation criteria have been heavily investigated in many studies. In contrast, the study on search engines has drawn little attentions. The heuristic search engines, especially the SFS and the SBS, and the stochastic search engines (e.g., Genetic algorithm) are pervasively employed for gene selection. Going through most literatures, it is noticed that almost all studies focused on the improvement of searching efficiency of stochastic algorithms. There are, however, very few attempts to modify heuristic search algorithms. In the stepwise strategy [16], that is, the floating (compound) search, selecting k features (genes) is followed by eliminating j "worst" selected ones, where j is less than k. Al-Ani et al. [17] use only the "elite" selected features to identify the important items from the unselected features. It is noted that these algorithms are designed totally in a discrete gene space and the results require the testing of more gene combinations. But the testing of more gene combinations usually leads to an increase of computational effort.

The example used in this chapter is the BD based SFS gene selection model. This model includes a greedy process to maximize BD. After assuming $f((x,c),s) = p_S(c|x)/(1-p_S(c|x))$, we have

$$BD = \sum_{all(x_i,c_i)} \log(f((x_i,c_i),S)). \qquad (9.7)$$

According to optimization theory, the steepest direction of adjusting S to maximize (9.7) is

$$\frac{\partial BD(S)}{\partial f((x,c),S)} = \sum_{all(x_i,c_i)} \frac{\partial BD(S)}{\partial f((x,c),S)} \frac{\partial f((x,c),S)}{\partial S}. \tag{9.8}$$

The above equation indicates that during the course of optimizing the *BD*(*S*), the updating of *S* depends on $\partial BD(S)/\partial f((x,c),S)$ and $\partial f((x,c),S)/\partial S$. The former one happens in a continuous domain, while the latter one is related to S and has to be tackled in a discrete gene domain. In this sense, (9.8) cannot be solved directly. In order to maximize the *BD*(*S*), many searching schemes have been designed. For example, the conventional SFS tests all combinations of *S* and one of unselected genes. It then identifies the one having the maximal *BD*. Clearly, the conventional SFS is conducted only in a gene domain. In other word, only the second part of (9.8) is considered by the conventional SFS. This means that the search direction of the conventional SFS cannot comply with the direction defined by (9.8), which is the steepest optimization direction. This shortcoming degrades the effectiveness of optimization.

To conduct gene selection along the optimization direction, equation (9.8) should be considered. To fix the second term of (9.8), which happens in a gene domain, the conventional SFS can be used. As to the first term of (9.8), it can be directly calculated in a way of

$$\frac{\partial BD(S)}{\partial f((x,c),S)} = \frac{1}{\partial f((x,c),S)} = \frac{1-p_S(c|x)}{p_S(c|x)}. \tag{9.9}$$

Given *S*, $\partial BD(S)/\partial f((x,c),S)$ is only related to x. With this observation, $\partial BD(S)/\partial f((x,c),S)$ can be regarded as a penalty weight to sample. Thus, the modified search engine is equivalent to a conventional SFS conducted on the weighted samples. The weights of samples are determined by (9.9).

Assuming that the weight assigned to the data sample (x_i,c_i) is w_i. With this weighted dataset, we adjust the criterion *BD* (9.2) as well as the probability estimations (9.3) and (9.4) accordingly. In details, we have

$$\mathrm{BD} = \frac{1}{N}\sum_{i=1}^{N} w_i \log \frac{p_s(c_i|x_i)}{p_s(\overline{c_i}|x_i)}, \tag{9.10}$$

where

$$p(x,c) = \sum_{x_k \in class\ c} \frac{w_k}{N} \kappa(x-x_k,h_k), \tag{9.11}$$

$$p(x) = \sum_{all c\ x_k} \frac{w_k}{N} \kappa(x-x_k,h_k). \tag{9.12}$$

Imagine that there is no a priori knowledge about a given dataset, it is natural that different samples exhibit different contributions to learning processes. Actually, most machine learning algorithms incorporate this idea. For example, when minimizing the mean square error $E = \sum_{(x_i,y_i)}(f(x_i,\Lambda)-y_i))^2$ through adjusting the parameter set Λ of f, a steepest decent algorithm [20] can be used to determine the direction of the updating Λ with

$$\frac{\partial E}{\partial \Lambda} = \sum_{all(x,y)} -\frac{\partial E}{\partial f}\frac{\partial f(x,\Lambda)}{\partial \Lambda} = \sum_{all(x,y)} -(f(x,\Lambda)-y)\frac{\partial f(x,\Lambda)}{\partial \Lambda} \tag{9.13}$$

It is noted that the contribution of (x,y) is penalized by $|f(x,\Lambda)-y|$. Another example is AdaBoosting [21], which is a typical boosting learning algorithm. During the course of learning, AdaBoosting repeats weighting the sample (x,y) with $we^{-yf(x)}$, where w is the current weight to (x,y). Also, the effect of overfitting should not be overlooked. To reduce the risk that the optimization may suffer from overfitting, one can add more weight on the negative samples, incorrectly-recognized ones, that exhibit more influence to the subsequent learning than on the positive ones. In such a way, the convergence rate can be speeded up, and the problem of overfitting can be alleviated [22]. AdaBoosting clearly can meet this expectation and exhibit good performance. Equation (9.13), however, indicates that the steepest decent algorithm fell short on tackling overfitting in a way that the correctly-recognized patterns still carry large weights. This fact has motivated modifications on the gradient-based algorithms [22]. Consider the above pattern-weighting strategy. As indicated by the equation (9.9), negative patterns (i.e., the ones with small values of BD) are assigned with larger weights than the positive ones. Obviously, this will be helpful in alleviating the problem of overfitting.

9.4 Point Injection for Gene Selection

Overfitting is a major issue affecting the performance of gene selection. In brief, overfitting means that learning results (i.e., the selected gene subsets in this case) may perform perfectly on the training data, but are unable to handle the testing data satisfactorily. Due to the nature of small pattern sets in most microarray-based data, the problem of overfitting exacerbates in performing gene selection. In order to alleviate this problem, the use of models with high capability of generalization, such as support vector machine and penalized Cox regression model, have been suggested [13, 18]. But purely relying on learning algorithms is not enough to tackle overfitting in most cases. Developing a specific strategy to solve the problem of overfitting appears to be an effective alternative. Zhou et al. [19] employed a bootstrap framework to obtain reliable mutual information estimates. However, its large computational requirement, which is arguably the main shortcoming of the approach, substantially restricts its application.

Given a dataset $D=\{X,C\}$ drawn from a distribution $\wp$ in the data domain $X\times C$. The real goal of the BD based gene selection process is to maximize $BD_{\wp}(S)$. Since $\wp$ is unknown in most cases, researchers substitute $BD_{\wp}(S)$ with the empirical estimate $BD_{(X,C)}(S)$ (simplified as $BD(S)$, as the above equations do). This substitution may have a bias because $BD(S)$ cannot always reflect $BD_{\wp}(S)$ correctly. The bias will also lead to overfitting, thereby the gene subset, which is selected to optimize

$BD(S)$, cannot optimize the real objective $BD_{\wp}(S)$. From the perspective of avoiding overfitting, it is preferable that $BD(S)$ varies smoothly around the whole data domain. Samples near each other should correspond to similar performance. This is the rationale behind the point injection technique. This technique is called the noise injection in many literatures. But the newly generated points are not expected to be real noise. In this chapter, in order to avoid the confusion, the term point injection, instead of noise injection, is used. There are two ways that the injected points participate in gene selection. With reference to the noise injection approaches for classification training [23], the injected points can be explored totally like the original samples. Using the original samples as well as the injected points, the probability estimation models required by the *BD* are built, and the quality of gene subsets is then evaluated. Alternatively, similar to the smooth error evaluation schemes [24], the injected points are employed only in the evaluation stage. On the other words, the probability estimators are built only upon the given samples. Subsequently, gene subsets are evaluated according to the *BD* values of the given samples and the injected points. Below, the latter mechanism is discussed.

Around a pattern x_i, a point injection technique adds v points which are generated from a distribution $b(x-x_i)$. v and $b(x-x_i)$ determine the performance of a point injection scheme [25] . v should be determined in a way to strike the balance between performance stability and computational efficiency. Also, it has been argued that, for the reasonable choices of v, such as $v = 8$, 10 or 20, the effect of point injection is slightly different [23, 25]. In the below experiments, the setting of $v=10$ is used. As for $b(x-x_i)$, its "width", which determines the variance of the injected points, is crucial. As point injection is used to test the properties of the region around x_i, a large width of $b(x-x_i)$ is not expected. On the other hand, a small width of $b(x-x_i)$ must bring an insignificant contribution. To determine an appropriate width, simulation based strategies can be used [23, 25]. Also, Sima et al. [24] developed an analytic approach to determine the width of $b(x-x_i)$. Aiming to reduce the bias intrinsic to the re-substitution error estimation as much as possible, this approach rely on the joint distribution (X,C) to determine the width of $b(x-x_i)$. This idea appears to be very effective and will be used in the further discussions.

Suppose that d_i is the distance of x_i to the nearest different-class samples, that is,

$$d_i = min\|x_i - x_j\|, x_j \notin class\ c_i. \tag{9.14}$$

Also, assuming that $d_i/2$ is the boundary of different classes, which means that, a point x' should have the same class label as x_i, if $\|x_i - x'\| \leq d_i/2$. With this assumption, several points around x_i can be generated from the Gaussian distribution $N(x_i, \sigma_i)$. σ_i is set as $d_i/6$ to guarantee that x' having $\|x_i - x'\| = d_i/2$ occurs with the close-zero probability.

Apart from $N(x_i, d_i/6)$, other distributions are investigated. For example, following Sima et al. [24], σ_i is set as d_i/α_i. Here d_i is defined in (9.14) and α_i is determined by $\alpha_i = F_D^{-1}(1/2)$, where F_D^{-1} is the cumulative distribution function of the distance of x_i to the points generated from $N(x_i, \sigma_i)$. Also the rectangular uniform distribution $b(x-x_i) = R(x_i - d_i/2, x_i + d_i/2)$ is tested. Compared with $N(x_i, d_i/6)$, these distributions produce either similar or inferior results.

9.5 Implementation

To give a concrete idea of the above modified search engine and point injection strategy, this section will show the implementation of these strategies. The conventional SFS, described in section 9.2, is used as demonstration example.

With the above point injection and weighting-sample strategies, the conventional SFS, described in section 9.2, will be modified as followings.

Step 1 Set the selected gene subset S empty. Assign a weight of 1 to each sample, that is, $w_i = 1, 1 \leq i \leq N$. Set the injected point set $\{X', C'\}$ empty.

Step 2 Repeat the followings until one have selected certain genes.

a) From the unselected genes, identify the gene g_m satisfying

$$BD_X(S+g_m) + BD_{X'}(S+g_m) = \max(BD_X(S+g) + BD_{X'}(g+S)). \quad (9.15)$$

$BD_X(S+g)$ is $BD(S+g)$ (9.10) of the given data $\{X, C\}$. $BD_{X'}(S+g)$ will be discussed in detail later.

b) (sample-weighting) Update the sample weights. Set w_i based on equation (9.9). Then normalize w_i as $w_i = w_i / \sum_k w_k$.

c) Point injection) Set $\{X', C'\}$ with empty. Around each pattern, say x_i, produce 10 points based on the distribution $N(x_i, d_i/6)$. Then place these points into $\{X', C'\}$. These injected points inherit the class label and the weight of x_i.

Since given samples cannot cover the whole data domain, the probability estimators built upon them are not able to describe every part of data space sufficiently. There are areas where points have small distribution probabilities, that is, $p(x, c_i)$ and $p(x)$ are all small. According to (9.9), given a point x, it is $| \frac{\partial BD(S)}{p_S(x,c)} | \propto | \frac{1}{p_S(x,c)} |$. It shows that, when $p(x, c_i)$ for all c_i are small, a very little shift of x may cause an extremely large change of $BD(S)$. It is better to avoid this uncontrollable condition, although it can be argued that the uncertain points equally affect the performance of different gene subsets. With this consideration, a simple strategy, called as maximal-probability-weighting-injected-point strategy, is designed. With this strategy, $BD_{X'}(g+S)$ is modified as

$$WBD_{X'}(S+g) = \frac{1}{|X'|} \sum_{all x_i \in X'} w'_i log \frac{p_{(S+g)}(c'_i|x'_i)}{p_{(S+g)}(\overline{c'_i}|x'_i)} \underbrace{max(p_{(S+g)}(x'_i|c'_i))}_{A}, \quad (9.16)$$

where $|X'|$ means the cardinality of $|X'|$. c'_i and w'_i are the weight and class of x'_i. Compared with the original $BD_{X'}(g+S)'$ (9.10), $WBD_{X'}(g+S)$ has a new term, i.e., term A. For uncertain points, all the conditional probabilities must be small. Thus, the term A of $WBD_{X'}(S+g)$ limits the impact of uncertain point, which is as expected.

9.6 Experimental Results

To evaluate the described strategies, the modified SFS depicted in the above section is compared with typical gene selection models. Below, SFS denotes the conventional BD based SFS, while our modified SFS is denoted as MSFS. The modified SFS using $WBD_{X'}(S+g)$ (9.16) rather than $BD_{X'}(S+g)$ is represented by the WMSFS. Besides these BD based SFSs, support vector machine based recursive feature (gene) elimination model (SVM RFE) [13] is also implemented. SVM RFE is a typical embedded gene selection method. Due the superior performance of SVM, SVM RFE is considered as a good gene selection model and has been used as evaluation baseline in gene selection study. To provide a detail comparative study on different methodologies, all the compared methods are applied to one synthetic dataset and three patient datasets.

9.6.1 Synthetic Data

The synthetic data is a 3-class and 8-gene dataset. The first four features are generated according to

- class 1 N((1,1, -1, -1), σ);
- class 2 N((-1, -1, 1, 1), σ);
- class 3 N((1, -1, 1, -1), σ).

The other four genes are randomly determined from normal distribution with zero means and unit variances. Thus, among 8 genes, the first 4 genes are equally relevant to this classification task, and the others are irrelevant. All the examined approaches are required to identify 4 genes. In this study, a selection result is considered correct only when it includes all relevant genes.

Also, this dataset has a small-sample set, including totally 9 samples among which 3 samples are from each of three classes. Different values of s are tested. Clearly, the smaller the s is, the simpler the classification problem is and the less likelihood overfitting occurs as a result. Thus, theoretically, the advantage of MSFS and WMSFS should become significant with the increase of s. For each s, 10,000 datasets are generated. SFS, MSFS and WMSFS run on these datasets, and the correct results are counted.

The obtained correctness rates are illustrated in Fig. 9.2. For a tested method under an experimental condition, the correctness rate means the percentage of the correct results obtained in all running trials. These results are consistent with the above theoretical analysis. When σ is small, there is unnoticeable performance difference between the compared methods. Actually, in the case of a small σ, on the same dataset, different methods basically produce the same results. When σ is large, however, MSFS and WMSFS illustrate the improved performance. Also, WMSFS is better than MSFS.

9.6.2 Patience Data

Different gene selection methods are compared on several cancer diagnosis datasets. Since, in these real datasets, no a priori knowledge is available, experimental classification results will be used to assess the quality of gene selection results. Using a selected gene subset, certain classifiers are constructed on training data that are also used for gene selection. Then, the built classifiers are evaluated on the

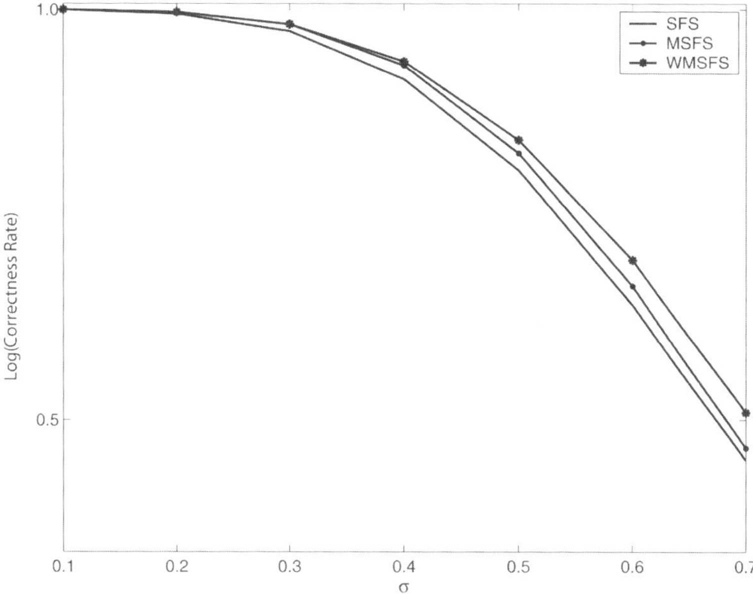

Fig. 9.2. Comparison between SFS, MSFS and WMSFS on synthetic data

testing dataset. Good classification results must indicate a respectable gene subset. Four typical classifiers are used. They are a multiply percepton model (MLP), two support vector machine models (SVM), and a 3-nearest neighbor rule classifier (3-NN). The MLP is available at http://www.ncrg.aston.ac.uk/netlab/, and six hidden neurons are used in all MLP models used. Also, in order to ease the problem of overfitting, each MLP model is trained through 100 training cycles. The other learning parameters are set with default values. The SVM models used are available at http://www.isis.ecs.soton.ac.uk /resources/svminfo. Two SVM models are the SVM with "Linear" kernel (SVM-L) and the SVM with "RBF" kernel (SVM-R), respectively.

The gene selection methods are compared on the following cancer diagnosis datasets.

Colon tumor classification This data contains 62 samples collected from colon-cancer patients. Among these samples, 40 samples are tumor, and 22 are labeled "normal". There are 2,000 genes selected based on the confidence in the measured expression levels. The 62 samples are split into two disjoint groups - one group with 40 samples for training and the other one with 22 samples for test. The investigations are repeated on 10 different sets of training and testing data to deliver reliable evaluations. The statistical results of 10 trials are presented. Also, in each training dataset, the original size ratio between two classes, that is, 40 tumor samples vs. 22 normal samples, is roughly remained.

Prostate cancer classification The objective of this task is to distinguish prostate cancer cases from non-cancer cases. The original raw data are published at http:// www.genome.wi.mit.edu/mpr/prostate. This dataset consists of 102 samples from the same experimental conditions. Among the 102 samples, there are 50 normal and 52 tumor samples. Each sample is described using 12600 genes. The 102 samples are divided into two disjoint groups - one group with 60 samples for training and the other with 42 samples for testing. Similar to the last example, the studies on this data are repeated on 10 different sets of training and testing data. The statistical results are summarized and presented in this chapter.

Leukemia subtype classification This data, which are available at http://www.broad.mit.edu/cgibin/cancer/datasets.cgi, are used for performing leukemia subtype classification. The given samples are labeled with ALL, MLL or AML. Training data contains 57 samples - 20 labeled with ALL, 17 with MLL and 20 with AML. In the test data, there are 15 samples - 4 ALL samples, 3 MLL ones and 8 AML ones. There are no SVM-related results in this example because the SVM classification models employed and the gene selection method SVM RFE are designed to deal with 2-class data only.

Analysis on the Stability of MSFS and WMSFS

There is an inherent randomness in the course of injecting points. It is thus necessary to investigate the stability of the MSFS and WMSFS. For this purpose, the MSFS and WMSFS run 10 times on the same training dataset. Then the obtained results are compared. Given a group of gene subsets of the same size, the appearance probability of each subset is calculated. The largest appearance probability, named LAP, can measure the likelihood of all the tested subsets being identical. LAP = 1 indicates that all given gene subsets totally match. LAP arrives at its minimum when the tested gene subsets are totally different from each other. In Fig. 9.3, the results obtained on three datasets are illustrated. It shows that, in most cases, LAP = 1. It means that the MSFS/WMSFS can deliver the same results in different runs using the same training dataset.

Comparisons of SFS, MSFS and WMSFS

To demonstrate the merits of the MSFS and WMSFS, these methods are compared with the SFS in terms of classification accuracy. As the MSFS and WMSFS have a stable performance, we run these schemes once on a given dataset. The comparative results are presented in Fig 9.4 (for colon cancer classification), Fig 9.5 (for prostate cancer classification) and Fig 9.6 (for leukimia subtype classification).

Due to the capability of SVM handling small sample sets, the results of SVM RFE are better than those of SFS a little bit. Also, it can be clearly noted that MSFS is much better than SFS and SVM RFE in most cases. This improvement can be contributed to the implementation of the point-injection and sample-weighting strategies. The performance can be further enhanced when the WMSFS is use. This enhancement is caused greatly by using the maximal-probability-weighting-injected-point strategy. For example, on the colon cancer dataset, the results achieved by

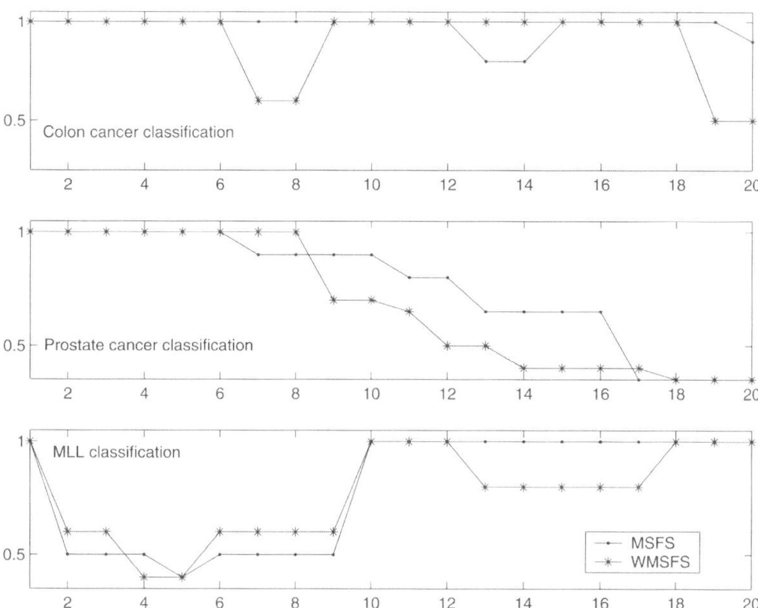

Fig. 9.3. Statistical results of MSFS and WMSFS. These results can show the stability of MSFS and WMSFS. The x-axes are the number of the selected genes. The y-axes are LAP

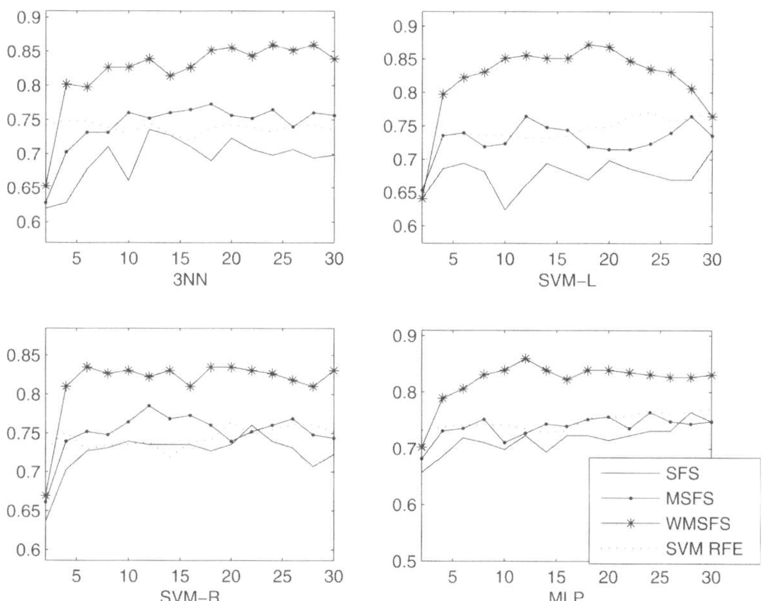

Fig. 9.4. Comparison on the colon cancer classification data. In these figures, the y-axes are the classification accuracy, and the x-axes are the number of the selected genes

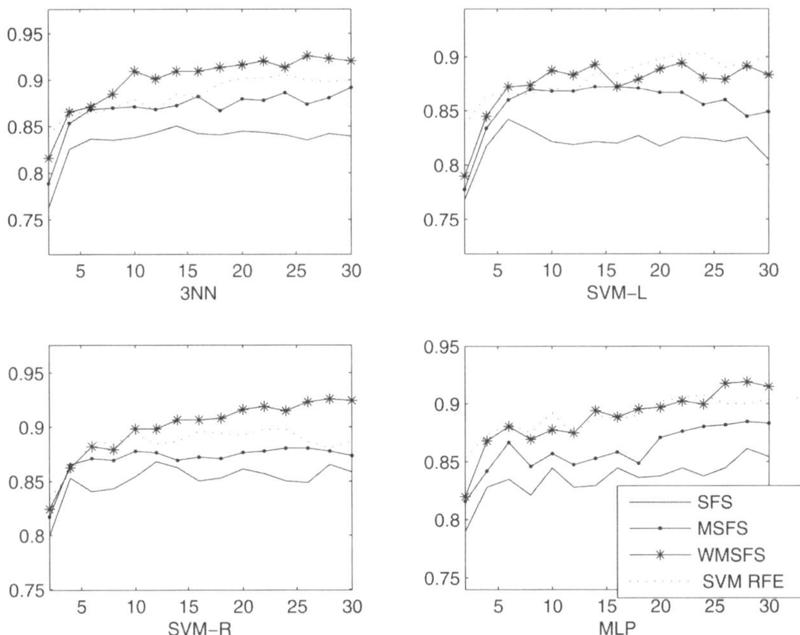

Fig. 9.5. Comparison on the prostate cancer classification data

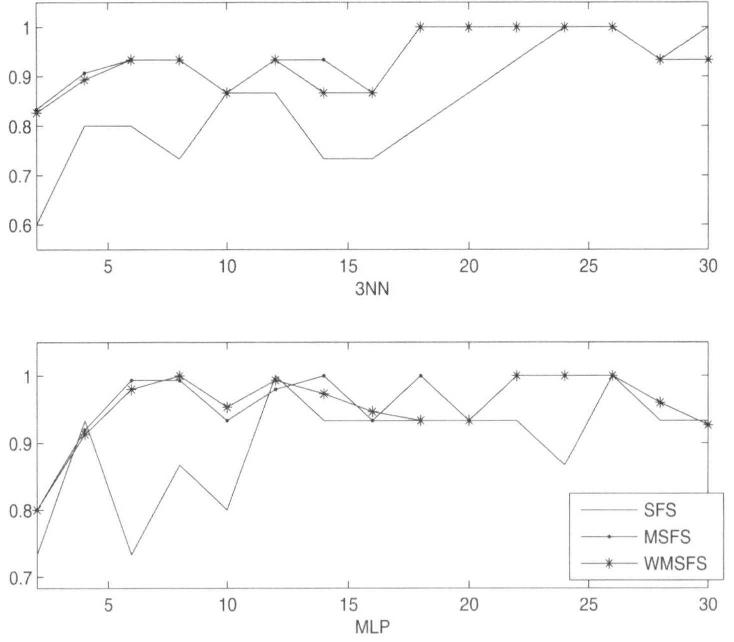

Fig. 9.6. Comparison on leukimia subtype classification data

WMSFS are much better, up to 10%, than those of SFS. Even compared with SVM RFE, a typical and effective embedded gene selection method, WMSFS can always deliver the improved result.

Detailed Results on Prostate Cancer

Given a set of selected genes, we must realize that good computational classification results are still unable to confirm whether or not the genes are biologically causative. Therefore, gene selection results have to be evaluated from biological perspective as well.

In Table 9.1, one gene result of the WMSFS are listed. The functions of these genes range from cell adhesion (VCL, NELL2) to immune response (DF, C7), from cellular transport (MRC2, RBP1) to regulation of transcription (LMO3), from protein kinase activity (ILK) to hormone activity (IGF1). It can be noted that almost all of these selected genes have been associated with development and diagnosis of prostate cancer - some of them are well-known prostate-cancer-associated genes, such as IGF1, GAGEC1, RBP1, DF, NELL2, ILK, etc., and others have been suggested to overexpress in prostate cancer samples, for example, C7, LMO3.

9.7 Conclusions

Gene selection aims to find the biomarkers of different diseases, such as prostrate cancer and lung cancer. From computational perspective, it has always been difficult to handle a huge dimensional data set like microarray data set. Reducing the gene space to manageable dimensions has become an essential issue. Gene selection models are categories into filter model and embedded model. This chapter focuses on the filter models. It is illustrated that search engine can be nicely modified through using a pattern-weighting idea. Also, since microarray data sets are usually in a small sample size, the problem of overfitting should be paid attentions on. It is shown that the point injection approach can be used address the problem of overfitting.

In the area of gene selection, there are several open issues. One of them is how to deliver a reliable result. It is known that, given different datasets, almost all gene selection methods rarely deliver identical outputs. To determine more robust result, a committee-voting scheme can be adopted. In this scheme, several gene selection approaches are used. Each approach is used to rank given genes. After summarizing the all gene-ranking results, the genes with the best quality are finally identified.

To determine the number of selected genes is another challenging issue. From the perspective of machine learning, this problem can be handled using the techniques of model selection or the intrinsic characteristics of a well-defined evaluation criterion. But it has to be realized that the biological nature of gene selection sometimes may prove to be too complex for a computational process to handle. Despite the mathematical sophistication of a gene selection algorithm, we know that we are impossible to claim that a gene-selection result obtained in machine-learning ways is totally reliable. The final resolution still largely depends upon the biological senses.

Table 9.1. The genes that are identified WMSFS to be related with the prostate cancer

Order of selection	Gene symbol	Gene title	Relation with prostate cancer
1	VCL	Vinculin	Vinculin, a cytoskeletal protein, can regulate the ability of cancer cell to move away from turmors. It may contribute to metastatic process of prostate cancer.
2	DF	D component alternative of complement (adipsin)	Adipsin, a member of the trypsin family of peptidases, is a component of the complement pathway playing an important part in humoral suppression of infectious agents. Uzma et al. [29] find out this gene up-regulates in the samples with prostate diseases, such as prostate cancer. Also, Chow et al. [30] suggest it a good cancer marker.
3	MRC2	Mannose receptor, C type 2	
4	NELL2	NEL-like 2 (chicken)	The close correlation of this gene to prostate cancer is also suggested in other studies [29][34].
5	RBP1	retinol binding protein 1, cellular	Retinoids are involved in cell growth, differentiation, and carcinogenesis. This gene has been found to overexpress in prostate carcinoma [31].
6	C7	complement component 7	This gene takes part in androgen-regulated processes that play important roles in malignant transformation of prostate gland [29].
7	IGF1	homeodomain interacting protein kinase 3	The role that this gene plays in prostate development and carcinogenesis has been well-recognized and widely examimed [32].
8	ILK	integrin-linked kinase	This gene overexpression can suppress anoikis, promote anchorage-independent cell cycle progression, and induce tumorigenesis and invasion [33].
9	GAGEC1	G antigen, family C, 1	The protein encoded in this gene is PAGE4, which is a Cytoplasmic protein and is prostate associated.
10	LMO3	LIM domain only 3 (rhombotin-like 2)	The protein encoded in this gene is a LIM-only protein (LMO), which is involved in cell fate determination. This gene has been noted to upregulate in the prostate cancer samples [29].

In summary, the advances in machine learning enable gene selection be reliably and efficiently conducted computationally. Gene selection can become more robust when more sophisticated computational technique is introduced. But owing to the nature of the problem, computational gene selection should work together with the biological means to confirm the disease-causing genes.

9.8 Acknowledgement

The authors would like to acknowledge the City University of Hong Kong for supporting this work through a SRG research project of Project No: 7001998570.

References

1. Ekins R, Chu FW (1999) Trends in Biotechnology 17:217–218.
2. Dudoit S, Fridlyand J, Speed TP (2002) Journal of the American Statistical Association 97(457):77–87.
3. Golub TR, Slonim DK, Tamayo P, et al. (1999) Science 286:531–537.
4. Li W, Yang Y (2002) How Many Genes Are Needed for a Discriminant Microarray Data Analysis? In: Lin SM, Johnson KF (eds) Methods of Microarray Data Analysis. Kluwer Academic, London.
5. Singh D, Febbo PG, Ross K, et al. (2002) Cancer Cell 1(2):203–209.
6. Liu H, Motoda H (1998) Feature Selection for Knowledge Discovery and Data Mining. Kluwer Academic, London.
7. Molina LC, Belanche L, Nebot A (2002) Feature Selection Algorithms: a Survey and Experimental Evaluation, Departament de Llenguatges i Sistemes Informtics, Universitat Politcnica de Catalunya, available at: http://www.lsi.upc.es/dept/techreps /html/R02-62.html, Technical Report.
8. Liu X, Krishnan A, Mondry A (2005) An Entropy-based gene selection method for cancer classification using microarray data, BMC Bioinformatics, 6:76.
9. Shah SC, Kusiak A (2004) Intelligence in Medicine 31(3):183–196.
10. Umpai TJ, Aitken S (2005) BMC Bioinformatics 6(148):1:11.
11. Xing EP, Jordan MI, Karp M (2001) Feature selection for high-dimensional genomic microarray data. In: Carla E. Brodley, Andrea Pohoreckyj Danyluk (eds) Proc. 18th Intl. Conf. On Machine Learning. San Francisco, Morgan Kaufmann.
12. Lee KE, Sha N, Dougherty ER, et al. (2003) Bioinformatics 19(1):90–97.
13. Guyon I, Weston J, Barnhill S, et al. (2002) Machine Learning 46:389–422.
14. Yeung K, Bumgarner RE, Raftery AE (2005) Bioinformatics 21(10):2394–2402.
15. Huang D, Chow TWS (2005) IEEE Trans. Circuits and Systems 52(4):785–793.
16. Pudil P, Novovicova J, Kittler J (1994) Pattern Recognition Letter 15:1119–1125.
17. Al-Ani A, Deriche M (2000) Optimal feature selection using information maximisation: case of biomedical data. In: Proc. of the 2000 IEEE Signal Processing Society Workshop, vol. 2.
18. Gui J, Li H (2005) Bioinformatics 21(13):3001–3008.
19. Zhou X, Wang X, Dougherty ER (2004) Journal of Biological Systems 12(3):371–386.
20. Bishop CM (1995) Neural Networks for Pattern Recognition. Oxford University Press, New York.

21. Hastie T, Tibshirani R, Friedman J (2001) The Elements of Statistical Learning. Springe, Berlin Heidelberg New York.
22. Lampariello F, Sciandrone M (2001) IEEE Trans. On Neural Networks 12(5):1235–1242.
23. Kim S, Dougherty ER, Barrera JY, et al. (2002) Journal of Computational Biology 9:127–146.
24. Sima C, Braga-Neto U, Dougherty ER (2005) Bioinformatics 21(7):1046–1054.
25. Skurichina M, Raudys S, Duin RP (2000) IEEE Trans. On Neural Networks 11(2):504–511.
26. Parzen E (1962) Ann. Math. Statistics 33:1064–1076.
27. Tibshirani R, Parker J, Hastie T, et al. (2003) Proc Natl Acad Sci USA 100(14):8418–8423.
28. Turashvili G, Bouchal J, Burkadze G, et al. (2005) Biomedical papers 149(1):63–68.
29. Uzma SS, Robert HG (2004) Journal of Cellular Biochemistry 91:161–169.
30. Chow ML, Moler EJ, Mian IS (2001) Physiol Genomics 5:99–111.
31. Jeronimo C, Henrique R, Oliveira J, et al. (2004) Journal of Clinical Pathology 57:872–876.
32. Cheng I, Stram DO, Penney KL, et al. (2006) Journal Natl. Cancer Inst. 98(2):123–124.
33. Graff JR, Deddens JA, Knoicek BW, et al. (1002) Clinical Cancer Research 7:1987–1991.
34. Zhang C, Li H, Fan J, et al. (2006) BMC Bioinformatics, 7:202–236.

10

Saccharomyces pombe and *Saccharomyces cerevisiae* Gene Regulatory Network Inference Using the Fuzzy Logic Network

Yingjun Cao[1], Paul P. Wang[2], and Alade Tokuta[1]

[1] Department of Mathematics and Computer Science, North Carolina Central University, 1801 Fayetteville Street, Durham, NC 27707.
{ycao,atokuta}@nccu.edu

[2] Department of Electrical and Computer Engineering, P.O. Box 90291, Duke University, Durham, NC 27708.
ppw@ee.duke.edu

Summary. In this chapter, a novel gene regulatory network inference algorithm based on the fuzzy logic network theory is proposed and tested. The key motivation for this algorithm is that genes with regulatory relationships may be modeled via fuzzy logic, and the strength of regulations may be represented as the length of accumulated distance during a period of time intervals. One unique feature of this algorithm is that it makes very limited a priori assumptions concerning the modeling. Hence the algorithm is categorized as a data-driven algorithm. With the theoretical guidelines to quantify the upper limits of parameters, the algorithm is implemented to infer gene regulatory networks for *Saccharomyces cerevisiae* and *Saccharomyces pombe*. The computation results not only prove the validity of the data-driven algorithm, but also offer a possible explanation concerning the difference of network stabilities between the budding yeast and the fission yeast.

10.1 Introduction

One of the most challenging problems in bioinformatics is to determine how genes inter-regulate in a systematic manner which results in various translated protein products and phenotypes. To find the causal pathways that control the complex biological functions, previous work have modeled gene regulatory mechanisms as a network topologically [1]. The importance of networking models is that normal regulatory pathways are composed of regulations resulting from many genes, RNAs, and transcription factors (TFs). The complicated inter-connections among these controlling chemical complexes are the driving forces in maintaining normal organism functions. A precise structural presentation of components should illustrate the key properties of the system.

Based on the network representation of gene regulations, a number of inference models have been proposed. They include Bayesian networks [2], hybrid Petri

Y. Cao et al.: *Saccharomyces pombe and Saccharomyces cerevisiae Gene Regulatory Network Inference Using the Fuzzy Logic Network*, Studies in Computational Intelligence (SCI) **94**, 237–256 (2008)
www.springerlink.com

net [3], growth network [4], genetic circuits [5], inductive logic programming [6], and hybrid networks [7]. These models have focused on different aspects of gene regulatory behaviors, and each model has contributed good inference results in certain aspects. The ongoing research on these models is focused on the challenges of data integration, non-linear data processing, noise tolerance, synchronization, and model over fitting [8].

Research on genome-wide gene regulations has used dynamic microarray data which quantify the genomic expression levels at each sample time. Given a series of microarray data, researchers have attempted to find the spatial and temporal modes of regulations regarding different conditions or different stages of cell cycles on different species [8]. But because of the hybridization process and the synchronization issues of time-series microarray, the data, very often contain missing, noisy, or unsynchronized data subsets. Thus data normalization and pre-processing techniques have become necessary to reduce the noise. Other techniques like SAGE [9], TFs mappings [10], and antibiotic arrays [11] have been designed to discover the regulatory mechanisms.

In this chapter, a novel network model, the fuzzy logic network (FLN), is proposed and thoroughly examined. The feasibility of applying this model to inferring gene regulatory networks is investigated. The FLN is a generalization of the Boolean network, but it is capable of overcoming the unrealistic constraints of Boolean values (ON/OFF symbolically). With distinctive properties in processing real life incomplete data and uncertainties, researchers have applied fuzzy logic to gene expression analysis, and by the use of specific scoring matric, *Saccharomyces cerevisiae* gene regulatory networks with biological verifications were inferred [12–14]. This chapter expands these previous research work.

The rest of the chapter is organized as follows: In Section 10.2, the definition of the FLN is introduced, and the critical connectivity of the FLN is deduced using the anneal approximation. Then, the structure of the inference algorithm is discussed in Section 10.3. Finally, in Section 10.4, the algorithm is used to infer the gene regulatory networks for *Saccharomyces cerevisiae* and *Saccharomyces pombe*. The inference results are compared and analyzed in this section. The chapter concludes in Section 10.5.

10.2 Fuzzy Logic Network Theory

The proposed FLN theory is based on theoretical deductions at the second level of fuzzy uncertainty. This means that the variables have been normalized into $[0,1]$ interval, and statistical methods are built on the basis of fuzzy variables as well as their relationships.

The FLN is defined as follows:

Given a set of N variables (genes), $\Sigma(t) = (\sigma_1(t), \sigma_2(t), \ldots, \sigma_N(t))$, $(\sigma_i(t) \in [0,1]$, $(i = 1, 2, \ldots, N)$, index t represents time), the variables are to be updated by means of the dynamic equations:

$$\sigma_i(t+1) = \Lambda_i(\sigma_{i_1}(t), \sigma_{i_2}(t), \ldots, \sigma_{i_K}(t)),\ (1 \le i \le N) \tag{10.1}$$

Table 10.1. Four commonly used fuzzy logical functions including their AND ($\wedge$), OR ($\vee$), and NOT ($\bar{\ }$).

Fuzzy Logical Functions	$a \wedge b$	$a \vee b$	$\bar{a}$
Max-Min	$\min(a,b)$	$\max(a,b)$	1-a
GC	$a \times b$	$\min(1, a+b)$	1-a
MV	$\max(0, a+b-1)$	$\min(1, a+b)$	1-a
Probabilistic	$a \times b$	$a+b-a \times b$	1-a

where Λ_i is a fuzzy logical function, and K represents the number of regulators for σ_i.

For an FLN, the logical functions may be constructed using the combinations of AND ($\wedge$), OR ($\vee$), and NOT ($\bar{\ }$). The total number of choices for fuzzy logical functions is determined only by the number of inputs. If a node has K ($1 \leq K \leq N$) inputs, then there are 2^K different logical functions. In the definition of the FLN, each node, $F_i(t)$, has K inputs. But this fixed connectivity will be relaxed later.

To apply the FLN to modeling gene regulatory networks, each fuzzy variable will represent a gene, and genetic regulatory relationships will be modeled as fuzzy logical functions. A fuzzy logical function is defined as a function $\Lambda : U \rightarrow [0,1]$ where $\Lambda(u)$ is the degree of the membership. Usually, it has to satisfy the requirement of the t-norm/t-co-norm, which is a binary operation that satisfies the identity, commutative, associative, and increasing properties [15]. Table 10.1 shows the commonly used fuzzy logical functions with distinctive dynamics [16].

Although the logical functions are expressed via simple algebraic expressions, they have their own distinctive properties. The Max-Min logical function which is closely related to Boolean logic, is one of the classical fuzzy logical functions. This logical function uses the maximum of two values to replace the Boolean OR, whereas the minimum replaces the Boolean AND. GC logical function is a combination of MV and Probabilistic logical functions. The MV logical function follows the trivalent logic whereas Probabilistic does not. In this chapter, all four fuzzy logical functions are tested on the *S. cerevisiae* dataset.

The critical connectivity of the FLN is crucial in the data-driven algorithm's application to gene regulatory network inference. It quantifies the algorithm's search strategy, and the computational complexity of the algorithm is determined by it. To study the detailed dynamics and the connectivity of the FLN, the annealed approximation [17, 18] has been used. Consider the following two FLN configurations at time t: $\Sigma(t)$ and $\widetilde{\Sigma}(t)$, where

$$\begin{aligned} \Sigma(t) &= \{\sigma_1(t), \sigma_2(t), \ldots, \sigma_N(t)\} \\ \widetilde{\Sigma(t)} &= \{\widetilde{\sigma_1}(t), \widetilde{\sigma_2}(t), \ldots, \widetilde{\sigma_N}(t)\} \end{aligned} \tag{10.2}$$

Assume that logical functions selected by the two configurations are not time variant throughout the dynamic process. Then the distance between the two configurations may be computed as the accumulated Hamming distance (AHD):

$$AHD(t) = \sum_{i=1}^{N} Hamming(\sigma_i(t), \widetilde{\sigma}_i(t)) \tag{10.3}$$

and

$$Hamming(\sigma_i(t), \widetilde{\sigma}_i(t)) = \begin{cases} 1 & \text{if } |\sigma_i(t) - \widetilde{\sigma}_i(t)| > \delta \\ 0 & \text{if } |\sigma_i(t) - \widetilde{\sigma}_i(t)| \leq \delta \end{cases} \tag{10.4}$$

The Hamming distance uses $\delta \in [0,1]$ (Hamming threshold) as a parameter to differentiate the closeness of two values. The distance between two Boolean values may also be computed using (10.4) with $\delta \equiv 0$. Thus, the AHD of the FLN is the extension of the Boolean distance. One may easily see that the maximum distance between $\Sigma(t)$ and $\widetilde{\Sigma}(t)$ is N, while the minimum distance is 0. In comparison with the distance, another quantity, $a_t \in [0,1]$, may be defined as the similarity of the two networks, i.e.,

$$a_t = 1 - \frac{AHD(t)}{N} \tag{10.5}$$

Suppose at time t, $\Sigma(t)$ and $\widetilde{\Sigma}(t)$ are at distance l_t. Then the probability of the two configurations having a distance l_{t+1} at time $t+1$ may be found. This change in distances represents the dynamic paths of the two configurations. Denote this probability as $P_t(l_{t+1}, l_t)$. Suppose $\Sigma(t)$ and $\widetilde{\Sigma}(t)$ have the same logical function selections for their corresponding variables but different initial values for each variable, and the variables in the two systems can select one out of S values (S is finite, $S \geq 2N$). The requirement of $S \geq 2N$ is to guarantee that different fuzzy logical functions may be used by the FLN [19]. The probability of selecting each of the S values is assumed to be the same, i.e. $\frac{1}{S}$.

Suppose A is the set of variables which are identical in $\Sigma(t)$ and $\widetilde{\Sigma}(t)$ at time t. Obviously, set A has $N - l_t$ variables. Define $Q(N_0)$ as the probability that N_0 variables have all their K parents from set A. Then, $Q(N_0)$ is a discrete random variable following the binomial distribution with parameter $\left(\frac{N-l_t}{N}\right)^K$. By definition,

$$\frac{N-l_t}{N} = 1 - \frac{l_t}{N} = a_t \tag{10.6}$$

so,

$$\begin{aligned} Q(N_0) &= \binom{N}{N_0} \left[\left(\frac{N-l_t}{N}\right)^K\right]^{N_0} \left[1 - \left(\frac{N-l_t}{N}\right)^K\right]^{N-N_0} \\ &= \binom{N}{N_0} \left[a_t^{\,K}\right]^{N_0} \left[1 - a_t^{\,K}\right]^{N-N_0} \end{aligned} \tag{10.7}$$

It is obvious that these N_0 variables will be the same at time $t+1$ in both $\Sigma(t+1)$ and $\widetilde{\Sigma}(t+1)$. For the remaining $N - N_0$ variables, since at least one of their parents will be different, there is a probability of $p = \frac{S(S-1)}{S^2}$ that a variable will be different in two networks at the next step, while $1-p$ is the probability that it will be the same.

More generally, let P be the probability that a function produces different values from different inputs. If a variable can take S values, and the probability of selecting one of these values is the same, then P may be expressed using the Bayesian rule.

$$P = p(S-1)\frac{1-p}{S-1} + (S-1)\frac{1-p}{S-1}(1-\frac{1-p}{S-1}) \tag{10.8}$$

Thus, through deductions using the annealed approximation, the following equation may be found:

$$P(l_{t+1}, l_t) = \frac{N!}{l_{t+1}!(N-l_{t+1})!}\left(P(1-a_t^K)\right)^{l_{t+1}} \left[1-P(1-a_t^K)\right]^{N-l_{t+1}} \tag{10.9}$$

As can be seen, (10.9) follows binomial distribution. Thus the possibility of the coverage at the next step will peak at the current mean. The dynamic recursive equation, then, may be expressed as

$$a_{t+1} = 1 - P(1-a_t{}^K) \tag{10.10}$$

A general situation is considered in which P is uniformly distributed. Then P can be computed as,

$$P = P(|\sigma_i(t) - \sigma_j(t)| \geq \delta) = (1-\delta)^2 \tag{10.11}$$

If the two networks converge, then the following marginal stability should be imposed:

$$\begin{aligned} &\tfrac{\partial a_{t+1}}{\partial a_t} < 1 \quad \text{(implies that the coverage does not decrease with time)} \\ &\lim_{t\to\infty} a_t = 1 \quad \text{(the condition for a full coverage in the steady state)} \end{aligned} \tag{10.12}$$

If a network does not have uniform connectivity for all nodes, we may assume that the nodes may have different number of parents with a discrete distribution ρ_k, where

$$\rho_k = Prob(\text{a node has } k \text{ parents}) \text{ and } \sum_{k=1}^{N} \rho_k = 1 \tag{10.13}$$

By applying (10.11), (10.12), and (10.13) to (10.10), the following relationship may be found.

$$\overline{K} = \frac{1}{(1-\delta)^2} \tag{10.14}$$

It has been found that, in yeast protein-protein networks, as well as in the Internet and social networks, the distribution of connectivity follows the Zipf's law [20], i.e.,

$$P((\text{number of inputs}) = K) \propto \frac{1}{K^\gamma}, 1 \leq K \leq N \tag{10.15}$$

where γ is a real number, usually between 2 and 3.

Hence, according to (10.15), the mean connectivity may be computed as

$$\overline{K} = \varepsilon \sum_{K=1}^{N} K \frac{1}{K^{\gamma}} = \varepsilon \sum_{K=1}^{N} \frac{1}{K^{\gamma-1}} \tag{10.16}$$

where ε is a constant to guarantee that the sum of distribution equals 1.

Then, define

$$H_N^{(\gamma)} = \sum_{i=1}^{N} \frac{1}{K^{\gamma}} \tag{10.17}$$

as the partial sum of the generalized harmonic series. It may be proved that

$$\varepsilon = \frac{1}{H_N^{\gamma}} \quad \text{and} \quad \overline{K} = \frac{H_N^{\gamma-1}}{H_N^{\gamma}} \tag{10.18}$$

Since there is no general formula for (10.17), approximations for the sum may be used if N is large enough, which is true for the application to gene regulatory network inference. The approximation of $H_N^{(\gamma)}$ is

$$H_N^{(\gamma)} \approx \begin{cases} \infty & \text{if } \gamma = 1 \\ \frac{\pi^2}{6} & \text{if } \gamma = 2 \\ 1.202 & \text{if } \gamma = 3 \\ \frac{\pi^4}{90} & \text{if } \gamma = 4 \\ 1.036 & \text{if } \gamma = 5 \\ \frac{\pi^6}{945} & \text{if } \gamma = 6 \end{cases} \tag{10.19}$$

By substituting (10.19) into (10.18), the mean connectivity of the network may be found as

$$\overline{K} \approx \begin{cases} \infty & \text{if } \gamma = 2 \\ 1.3685 & \text{if } \gamma = 3 \\ 1.1106 & \text{if } \gamma = 4 \\ 1.0447 & \text{if } \gamma = 5 \\ 1.0183 & \text{if } \gamma = 6 \end{cases} \tag{10.20}$$

By applying (10.8), (10.12), and (10.13) to (10.10), the relationship between S and $\overline{K}$ is

$$S \geq \frac{\overline{K}}{\overline{K} - 1} \tag{10.21}$$

Therefore, by substituting $\overline{K}$ in (10.20) to (10.21), the value of S may be found to have a lower bound, i.e.

$$S > \begin{cases} 1 & \text{if } \gamma = 2 \\ 3.7137 & \text{if } \gamma = 3 \\ 10.0416 & \text{if } \gamma = 4 \\ 23.3714 & \text{if } \gamma = 5 \\ 55.6448 & \text{if } \gamma = 6 \end{cases} \tag{10.22}$$

In general, the connectivity of a real network should be greater than the critical connectivity because real networks are usually much more complicated. In other words, the critical connectivity only serves as a lower bound. In addition, more relaxed criteria may find more possible regulations, and provide a much smaller search space for further investigations. Thus an S with $\gamma \geq 3$ should be chosen in initial searches, which means S should be more than 2. This triggers the question of whether a crisp Boolean network is powerful enough to infer the genetic network structure.

If (10.20) is substituted into (10.14), the requirement that δ must satisfy may be found as

$$\delta > \begin{cases} 0.1452 & \text{if } \gamma = 3 \\ 0.0511 & \text{if } \gamma = 4 \\ 0.0216 & \text{if } \gamma = 5 \\ 0.0090 & \text{if } \gamma = 6 \end{cases} \tag{10.23}$$

When γ increases, the network has to adjust itself by adopting stricter criteria. The result also agrees with the relationship between S and γ from (10.22). However, when the FLN is used on a real dataset, δ must also be increased to account for the noise inside.

10.3 Algorithm

Let $\boldsymbol{G} \in R^{n\times m}$ be the time-series microarray data where n is the number of genes in the data and m is the number of time slots in the microarray set. The algorithm will first randomly select $G_r = (G_{r_1}, G_{r_2}, ..., G_{r_i}, ...G_{r_K})$, a group of regulators that regulates G_t $(G_{r_i}, G_t \in R^{1\times m}, t \neq r_1, r_2, ...r_K)$. Then the algorithm will filter the regulators through a fuzzy logic mask, $FLogic$[1], to generate a pseudo-gene-time-series, $G_s \in R^{1\times m}$ where

$$G_s^j = Flogic(G_{r_1}^j, G_{r_2}^j, ..., G_{r_i}^j, ...G_{r_K}^j), j \in \underline{m} \tag{10.24}$$

The distance between G_r and G_t is then computed as:

$$Distance(G_r, G_t) = AHD(G_s, G_t) = \sum_{j=1}^{m-1} Hamming(G_t^{j+1}, G_s^j) \tag{10.25}$$

where the Hamming distance is computed according to (10.4). The value of δ should base on (10.23), on the noise level, and on the data completeness. As shown in (10.25), the AHD between G_t and G_s is computed with a time shift throughout the time series, which is a reasonable assumption that regulations happen with one time delay.

For each group of possible regulators and the regulated gene, the algorithm determines its AHD, and records it. In the end, the algorithm will infer regulatory groups

[1] $Flogic$ is one of the possible fuzzy logical functions that is applied on K variables

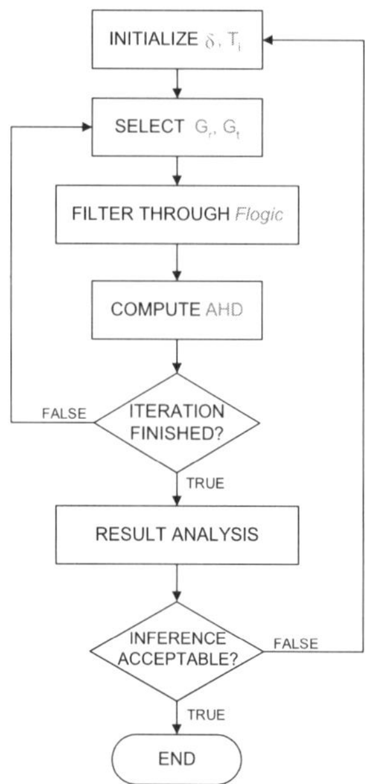

Fig. 10.1. Algorithm flowchart

with $\frac{AHD}{m}$ less than another threshold, $T_i, (T_i \in [0,1])$, which is the inference threshold. This threshold is defined as the percentage of time periods that a regulation persists with respect to the total time slots in the microarray. The flow chart of the algorithm is depicted in Fig.10.1

The complexity of the algorithm is $O(n^{\overline{K}})$ where n is the number of genes, and $\overline{K}$ is the critical connectivity of the FLN. As shown in (10.20), the mean connectivity should be less than 1.3685. Thus the complexity of the algorithm is almost linear with the number of genes. Based on the result shown in (10.20) and [21], we assume the maximum number of regulators for each inference group is 2. In this scenario, there may be more than 2 regulators for a regulated gene but we assume that regulators are correlated in a pair-wise manner. This assumption, which assumes that two regulators are more efficient to deliver regulatory mechanisms, ensures the critical connectivity. One should note that the total number of regulators is not limited by this assumption (any gene in the network may have N regulators maximally).

10.4 Applications to Yeast Datasets

To test the algorithm, the α-factor arrest subset of *S. cerevisiae* cell cycle dataset [22] and the *S. pombe* cell cycle dataset [23] are selected. Although the budding yeast, *S. cerevisiae*, and the fission yeast, *S. pombe*, are generally named as yeast, they diverged about 1.1 billion years ago [24]. It has been shown that these two fungi have different chromosome structures, cell cycle synergy, and different gene functions. In addition, it has been reported that the S and M phases of *S. cerevisiae* are very hard to disrupt while the disrupt of *S. pombe* cell cycles can be relatively easier [25]. The inference results shown later in this chapter confirm the differences of gene regulation strategies in the two yeast, and offer a possible explanation on why the *S. cerevisiae* cell cycle is more stable.

10.4.1 Data Pre-processing

The *S. cerevisiae* dataset includes 6178 commonly accepted open reading frames, and for the α-arrest subset there are 18 time slots. The *S. pombe* dataset includes 99.5 percent of the total open reading frames, but we choose to use the 407 genes that have been inferred to be cell-cycle regulated [23]. For the *S. pombe* dataset, three *elutriation* and two *cdc*25 block-release subsets are combined to produce a series of 115 time slots for each gene. To reduce errors introduced by noise, and to reduce computational complexity, both datasets are pre-processed with 3 criteria. Genes that do not satisfy all three criteria are deleted. The criteria are stated as follows:

- Only genes having more than two thirds valid time slots, with respect to the entire time span in the microarray, are considered above the noise level. The reason for this requirement is that some genes do not have expression strong enough to counter the background noise at some time slots. Totally 115 genes do not fit this criterion in the *S. pombe* dataset. The number of genes deleted in the *S. cerevisiae* dataset is 125.
- For *S. cerevisiae*, the maximum value of each gene's expression must be at least three times greater than its minimum value in the time series. For *S. pombe*, the ratio is limited to 1. If not, the gene is excluded from the dataset. This requirement guarantees that genes running inside the algorithm have a dynamic range of expression. Thus it reduces the computational time by limiting the search space. *S. cerevisiae* dataset has 5366 genes deleted, but no gene is deleted from *S. pombe* dataset due to this criterion.
- Genes with spikes in the time series are not included. The signal-to-noise ratio of the spike is defined as five. 4 genes in the *S. pombe* dataset and 8 genes in the *S. cerevisiae* dataset have spikes, and are deleted.

After the pre-processing and filtering, 680 genes in the *S. cerevisiae* subset are found to satisfy all three criteria, and, as far as the *S. pombe* dataset is concerned, 286 genes have survived the cut. The values of gene expression are then normalized into $[0,1]$ interval throughout the time series. For *S. cerevisiae*, the values of the dataset are changed from log-odds into true values. Then, the maximum value of each gene series is found, and used to divide the expression of that gene in the series. For *S. pombe*, every gene series in each subset is normalized to have zero median. In the next step, the maximum value of a gene's expression inside each subset is found, and used to divide the values of that gene's expression in the same subset (the five subsets are normalized separately). After these steps, the values of each gene in the dataset have been normalized into $[0,1]$ interval.

10.4.2 Inference Results of the *S. cerevisiae* Dataset

We investigate the inference results when regulators apply controls to the regulated gene in pairs. All four logical functions in Table 10.1 are tested with same parameter settings ($\delta = 0.01, T_i = 21\%$), and they have inferred four different regulatory networks. The MV logical function do not introduce as many false positives as that from using other commonly used fuzzy logical functions. Furthermore, MV logical function causes the algorithm to be less sensitive to small variations of parameters i.e., δ from (10.4) and the inference threshold of the algorithm, T_i. The inferred network based on the MV logical function is shown in Fig. 10.2 and Table 10.2.

Out of 4.3×10^9 possible groups, the algorithm locates 51 regulatory groups (32 regulatory pairs) involving 21 genes with average connectivity of 1.5238. There are 17 verified regulations, 5 unknown regulations, and 10 dubious regulations. Table 10.2 shows the verified regulations with functions of the regulator and regulated gene. One interesting finding is that 15 out of the 21 genes in the network have been proved to be involved in yeast mating or the cell cycle, and most of them are downstream mating regulatory genes. In addition, the backbone of the network (nodes with high connectivities) is made up of 9 out of these 15 genes. The clustering attribute in the result, although unexpected, may explain why 14 out of the 17 verified regulations are based on close relationships.

The inferred network also shows network motifs. The network includes seven feed-forward loops, three single-input modules, as well as the dense overlapping and bi-fan modules [33, 34]. Through comparative studies on complex biological, technological, and sociological networks, it has been shown that these modules share different evolutionary properties and advantages [35, 36]. The feed-forward loop is believed to play a functional role in information processing. This motif may serve as a circuit-like function to activate output only if the input signal is persistent, and allows a rapid deactivation if the input signal is off. Further, the bi-fan structure of (*PRM1*, *FUS2*) and (*FIG1*, *ASG7*) are coupled with a number of feed-forward motifs. The inferred network also includes two internal cycles (*FIG1* $\longleftrightarrow$ *PRM1*, *FIG1* $\longleftrightarrow$ *FIG2*) and one feedback loop among *FIG1*, *FIG2* and *ASG7*. All the genes in the cycles or feedback loop are involved in the signaling for yeast mating, and the close regulations among them are integral to yeast mating. Although network motif studies on *E. Coli* have not found cyclic structures [34, 37], the feedback loop is believed to be the most important regulatory mechanism for cells to adapt to new environments. The inferred network shows that while preserving specific regulatory strategies, different species share a striking similarity of regulation mechanisms.

10.4.3 Inference Results of the *S. pombe* Dataset

The algorithmic parameters for the *S. pombe* dataset are set as $\delta = 0.018$ because the network is selected to be a more general network according to (10.23), and $T_i = 71\%$ because the combination of five different subsets lowers the percentage of time that a regulation may persist. The quantifications of the two thresholds are also based on previous investigations on the algorithm's behaviors [12]. The algorithm uses the

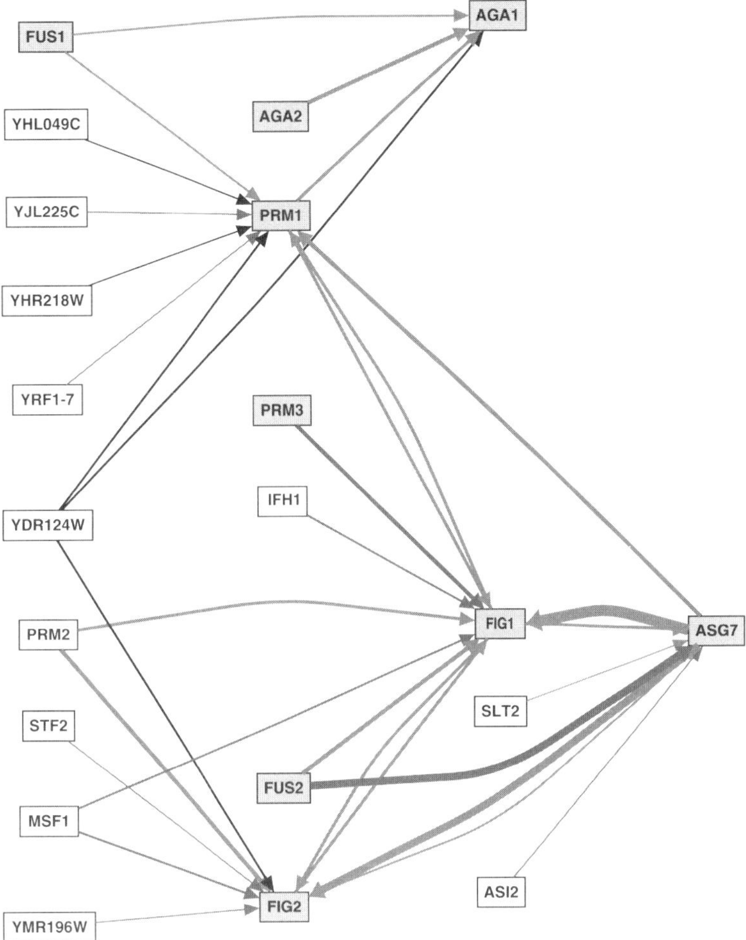

Fig. 10.2. The α-factor gene regulatory network using the MV fuzzy logical function. In the network, there are 21 genes and 32 regulatory arcs. Nodes with high connectivities are green. The colors of arcs are coded as: functionally verified regulations (red), regulations involving genes with unknown functions (black), and dubious regulations (blue). The width of the arcs represents the regulatory strength between the regulator and regulated gene, and it is calculated as $(m - AHD)$

MV logical function to find 105 regulations (125 regulatory pairs) among 108 genes. The regulatory network is shown in Fig. 10.3.

In the network, there are 108 genes and 125 regulatory arcs. The colors of arcs are coded as: functionally verified regulations (red), regulations involving genes with unknown functions (black), and dubious regulations (blue). The width of the arcs represents the regulatory strength between the regulator and regulated gene, and it is calculated as $(m - AHD)$. The network includes 59 functionally verified regulations,

Table 10.2. Functionally verified regulations in the inferred *S. cerevisiae* gene regulatory network. The regulations are grouped by regulators and the criteria of verifications are categorized as functional (verified by gene functions from Saccharomyces Genome Database, 4 regulations), close relationship (regulators and regulated genes are usually co-expressed, co-regulated, 14 regulations). Some of the verifications are also based on the included references

Regulator	Regulated gene	Verification
AGA2: adhesion subunit of a-agglutinin of a-cells	**AGA1**: anchorage subunit of a-agglutinin of a-cells	close relationship [26]
ASG7: regulates signaling from Ste4p	**FIG1**: integral membrane protein for efficient mating	close relationship [27]
	FIG2: cell wall adhesin specifically for mating	functional
	PRM1: SUN family gene involved in cell separation	close relationship [28]
FIG1: integral membrane protein for efficient mating	**ASG7**: regulates signaling from Ste4p	close relationship [27]
	FIG2: cell wall adhesin specifically for mating	close relationship [29]
	PRM1: pheromone-regulated protein for membrane fusion during mating	close relationship [27]
FIG2: cell wall adhesin specifically for mating	**FIG1**: integral membrane protein for efficient mating	close relationship [29]
	ASG7: regulates signaling from Ste4p	functional
FUS1: membrane protein required for cell fusion	**AGA1**: anchorage subunit of a-agglutinin of a-cells	close relationship [30]
	PRM1: pheromone-regulated protein for membrane fusion during mating	close relationship [31]
PRM1: pheromone-regulated protein for membrane fusion during mating	**AGA1**: anchorage subunit of a-agglutinin of a-cells	close relationship [30]
	FIG1: integral membrane protein for efficient mating	close relationship [27]
PRM2: pheromone-regulated protein regulated by Ste12p	**FIG1**: integral membrane protein for efficient mating	close relationship [27]
	FIG2: cell wall adhesin specifically for mating	close relationship [27]
SLT2: suppressor of lyt2	**ASG7**: regulates signaling from Ste4p	close relationship [32]
FUS2: cytoplasmic protein for the alignment of parental nuclei before nuclear fusion	**FIG1**: integral membrane protein for efficient mating	close relationship

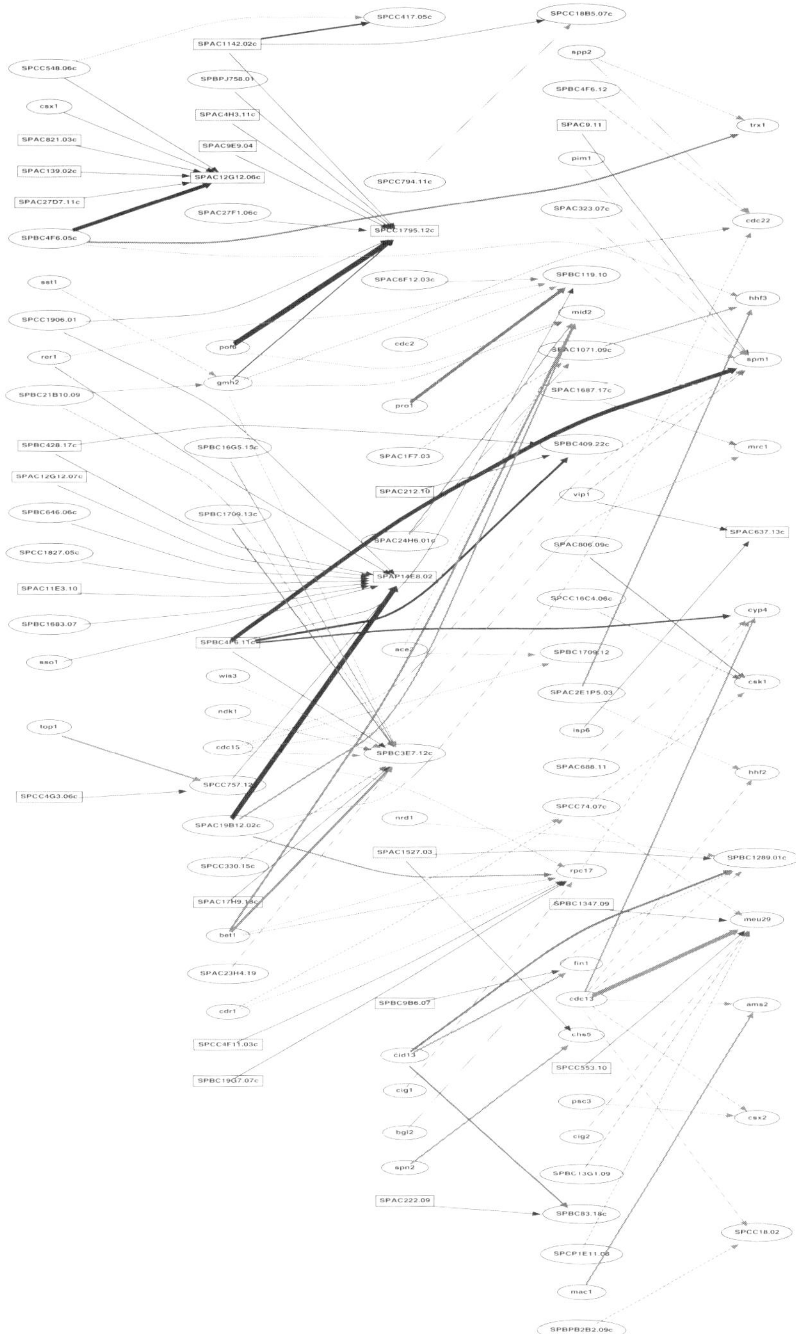

Fig. 10.3. The *S. pombe* gene regulatory network using the MV fuzzy logical function

47 regulations involving genes with unknown functions, and 19 dubious regulations. The 59 functionally verified regulations are listed in Table 10.3 with the functions of regulators and regulated genes.

Take the *ace2* - *mid2* and *cdc15* - *spm1* regulatory pairs as examples. Gene *ace2* is a transcription factor that regulates the transcription of genes required for cell separation; *mid2* is essential for the central positioning of the division septum before the cell divides and in organizing the septin ring during late mitosis. Gene *cdc15* mediates cytoskeletal rearrangements required for cytokinesis on the onset of mitosis, and *spm1* involves in the regulation of cell wall structure. The average connectivity of the inferred network is 1.157 which fits the requirement of the Zipf's law as shown in (10.20) with γ between 2 and 3.

The regulations among genes in *S. pombe* are different from *S. cerevisiae* largely because the regulatory network of *S. pombe* does not include any feed-forward structure or cycle. As shown before, the feed-forward loop is a stable motif for the network, and this might offer an explanation why *S. pombe* cell cycles are less stable. We also found that the regulatory network of *S. pombe* is sparser than that of the *S. cerevisiae*. As for the regulatory logic in the two networks, we found striking similarities between the two yeast. For *S. pombe*, 75 percent of regulations are OR logic while the majority of the remaining 25 percent are single-regulator situations. For *S. cerevisiae*, 63 percent are OR logic and 17 percent are single-regulator scenarios.

10.5 Conclusions and Discussion

The biologically plausible results from the applications of our FLN algorithm to the *S. cerevisiae* and *S. pombe* datasets suggest that the data-driven algorithm is potentially a powerful tool to decipher the causal pathways involved in complex biological systems. In this work, the focus has been on the theoretical deduction of the FLN's dynamic behavior and on the computational aspects of the inference algorithm. The theory of the FLN not only provides a sound theoretical guarantee for algorithmic parameter adjustments, but also is also a novel proposal for a new network model with potentially broad applications in modeling complex networks. From the computation results, the algorithm has provided detailed and insightful causal relationships among various genes. Thus, we believe that, given less noisy data, the FLN algorithm may be applied to a large range of biological systems having different spatial or temporal scales.

Regarding future research on the theoretical aspects of the FLN, we think that the dynamics and the steady-state properties of the FLN are important. Further effort should also focus on the effect of other distance metrics as means of comparing the performance of the modeling. It is also our belief that there is still room for improvement in computational complexity via heuristic search. Although the theory of the FLN is still in its infancy, particularly with respect to the details of network evolution, we think that the FLN, in the future, can model the real world uncertainty and mimic the behaviors of complex systems.

Table 10.3. Functionally verified regulations in the inferred *S. pombe* gene regulatory network. The regulations are grouped by regulators and the criteria of verifications are categorized as functional (verified by gene functions from GeneDB database, 49 regulations), close relationship (regulators and regulated genes are usually co-expressed or co-regulated, 9 regulations), and homolog (1 homolog). Some of the verifications are also based on the included references

Regulator	Regulated gene	Verification
ace2: regulator of cell separation genes	**mid2**: positioning of the division septum before the cell divides **SPBC1709.12**: Rho GTPase binding signaled by cell cycle	functional functional
bet1: controls intracellular protein transport and cell wall formation	**mid2**: positioning of the division septum before the cell divides **SPBC3E7.12c**: chitin biosynthesis **SPCC74.07c**: SUN family gene involved in cell separation	close relationship functional functional
bgl2: regulates cell expansion during growth and cell-cell fusion during mating	**SPBC1289.01c**: involves in septum formation	functional [38]
cdc13: controls the cell cycle at the G2/M (mitosis) transition	**ams2**: required for proper chromosome segregation **csx2**: involves in cell-cycle regulated ADP-ribosylation **cyp4**: peptidyl-prolyl cistrans isomerase involved in mitosis **hhf2**: histone H4 **meu29**: up-regulated in meiotic **SPBC1289.01c**: cell wall chitin protein **SPCC18.02**: involves hydrogen anti-porter activity in cell cycle	functional functional [39] functional [40] functional functional functional functional
cdc15: mediates cytoskeletal rearrangements required for cytokinesis	**mrc1**: mediator of replication checkpoint 1 **rpc17**: RNA polymerase **SPBC1709.12**: Rho GTPase binding signaled by cell cycle **SPBC3E7.12c**: chitin biosynthesis **spm1**: involves in cell separation	functional functional functional [41] functional functional
cdc2: controls the eukaryotic cell cycle	**SPBC119.10**: asparagine synthase involved in glutamine-hydrolyzing	functional
cdr1: mitotic inducer	**rpc17**: RNA polymerase **SPCC74.07c**: involve in beta-glucosidase activity at cell separation	functional functional
cid13: creates the 3' poly(A) tail of suc22 mRNA affecting DNA replication	**fin1**: Promotes chromosome condensation and nuclear envelope dynamics during mitosis	functional [42]

(*continued*)

Table 10.3 (Continued)

Regulator	Regulated gene	Verification
cig1: cyclin regulating G1/S transition	**rpc17**: RNA polymerase	functional
cig2: cyclin regulating G2/M transition	**meu29**: up-regulated in meiotic	functional
gmh2: affects Golgi membrane and chitin synthase	**SPBC3E7.12c**: chitin biosynthesis	functional [43]
mac1: required for cell separation	**ams2**: required for proper chromosome segregation	functional
mid2: positioning of the division septum before the cell divides	**spm1**: involves in cell separation	functional
nrd1: negative regulator of sexual differentiation affecting chitin biosynthesis	**SPBC1289.01c**: cell wall chitin protein	functional [44]
pim1: involves in the control of mitosis	**spm1**: involves in cell separation	functional
pof6: involves in cell division	**mid2**: positioning of the division septum before the cell divides	functional
psc3: required for normal mitosis	**csx2**: involves in cell-cycle regulated ADP-ribosylation	functional
rer1: COPI-coated vesicle	**SPBC119.10**: asparagine synthase	close relationship [45]
rpc17: RNA polymerase	**cyp4**: PPIases to accelerate the folding of proteins	close relationship [46]
SPAC1071.09c: contains a DnaJ domain which mediates interactions with histone-modifying heat shock proteins	**hhf3**: histone 4	functional
SPAC19B12.02c: high similarity to 1,3-beta-glucanosyl transferase	**cdc22**: provides the precursors necessary for DNA synthesis **mid2**: positioning of the division septum before the cell divides	close relationship [47] close relationship
SPAC1F7.03: involves in calcium transport that affects heat shock genes	**SPAC1071.09c**: interacts with heat shock proteins	functional [48]
SPAC23H4.19: putative cell wall biogenesis protein	**SPBC3E7.12c**: chitin biosynthesis	functional
SPAC24H6.01c: involved in phospholipid biosynthesis affecting cell division	**mid2**: positioning of the division septum before the cell divides **SPAC1071.09c**: interacts with heat shock proteins	functional [49] functional [50]

(*continued*)

Table 10.3 (Continued)

Regulator	Regulated gene	Verification
SPAC2E1P5.03: mediates interaction with heat shock proteins that interacts with histones	**hhf2**: histone 4 **hhf3**: histone 4	functional [51] functional [51]
SPAC323.07c: member of the MatE family of integral membrane proteins	**spm1**: involves in cell separation	close relationship
SPAC688.11: required for hyphal growth	**cyp4**: peptidyl-prolyl cis-trans isomerase involved in mitosis	close relationship [52]
SPBC16G5.15c: required for the correct timing, positioning and contraction of the division septum	**SPBC3E7.12c**: involves chitin biosynthesis	functional
SPBC4F6.05c: involves in sugar biding that affects histones	**hhf3**: histone 4	functional [53]
SPBC4F6.12: regulates integrin or growth factor-mediated responses	**cdc22**: provides the precursors necessary for DNA synthesis	functional
SPBPB2B2.09c: involves in thiamine biosynthesis	**SPCC18.02**: involves hydrogen anti-porter activity	close relationship [54]
SPCC548.06c: involves in glucose transport that affects chitin synthesis	**SPCC417.05c**: stimulates chitin synthase III activity	functional [55]
SPCC74.07c: involves in beta-glucosidase activity at cell separation	**csk1**: cell cycle kinase **meu29**: up-regulated in meiotic	functional functional
SPCC794.11c: involves in formation of clathrin coats at the Golgi and endosomes	**SPCC18B5.07c**: nuclear pore protein (nucleoporin)	homolog [56]
SPCP1E11.08: nuclear protein involved in ribosome biogenesis	**meu29**: up-regulated in meiotic	close relationship
spn2: septin involved in cell separation	**chs5**: involves in chitin synthesis and also required for mating	functional
spp2: DNA primase, large (non-catalytic) subunit	**cdc22**: provides the precursors necessary for DNA synthesis **trx1**: putative thioredoxin that affects DNA primase	functional functional
sst1: member of sodium or calcium exchanger protein family of membrane transporters	**gmh2**: affects Golgi membrane and chitin synthase	functional [57]
top1: DNA topoisomerase I, involved in chromatin organization	**SPCC757.12**: alpha-amylase with special chromatin structure	functional [58]
wis3: regulates cell cycle progression	**SPBC3E7.12c**: involves in chitin biosynthesis	functional

10.6 Acknowledgment

The authors want to thank the Pittsburgh Supercomputing Center for providing high-speed computer clusters for this work.

References

1. Strogatz S (2001) Exploring complex networks. Nature 410:268–276
2. Nachman I, Regev A, Friedman N (2004) Inferring quantitative models of regulatory networks from expression data. Bioinformatics 20:i248–i256
3. Matsuno H, Doi A, Nagasaki M, Miyano S (2000) Hybrid Petri net representation of gene regulatory network. Pacific Symposium on Biocomputing 5:341–352
4. TeichMann S, Babu M (2004) Gene regulatory network growth by duplication. Nature Genetics 36:492–496
5. Sprinzak D, Elowitz M (2005) Reconstruction of genetic circuits. Nature 438:443–448
6. Ong I, Page D, and Costa V (2006) Inferring Regulatory Networks from Time Series Expression Data and Relational Data via Inductive Logic Programming. In: Proc. 16th International Conference on Inductive Logic Programming, Spain
7. Kwon A, Hoos H, Ng R (2003) Inference of transcriptional regulation relationships from gene expression data. Bioinformatics 19:905–912
8. Joseph Z (2004) Analyzing time series gene expression data. Bioinformatics 20:2493–2503
9. Georgantas R, Tanadve V, Malehorn M, Heimfeld S, Chen C, Carr L, Martinez-Murillo F, Riggins G, Kowalski J, Civin C (2004) Microarray and serial analysis of gene expression analyses indentify known and novel transcripts overexpressed in hematopoietic stem cells. Cancer Research 64:4434–4441
10. Lee T, Rinaldi N, Robert F, Odom D, Joseph Z, Gerber G, Hannettt N, Harbinson CT, Thompson C.M, Simon I, Zeitlinger J, Jennings E, Murray H, Gordon DB, Ren B, Wyrick J, Tagne J, Volkert T, Fraenkel E, Gifford D, Young R (2002) Transcriptional regulatory networks in *Saccharomyces cerevisiae*. Science 298:799–804
11. Fung E, Thulasiraman V, Weinberger SR, Dalmasso E (2001) Protein biochips for differential profiling. Current Opinion in Biotechnology 12:65–69
12. Cao Y, Wang P, Tokuta A (2005) 3. In: Gene regulating network discovery. Volume 5 of Studies in Computational Intelligence. Springer-Verlag GmbH 49–78
13. Woolf P, Wang Y (2000) A fuzzy logic approach to analyzing gene expression data. Physiological Genomics 3:9–15
14. Cao Y, Wang P, Tokuta A (2006) *S. pombe* regulatory network construction using the fuzzy logic network. Poster, LSS Computational Systems Bioinformatics Conference, Stanford University
15. Hajek P (1998) Metamathematics of fuzzy logic. Volume 4 of Trends in Logic. Kluwer academic publishers, Boston
16. Reiter C (2002) Fuzzy automata and life. Complexity 3:19–29
17. Derrida B, Pomeau Y (1986) Random networks of automata: A simple annealed approximation. Europhysics letters 1:45–49
18. Sole R, Luque B, Kauffman S (2000) Phase transition in random network with multiple states. Technical report, Santa Fe Institute
19. Resconi G, Cao Y, Wang P (2006) Fuzzy biology. In: Proc. 5th International Symposium on Intelligent Manufacturing Systems, Sajarya, Turkey 29–31

20. Kauffman S, Peterson C, Samuelsson B, Troein C (2004) Genetic networks with canalyzing Boolean rules are always stable. PNAS 101:17102–17107
21. Cao Y (2006) Fuzzy Logic Network Theory with Applications to Gene Regulatory Networks. PhD thesis, Duke University
22. Spellman P, Sherlock G, Zhang M, Iyer V, Anders K, Eisen M, Brown P, Botstein D, Futcher B (1998) Comprehensive identification of cell cycle regulated genes of the yeast saccharomyces cerevisiae by microarray hybridization, molecular biology of the cell. Molecular Biology of the Cell 9:3273–3297
23. Rustici G, Mata J, Kivinen K, Lio P, Penkett C, Burns G, Hayles J, Brazma A, Nurse P, Bahler J (2004) Periodic gene expression program of the fission yeast cell cycle. Nature genetics 36:809–817
24. Heckman D, Geiser D, Eidell B, Stauffer R, Kardos N, Hedges S (2001) Molecular evidence for the early colonization of land by fungi and plants. Science 293:1129–1133
25. Forsburg S (1999) The best yeast. Trends in genetics 15:340–344
26. Cappellaro C, Baldermann C, Rachel R, Tanner W (1994) Mating type-specific cell-cell recognition of saccharomyces cerevisiae: cell wall attachment and active sites of a- and alpha-agglutinin. EMBO Journal 13:4737–4744
27. Lockhart S, Zhao R, Daniels K, Soll D (2003) Alpha-pheromone-induced "shmooing" and gene regulation require white-opaque switching during candida albicans mating. Eukaryotic cell 2:847–855
28. Stone E, Heun P, Laroche T, Pillus L, Gasser S (2000) Map kinase signaling induces nuclear reorganization in budding yeast. Current Biology 10:373–382
29. Erdman S, Lin L, Malczynski M, Snyder M (1998) Pheromone-regulated genes required for yeast mating differentiation. Journal of Cell Biology 140:461–483
30. Oehlen L, McKinney J, Cross F (1996) Ste12 and mcm1 regulate cell cycle-dependent transcription of far1. Molecular Cell Biology 16:2830–2837
31. Jin H, Carlile C, Nolan S, Grote E (2004) Prm1 prevents contact-dependent lysis of yeast mating pairs. Eukaryotic Cell 3:1664–1673
32. Wang K, Vavassori S, Schweizer L, Schweizer M (2004) Impaired prpp-synthesizing capacity compromises cell integrity signalling in saccharomyces cerevisiae. Microbiology 150:3327–3339
33. Milo R, Shen-Orr S, Itzkovitz S, Kashtan N, Chklovskii D, Alon U (2002) Network motifs: Simple building blocks of complex networks. Science 298:824–827
34. Shen-Orr S, Milo R, Mangan S, Alon U (2002) Network motifs in the transcriptional regulation network of Escherichia coli. Nature Genetics 31:64–68
35. Thieffry D, Romero D (1999) The modularity of biological regulatory networks. Biosystems 50:49–59
36. Milo R, Itzkovitz S, Kashtan N, Levitt R, Shen-Orr S, Ayzenshtat I, Sheffer M, Alon U (2004) Superfamilies of evolved and designed networks. Science 303:1538–1542
37. Thieffry D, Huerta A, Perez-Rueda E, Collado-Vides J (1998) From specific gene regulation to genomic networks: A global analysis of transcriptional regulation in Escherichia coli. BioEssays 20:433–440
38. Kapteyn J, Montijn R, Vink E, de la Cruz J, Llobell A, Douwes J, Shimoi H, Lipke P, Klis F (1996) Retention of *Saccharomyces cerevisiae* cell wall proteins through a phosphodiester-linked beta-1,3-/beta-1,6-glucan heteropolymer. Glycobiology 6:337–345
39. Jacobson E, Meadows R, Measel J (1985) Cell cycle perturbations following DNA damage in the presence of ADP-ribosylation inhibitors. Carcinogenesis 6:711–714

40. Morris D, Phatnani H, Greenleaf A (1999) Phospho-carboxyl-terminal domain binding and the role of a prolyl isomerase in pre-mRNA 3'-end formation. Journal of Biological Chemistry 274:31583–31587
41. Richnau N, Aspenstrom P (2001) RICH, a rho GTPase-activating protein domain-containing protein involved in signaling by cdc42 and rac1. Journal of Biological Chemistry 276:35060–35070
42. Read R, Martinho R, Wang S, Carr A, Norbury C (2002) Cytoplasmic poly(A) polymerases mediate cellular responses to S phase arrest. PNAS 99:12079–12084
43. Trautwein M, Schindler C, Gauss R, Dengjel J (2006) Arf1p, Chs5p and the CHAPs are required for export of specialized cargo from the Golgi. the EMBO Journal 25:943–954
44. Lee J, Choi J, Park B, Park Y, Lee M, Park H, Maeng P (2004) Differential expression of the chitin synthase genes of Aspergillus nidulans, chsA, chsB, and chsC, in response to developmental status and environmental factors. Fungal Genetics and Biology 41:635–646
45. Moussalli M, Pipe S, Hauri H, Nichols W, Ginsburg D, Kauffman R (1999) Mannose-dependent endoplasmic reticulum (ER)-golgi intermediate compartment-53-mediated ER to Golgi trafficking of coagulation factors V and VIII. Journal of Biological Chemistry 274:32539–32542
46. Chao S, Greenleaf A, Price D (2001) Juglone, an inhibitor of the peptidyl-prolyl isomerase pin1, also directly blocks transcription. Nucleic Acids Research 29:767–773
47. Mouyna I, Hartland R, Fontaine T, Diaquin M, Simenel C, Delepierre M, Henrissat B, Latge J (1998) A 1,3-beta-glucanosyltransferase isolated from the cell wall of Aspergillus fumigatus is a homologue of the yeast Bgl2p. Microbiology 144:3171–3180
48. Mosser D, Kotzbauer P, Sarge K, Morimoto R (1990) In vitro activation of heat shock transcription factor DNA-binding by calcium and biochemical conditions that affect protein conformation. PNAS 87:3748–3752
49. Pierucci O, Rickert M (1985) Duplication of Escherichia coli during inhibition of net phospholipid synthesis. Journal of Bacteriology 162:374–382
50. Glonek J, Lipinska B, Krzewski K, Zolese G, Bertoli E, Tanfani F (1997) HtrA heat shock protease interacts with phospholipid membranes and undergoes conformational changes. Journal of Biolgocial chemistry 272:8974–8982
51. Solomon M, Larsen P, Varshavsky A (1988) Mapping protein-DNA interactions in vivo with formaldehyde: evidence that histone H4 is retained on a highly transcribed gene. Cell 53:937–947
52. Derkx P, Madrid S (2001) The Aspergillus niger cypA gene encodes a cyclophilin that mediates sensitivity to the immunosuppressant cyclosporin A. Molecular Genetics and Genomics 266:527–536
53. Riou-Khamlichi C, Menges M, Healy J, Murray J (2000) Sugar control of the plant cell cycle: Differential regulation of Arabidopsis D-type cyclin gene expression. Molecular and Cellular Biology 20:4513–4521
54. Gastaldi G, Cova E, Verri A, Laforenza U, Faelli A, Rindi G (2000) Transport of thiamin in rat renal brush border membrane vesicles. Kidney International 57:2043–2054
55. Bulik D, Olczak M, Lucero H, Osmond B, Robins P, Specht C (2003) Chitin synthesis in *Saccharomyces cerevisiae* in response to supplementation of growth medium with glucosamine and cell wall stress. Eukaryotic Cell 2:886–900
56. Jekely G, Arendt D (2006) Evolution of intraflagellar transport from coated vesicles and autogenous origin of the eukaryotic cilium. BioEssays 28:191–198
57. Corven E, Os C, Mircheff A (1986) Subcellular distribution of ATP-dependent calcium transport in rat duodenal epithelium. Biochimica et biophysica acta 861:267–276
58. Levy-Wilson B (1983) Chromatin structure and expression of amylase genes in rat pituitary tumor cells. DNA 2:9–13

11

Multivariate Regression Applied to Gene Expression Dynamics

Olli Haavisto and Heikki Hyötyniemi

Helsinki University of Technology, Control Engineering Laboratory, P.O.Box 5500, FI-02015 TKK, Finland
olli.haavisto@tkk.fi, heikki.hyotyniemi@tkk.fi

Summary. Linear multivariate regression tools developed on the basis of the traditional statistical theory are naturally suitable for high-dimensional data analysis. In this article, these methods are applied to microarray gene expression data. At first, a short introduction to dimension reduction techniques in both static and dynamic cases is given. After that, two examples, yeast cell response to environmental changes and expression during the cell cycle, are used to demonstrate the presented subspace identification method for data-based modeling of genome dynamics. The results show that the method is able to capture the relevant, higher level dynamical properties of the whole genome and can thus provide useful tools for intelligent data analysis. Especially the simplicity of the model structures leads to an easy interpretation of the obtained results.

11.1 Introduction

The gene expression network of a biological cell can be interpreted as a dynamical complex system, which consists of a large number of variables (gene activities) and connections between them. Because of the large dimension of the system the precise analysis of the structure and functioning of the network is a difficult problem. Microarray measuring technology developed during the last decades has, however, increased the possibilities to analyse genetic networks by enabling the use of data-based approaches and utilization of the increasing computational capacity.

A vast range of different methods like artificial neural networks, support vector machines and fuzzy systems have been applied to gene expression data modeling (see e.g. [2, 13, 15]). However, in addition to these soft computing methods particularly designed for high dimensional data mining, also the more traditional statistical multiregression tools can provide means to analyse gene expression data sets. One advantage of these methods is that the obtained models typically remain quite simple; usually linearity is assumed, which allows the precise analysis of the models as opposed to the more complex neural networks.

This chapter proposes the usage of a linear state space model to describe the overall gene expression dynamics of yeast *Saccharomyces cerevisiae*. The aim of

O. Haavisto and H. Hyötyniemi: *Multivariate Regression Applied to Gene Expression Dynamics*, Studies in Computational Intelligence (SCI) **94**, 257–275 (2008)
www.springerlink.com

the modeling differs from the traditional approaches where the idea often is to form a connection network for a small subset of genes [16, 17]. Instead, the goal here is to include all the genes of the organism in the analysis and to capture the main *dynamical* behaviour of the genome. It is assumed here that the important activity variations of the genome are highly redundant and can thus be compressed by using a low dimensional latent variable vector. This assumption also makes biologically sense instead of being just a mathematical trick to simplify the modeling. The obtained dynamical models can be applied to the simulation and prediction of the gene expression values.

The presented dimension reduction methods are demonstrated by two examples: response of yeast cells to environmental changes and yeast gene activation during cell cycle. In the first case, the changes in the activation levels of yeast genes are modeled when a sudden environmental change is applied to the cultivation media. Since the yeast cells are disturbed in these time series, they are in the following referred to as yeast stress experiments. Some of these results are initially published in [8]. The second example concentrates on the natural cyclic alternation of a group of yeast genes during the cell cycle.

11.2 Approach

This section discusses the principles of the modeling approach further elaborated in this study. First, some justification for the selection of a linear model structure is given. Then, the degrees of freedom based data analysis method is presented.

11.2.1 Linearity

In the case of a biological cell, the linearity of the model structures can be argued by the properties of the cell. A cell can be assumed to have a high number of internal state variables (i.e. gene activation levels and chemical concentrations) which are highly connected to each other. It is known that the regulatory networks in a cell are powerful buffers for external disturbances. This means, that the feedback connections between the state variables are capable of keeping the vital processes going in different environments, thus enabling the cell to survive in changing external conditions.

Under these assumptions, a nice analogy between an elastic mechanical object and a biological cell can be found [7]. In the case of an external disturbance, whether it is an external force applied to an elastic object or a change in the environmental conditions of a cell, the elastic system can adapt to the new situation by changing its internal state. Provided that the disturbance is small enough, after a transient the system can find a new balance and stay unbroken (or alive). When the disturbance is removed, the system recovers to the original internal state.

If the biological cell is seen as an elastic system, a certain level of smoothness in its behaviour can be assumed. This, on the other hand, implies that the system is *locally linearizable*, i.e. around a fixed nominal operating point, the minor changes can be modeled using a linear model structure.

11.2.2 Degrees of Freedom

When dealing with dynamical systems containing a large number of variables, the traditional modeling approaches have been based on the *constraints* which determine the relations of the system variables. For example, *differential equations* are a typical choice when modeling the dynamics: Each individual equation covers one relation or constraint between the investigated variables. To unambiguously describe the whole system, as many differential equations are required as there are system variables. However, even in the case of a "simple" biological cell like yeast, the gene regulation network includes a huge number of genes (over 6000) which can be connected to each other. Even though the gene regulation networks are typically assumed to be sparse instead of being completely connected, this means that a large number of constraints is required to define the system dynamics. When the system dimension is high, this approach is not feasible anymore.

The opposite way to analyse a multivariate system is to collect data from the system and use them to find out the main directions of variation or *degrees of freedom* present in the system. If the system variables are highly connected and the dynamics are restricted by many constraints, the actual number of meaningful degrees of freedom remains low. Accordingly, it may turn out that the dynamics can be described with a much lower number of variables. That is because each individual constraint reduces the number of possible variation directions by one, thus hopefully resulting to a low dimensional space of degrees of freedom [12].

One does not need to know the exact outlook of the constraints. As long as these underlying structures keep the system in balance and maintain its integrity, they can even be ignored. From the system theoretical point of view this is a relief: one does not need to tackle with the problems characteristic to closed-loop identification (see e.g. Sect. 13.4 in [14]).

The main assumption in this study is that the gene expression network in such a case actually has few possible degrees of freedom, that is, all the relevant variable combinations can be described by only a small number of latent variables. This enables the use of a low-dimensional state vector to comprise the core of the system dynamics, whereas the actual gene expression values are produced as a linear combination of these state variables or "functional modes". It has already been shown in a few publications that this assumption is quite well justified; the latent variables can be calculated for example using principal component analysis (or singular value decomposition) [9, 18, 22].

11.3 Multivariate Regression Tools

There exist various data analysis tools for static data compression and multivariate regression [11], and the principles utilized in these methods can further be extended to cover also dynamical systems.

11.3.1 Principal Component Analysis

Principal component analysis (PCA) (see e.g. [1]) is the standard multivariate method to analyse high dimensional static data with correlating variables. The idea of PCA is to find the orthogonal directions of highest variance in the data space. These directions are called the principal component directions, and the most important of them can be used as a lower dimensional latent basis to which the original data is mapped. Because of the dimensionality reduction in the mapping, data compression takes place. Assuming that variance in the data carries information, the PCA compression is optimal in the sense of preserving this information.

Let us assume that for example microarray time series data are collected in matrix $\mathbf{Y} \in \mathbb{R}^{l \times N}$, where each column represents one time point and each row an individual gene. It is possible to utilize PCA to the data in two ways; either to reduce the number of genes or the number of time points [18]. If the purpose is to reduce the number of genes, one should interpret them as variables, whereas different time series represent the available samples. In that case, PCA mapping for a (column) data vector $\mathbf{y}(k) \in \mathbb{R}^{l \times 1}$ is defined as

$$\mathbf{y}(k) = \mathbf{W}\mathbf{x}(k) + \mathbf{e}(k), \tag{11.1}$$

where $\mathbf{x}(k) \in \mathbb{R}^{n \times 1}$ is the k:th n-dimensional *score vector* and $\mathbf{W} \in \mathbb{R}^{l \times n}$ contains the principal component vectors as columns. If the amount n of latent variables in x is smaller than the original data dimension l, due to the dimension reduction the original data cannot be exactly reproduced, and the error term $\mathbf{e}(k)$ differs from zero. To conveniently map all the data, (11.1) can be presented in the matrix form:

$$\mathbf{Y} = \mathbf{W}\mathbf{X} + \mathbf{E}, \tag{11.2}$$

where $\mathbf{X} = [\mathbf{x}(1), \mathbf{x}(2), \ldots, \mathbf{x}(N)] \in \mathbb{R}^{n \times N}$ contains the score vectors for all samples, and $\mathbf{E} = [\mathbf{e}(1), \mathbf{e}(2), \ldots, \mathbf{e}(N)] \in \mathbb{R}^{l \times N}$.

Calculation of PCA can be done by utilizing the sample correlation matrix estimated from the data:

$$\hat{\mathbf{R}}_{\mathbf{yy}} = \frac{1}{N}\mathbf{Y}\mathbf{Y}^T. \tag{11.3}$$

The eigenvalue decomposition of $\hat{\mathbf{R}}_{\mathbf{yy}}$ is

$$\hat{\mathbf{R}}_{\mathbf{yy}} = \mathbf{\Phi}\mathbf{\Lambda}\mathbf{\Phi}^T, \tag{11.4}$$

where $\mathbf{\Phi} = [\phi_1, \phi_2, \ldots, \phi_l]$ contains the eigenvectors as columns and $\mathbf{\Lambda}$ is a diagonal matrix with eigenvalues λ_j on the diagonal. It is convenient to assume that the eigenvectors are sorted according to the corresponding eigenvalues starting from the largest one. Now the eigenvectors are directly all the principal component vectors of the data, whereas the eigenvalues represent the variance captured by each principal component. Thus, by combining the n eigenvectors corresponding to the n largest eigenvalues, the principal component mapping can be defined:

$$\mathbf{W} = [\phi_1, \phi_2, \ldots, \phi_n]. \tag{11.5}$$

The idea of PCA is to collect all the relevant correlations of the data in the latent variables, whereas the unimportant variations (hopefully mainly noise) are left to the error term **E**. This way the original high-dimensional dependencies in the data can be coded as efficiently as possible. Assuming that the gene expression system that produced the data in **Y** is redundant, the principal components in **W** can be interpreted as characteristic modes or "eigengenes", which together can explain most of the variations in the original data. The score values in **X** now contain the activation levels of these eigengenes as a function of time, and the weights in **W** describe how much each eigengene participates in explaining each of the original genes. This scheme has been investigated for example in [10], where it is shown that already a couple of main characteristic modes, i.e. principal components, can reproduce the data quite accurately.

Since the case of gene expression modeling is clearly a dynamic problem and time series data are available, the dynamics have to be taken into account in the modeling. One attempt to this direction was taken in [9], where a time translation matrix for the captured characteristic modes was calculated. However, the approach was rather elementary and was applicable only to a fixed number of eigengenes.

11.3.2 Subspace Identification

A quite recent extension of dimension reduction techniques to cover dynamical problems, *subspace identification* (SSI) [21], is demonstrated here. Subspace identification provides natural means to analyse multivariate dynamical systems either as deterministic-stochastic systems with separate (multidimensional) input signal or as purely stochastic systems, where the dynamics are driven by white noise signals.

The dynamics of the system are now assumed to be captured by the standard linear state space structure

$$\begin{cases} \mathbf{x}(k+1) = \mathbf{A}\mathbf{x}(k) + \mathbf{B}\mathbf{u}(k) + \mathbf{w}(k) \\ \mathbf{y}(k) = \mathbf{C}\mathbf{x}(k) + \mathbf{D}\mathbf{u}(k) + \mathbf{v}(k), \end{cases} \tag{11.6}$$

where the input signal for the system at the discrete time instant k is denoted as $\mathbf{u}(k) \in \mathbb{R}^m$ and the output of the system is $\mathbf{y}(k) \in \mathbb{R}^l$. The core of the dynamical behaviour is captured by the state variables in vector $\mathbf{x}(k) \in \mathbb{R}^n$, whereas $\mathbf{w}(k)$ and $\mathbf{v}(k)$ are assumed to be white noise signals of the appropriate dimension. In the purely stochastic case the input signal terms of the both equations are omitted. Based on a given set of time series samples of output $\{\mathbf{y}(1), \mathbf{y}(2), \ldots, \mathbf{y}(N)\}$ (and possibly of input $\{\mathbf{u}(1), \mathbf{u}(2), \ldots, \mathbf{u}(N)\}$ in the deterministic-stochastic case) collected from the system in question, the aim of the modeling is to determine the constant parameter matrices **A** and **C** (and **B** and **D**) so that the stochastic properties of system (11.6) as well as possible correspond to the stochastic properties of the given data.

The starting point of SSI is to interpret the original dynamic time series data in a static form. To obtain this, i consecutive time series data points are combined to one high-dimensional "static" data point:

$$\begin{pmatrix} \mathbf{y}(0) \\ \mathbf{y}(1) \\ \vdots \\ \mathbf{y}(i-1) \end{pmatrix}. \tag{11.7}$$

It is assumed that the length i of this data window is selected so that the static data point captures the dynamical behaviour of the system between samples $\mathbf{y}(0)$ and $\mathbf{y}(i-1)$. All possible static data points are further collected in the columns of a *block Hankel matrix*. For example, the block Hankel matrix for the output data is defined as:

$$\mathbf{Y} = \left(\frac{\mathbf{Y}_p}{\mathbf{Y}_f}\right) = \left(\begin{array}{cccc} \mathbf{y}(0) & \mathbf{y}(1) & \dots & \mathbf{y}(j-1) \\ \mathbf{y}(1) & \mathbf{y}(2) & \dots & \mathbf{y}(j) \\ \vdots & \vdots & \ddots & \vdots \\ \mathbf{y}(i-1) & \mathbf{y}(i) & \dots & \mathbf{y}(i+j-2) \\ \hline \mathbf{y}(i) & \mathbf{y}(i+1) & \dots & \mathbf{y}(i+j-1) \\ \mathbf{y}(i+1) & \mathbf{y}(i+2) & \dots & \mathbf{y}(i+j) \\ \vdots & \vdots & \ddots & \vdots \\ \mathbf{y}(2i-1) & \mathbf{y}(2i) & \dots & \mathbf{y}(2i+j-2) \end{array}\right). \tag{11.8}$$

Each column of the upper part ($\mathbf{Y}_p$) of the matrix is one static data point, formed by piling i consecutive output data points (column vectors) one on the other. These data are referred as the "past" static data points whereas the static data points in the lower part of the matrix ($\mathbf{Y}_f$) form the set of "future" data points. The number of columns, j, is the number of these new static data point pairs. Given that we have N original data points $\mathbf{y}(k)$, that is, $k = 0, 1, \dots, N-1$, the maximum number of columns is $j = N - 2i + 1$.

For the input data, the block Hankel matrix $\mathbf{U}$ containing the submatrices $\mathbf{U}_p$ and $\mathbf{U}_f$ can be defined accordingly.

Stochastic SSI Algorithm

The new data structures $\mathbf{Y}_p, \mathbf{Y}_f, \mathbf{U}_p$ and $\mathbf{U}_f$ can now be utilized for the actual model calculation. Let us first consider the purely stochastic estimation, where no deterministic input signal $\mathbf{u}(k)$ is allowed. The common approach to all SSI algorithms is that at first, estimate for the sequence of states $\hat{\mathbf{X}}_i = [\hat{\mathbf{x}}^T(i), \hat{\mathbf{x}}^T(i+1), \dots, \hat{\mathbf{x}}^T(i+j-1)]^T$ is calculated. From that, the actual system matrices can be estimated.

As the name subspace identification suggests, the key part of the algorithm deals with spaces spanned by the row vectors of the data matrices. In the stochastic case, the main result given and proved in [21] (Sect. 13.3) states that the row vectors of the orthogonal projection

$$\mathscr{O}_i = \mathbf{Y}_f / \mathbf{Y}_p = \mathbf{Y}_p^T \left(\mathbf{Y}_p \mathbf{Y}_p^T\right)^{\dagger} \mathbf{Y}_p \mathbf{Y}_f \tag{11.9}$$

span the same space as the rows of $\hat{\mathbf{X}}_i$. In (11.9) the row space of the future outputs $\mathbf{Y}_f$ is actually projected to the row space of the past outputs $\mathbf{Y}_p$ (the superscript $(\cdot)^{\dagger}$ denotes the Moore-Penrose pseudoinverse, i.e. $\mathbf{Z}^{\dagger} = \left(\mathbf{Z}\mathbf{Z}^T\right)^{-1}\mathbf{Z}$). By utilizing singular value decomposition (SVD) to the projection $\mathscr{O}_i$, the estimate $\hat{\mathbf{X}}_i$ is obtained.

When the line between past and future in (11.8) is shifted one step towards the future, two additional matrices can be defined:

$$\mathbf{Y} = \begin{pmatrix} \mathbf{Y}_p^+ \\ \hline \mathbf{Y}_f^- \end{pmatrix} = \left(\begin{array}{cccc} \mathbf{y}(0) & \mathbf{y}(1) & \dots & \mathbf{y}(j-1) \\ \mathbf{y}(1) & \mathbf{y}(2) & \dots & \mathbf{y}(j) \\ \vdots & \vdots & \ddots & \vdots \\ \mathbf{y}(i-1) & \mathbf{y}(i) & \dots & \mathbf{y}(i+j-2) \\ \mathbf{y}(i) & \mathbf{y}(i+1) & \dots & \mathbf{y}(i+j-1) \\ \hline \mathbf{y}(i+1) & \mathbf{y}(i+2) & \dots & \mathbf{y}(i+j) \\ \vdots & \vdots & \ddots & \vdots \\ \mathbf{y}(2i-1) & \mathbf{y}(2i) & \dots & \mathbf{y}(2i+j-2) \end{array}\right). \tag{11.10}$$

If the same projection as (11.9) of these new matrices, $\mathbf{Y}_p^+$ and $\mathbf{Y}_f^-$ is calculated, the "next" state values $\hat{\mathbf{X}}_{i+1}$ are obtained using again the singular value decomposition. According to the original state space model (11.6) with no inputs, it now holds

$$\begin{pmatrix} \hat{\mathbf{X}}_{i+1} \\ \mathbf{Y}_i \end{pmatrix} = \begin{pmatrix} \mathbf{A} \\ \mathbf{C} \end{pmatrix} \left(\hat{\mathbf{X}}_i \right) + \begin{pmatrix} \rho_w \\ \rho_v \end{pmatrix}, \tag{11.11}$$

where $\mathbf{Y}_i$ is the ith column of $\mathbf{Y}$ and the last term contains the residuals ρ_w and ρ_v. Assuming that the residuals are uncorrelated with the state sequence, a least squares solution can be used to obtain the matrices $\mathbf{A}$ and $\mathbf{C}$:

$$\begin{pmatrix} \mathbf{A} \\ \mathbf{C} \end{pmatrix} = \begin{pmatrix} \hat{\mathbf{X}}_{i+1} \\ \mathbf{Y}_i \end{pmatrix} \left(\hat{\mathbf{X}}_i^\dagger \right). \tag{11.12}$$

Combined Deterministic-Stochastic SSI Algorithm

In order to include the external, deterministic input signal $\mathbf{u}(k)$ in the state space model, it is convenient to combine all the information of the "past" in one matrix:

$$\mathbf{W}_p = \begin{pmatrix} \mathbf{U}_p \\ \mathbf{Y}_p \end{pmatrix}. \tag{11.13}$$

Additionally, the "past" and "current" information is combined by defining matrix

$$\mathbf{W}_p^+ = \begin{pmatrix} \mathbf{U}_p^+ \\ \mathbf{Y}_p^+ \end{pmatrix}, \tag{11.14}$$

where $\mathbf{U}_p^+$ is of the similar form as $\mathbf{Y}_p^+$ in (11.10).

Now the projection $\mathscr{O}_i$ must include also the input variables. To obtain a similar result as in the stochastic case, an *oblique projection* must be used:

$$\mathscr{O}_i = \mathbf{Y}_f /_{\mathbf{U}_f} \mathbf{W}_p = (\mathbf{Y}_f / \mathbf{U}_f^\perp)(\mathbf{W}_p / \mathbf{U}_f^\perp)^\dagger \mathbf{C}, \tag{11.15}$$

where $\mathbf{U}_f^\perp$ is the orthogonal complement of $\mathbf{U}_f$ (i.e. the row space of $\mathbf{U}_f^\perp$ is orthogonal to the row space of $\mathbf{U}_f$ and together the rows of $\mathbf{U}_f^\perp$ and $\mathbf{U}_f$ span the whole space). This operation actually projects the future outputs $\mathbf{Y}_f$ *along* the future inputs $\mathbf{U}_f$ to the row space of the past information $\mathbf{W}_p$. It can be shown (see [21], Sect. 13.4) that the row space of the projection $\mathscr{O}_i$ equals the row space of $\tilde{\mathbf{X}}_i$, which is an estimate of the state sequence of the deterministic-stochastic state space model. By utilizing SVD, it is further possible to extract $\tilde{\mathbf{X}}_i$ from the projection. Identically with the previous case, also the "next" state values $\tilde{\mathbf{X}}_i$ are calculated from the projection

$$\mathscr{O}_{i+1} = \mathbf{Y}_f^- /_{\mathbf{U}_f^-} \mathbf{W}_p^+ . \tag{11.16}$$

When the state sequences are known, it is a simple task to solve the parameter matrices $\mathbf{A}, \mathbf{B}, \mathbf{C}$ and $\mathbf{D}$ from the equation

$$\begin{pmatrix} \tilde{\mathbf{X}}_{i+1} \\ \mathbf{Y}_i \end{pmatrix} = \begin{pmatrix} \mathbf{A} & \mathbf{B} \\ \mathbf{C} & \mathbf{D} \end{pmatrix} \begin{pmatrix} \tilde{\mathbf{X}}_i \\ \mathbf{U}_i \end{pmatrix} + \begin{pmatrix} \rho_w \\ \rho_v \end{pmatrix} . \tag{11.17}$$

Matrices $\mathbf{U}_i$ and $\mathbf{Y}_i$ consist of the i:th block rows of matrices $\mathbf{U}$ and $\mathbf{Y}$, respectively, thus containing once all the original data points. If the residuals ρ_w and ρ_v are assumed to be uncorrelated with the state sequence and the input data, the least squares solution is again a natural choice to solve the combined parameter matrix:

$$\begin{pmatrix} \mathbf{A} & \mathbf{B} \\ \mathbf{C} & \mathbf{D} \end{pmatrix} = \begin{pmatrix} \tilde{\mathbf{X}}_{i+1} \\ \mathbf{Y}_i \end{pmatrix} \begin{pmatrix} \tilde{\mathbf{X}}_i \\ \mathbf{U}_i \end{pmatrix}^{\dagger} . \tag{11.18}$$

Extensions and Modifications

In the data-based analysis of gene expression networks, an additional difficulty is introduced by the low amount of available time series data points with respect to the system dimension. In order to be able to apply the SSI algorithms to this kind of data, some changes to the implementation of the algorithms had to be made when compared with the standard algorithms provided in [21]. Basically, the faster calculation of the model by using the QR decomposition had to be replaced by the actual matrix operations to obtain the defined matrix projections.

It is normally assumed that all the N data points are from a single, long and continuous time series. However, because of the nature of microarray measurements, this time a group of relative short time series formed the complete data set. A separate model was estimated for yeast stress behaviour and cell cycle, but inside these groups, each time series described the same dynamical system. To be able to use all the data points efficiently in the identification, two additional operations were performed for the data: 1) *padding* of the short time series in the stress case, and 2) combination of the static data points from different time series.

In padding each short time series (jth series containing originally N_j data points) was extrapolated using the assumptions of the system steady states: Assuming that *before* the environmental change the gene expression values remain constant, we can say that for all negative time indices the values of the output are equal to the time zero value, i.e. $\mathbf{y}(k) = \mathbf{y}(0)$, when $k < 0$. Correspondingly, if the system has reached the final state at the end of the time series ($k = N_j - 1$), gene expression remains constant *after* that, i.e. $\mathbf{y}(k) = \mathbf{y}(N_j - 1)$, when $k > N_j - 1$. This way it was possible to include in the identification also the static data points which contain some original data points outside of the range $k = 0, 1, ..., N_j - 1$ and thus increase the number of static data points.

11.4 Examples

The aim of the following examples is to demonstrate the usage of multilinear dimension reduction techniques and especially subspace identification for dynamical gene expression data analysis. As opposed to typical methods to model gene expression dynamics, SSI is able to include a very large number of genes in the modeling.

11.4.1 Data

The stress response of the wild type yeast *Saccharomyces cerevisiae* was modeled using in total 20 time series selected from the two public data sets [3, 5]. In each series, the environmental conditions of the yeast cultivation were changed at the beginning of the experiment, and microarray measurements of the resulting gene activities were performed. Before modeling, the data were normalized and linearly resampled to a constant sampling interval. Additionally, the missing data points were estimated using the k nearest neighbor method, as suggested in [20]. The genes with more than 12% of data points missing were excluded from the further analysis, thus leaving 4176 genes in the final data set. The included stress experiments are listed in the upper part of Table 11.1. Since several experiments were conducted using the same stress factors, only ten individual environmental conditions could be separated from the data. These are listed in Table 11.2.

Cell cycle microarray data for the second study were obtained from the publication by Spellman et al. [19], which contains four cell cycle data sets measured using the cDNA technology. In each data set, a different synchronization method is used to arrest the yeast cells. These methods are α-factor based synchronization, size-based synchronization (or elutriation), and two methods based on modified yeast strains (cdc15 and cdc28). Originally, the cdc28 experiment was published by Cho et al. [4]. The data were also normalized and missing values were estimated as above.

Originally, the alpha, cdc28, and elutriation data sets each contained data with constant sampling interval, whereas the cdc15 data set had five missing timepoints but otherwise a constant sample time. These values were filled using linear interpolation separately for each ORF. The sampling interval was, however, varying between the time series, and also the cell cycle length in the experiments was different. For our modeling purposes, data points with a constant and common sampling interval were required. Additionally, in order to make the experiments comparable, the cell cycle lengths should have been equal. To fulfill these conditions, we first estimated the cell

Table 11.1. Time series experiments included in the study and the number of samples in each series after data preprocessing

Type	Publication	Experiment	Data points
Stress	Gasch et al.	Heat (hs-1)	6
		Heat (hs-2)	5
		Heat 37 → 25°C	7
		Heat 29 → 33°C	3
		Heat 29 → 33°C (in sorbitol)	3
		Heat 29 → 33°C (sorbitol 1 M → 0 M)	3
		H2O2 treatment	10
		Menadione exposure	10
		DTT exposure (dtt-1)	10
		Diamide treatment	7
		Hyper-osmotic shock (Sorbitol)	9
		Hypo-osmotic shock (Sorbitol)	5
		Amino acid starvation	10
		Nitrogen depletion	10
	Causton et al.	Heat shock	9
		Acid	7
		Alkali	7
		H2O2 treatment	9
		Salt	9
		Sorbitol	9
Cell cycle	Spellman et al.	α-factor	18
		Elutriation	8
		cdc15	22
	Cho et al.	cdc28	17

Table 11.2. Stress factors (environmental variables) in the yeast stress example

Number	Quantity
1.	Temperature
2.	pH
3.	H_2O_2 concentration
4.	Menadione bisulfate concentration
5.	Diamide concentration
6.	Sorbitol concentration
7.	Amino acids concentration
8.	Ammonium sulfate concentration
9.	Dithiothrietol (DTT) concentration
10.	Sodium chlorine concentration

cycle length in each series using discrete Fourier transformation for all the cell-cycle related genes detected by Spellman *et al.* [19]. The power spectra of these genes in a time series were averaged and the frequency corresponding to the maximum of the averaged spectrum was interpreted as the approximative cell cycle frequency in that experiment. This lead to the following cell cycle lengths for the experiments: 63 min (α-factor), 116 min (cdc15), 88 min (cdc28) and 465 min (elutriation). Using these values and the known original sampling intervals it was possible to scale the time values of the experiments so that the cell cycle lengths become equal. After that, the data were linearly interpolated to have equal sample time in each experiment. The new sampling interval was chosen to match with the original sample time of the α-factor series.

Table 11.1 summarizes the experiments used in both examples and the number of data points after the preprocessing. The stress data contained altogether 148 time points, whereas the total length of the cell cycle data was 65 time points.

11.4.2 Results

In the following, the modeling of both examples, yeast stress experiments and cell cycle, are explained and the obtained results are presented.

Stress Response

The idea of modeling the stress experiments was to estimate a model for the dynamics of the yeast genome after a sudden environmental change. Since there existed a well-defined input signal for the system, a deterministic-stochastic state space model (11.6) was suitable for this modeling case. The stress factors (see Table 11.2) in the experiments were treated as components of the input signal vector $\mathbf{u}(k)$ for the model. Additionally, it is known that the yeast genome contains general stress genes, i.e. genes whose activation is increased during almost any environmental change. Since the model structure is linear, also the squares of the input factors were included in the input vector so that it could be possible to model these genes properly. As a result, the input dimension of the model was twenty. The gene expression values of the individual genes were collected as the elements of the output vector $\mathbf{y}(k)$. After the data preprocessing stage, the dimension of the output vector (number of genes in the model) was 4176.

Stochastic-deterministic SSI with padding was applied to the data as explained in Sect. 11.3. It was assumed that five consecutive samples were enough to capture the dynamics of the genome, so each static data point contained five original samples (i.e. $i = 5$). This assumption was partly due to the restriction introduced by the length of the shortest time series in the data set. By investigating the singular values of the oblique projection, it was detected that a suitable system dimension n was four. This is in line with the results in [10, 18]. As a result of the identification, the parameter matrices of the dynamical state space model for yeast gene expression in stress situations were obtained.

Once the model parameters were known, the model could be used to simulate the gene expression behaviour of a yeast cultivation when a change in one (or several) of the input variables takes place. Because of the lack of proper and independent validation data, the model was evaluated by simulating the same stress experiments

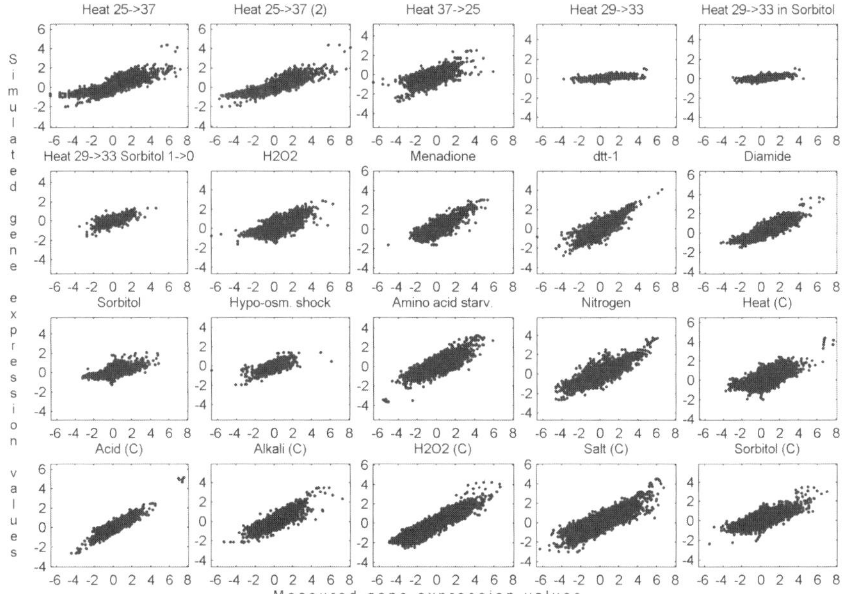

Fig. 11.1. The scatter plots of the measured and estimated stress experiment values (original figure published in [8])

and comparing the simulated and measured gene expressions. The correlation coefficients between these values in the altogether twenty experiment time series were ranging from 0.45 to 0.90 with the mean of 0.75, thus indicating that the model was able to reproduce the experiments quite well. Figure 11.1 shows the scatter plots of the measured and simulated gene expression values for each series. Especially the experiments with high activation values are modeled with more accuracy.

The model performance can also be analyzed by visualizing the measured and simulated gene expression values individually for each experiment. As an example, Fig. 11.2 shows the measured and simulated responses of a group of yeast stress genes when dithiothrietol (dtt) was added to the cultivation medium. The genes are selected to be stress-related according to the GO-slim annotation. The number of genes is restricted only because of visualization reasons; the model was able to simulate the expression of all the genes included in the model. Clearly the simulated response is well reproducing the measured values, even though the number of latent variables is limited to four. Especially the general and slower expression changes are accurately modeled, whereas the more rapid alternations are filtered. This is due to the strong dimension reduction in the modeling phase.

Because of the low number of state variables in the model, they can be analysed individually for each experiment. In Fig. 11.3, the state sequences for two experiments, addition of dtt and alkali, are shown. It seems that while dtt is affecting the gene activation gradually and permanently, the pH shock rapidly activates (or deactivates) the related genes. As the cells adapt to the new pH level, the activation levels

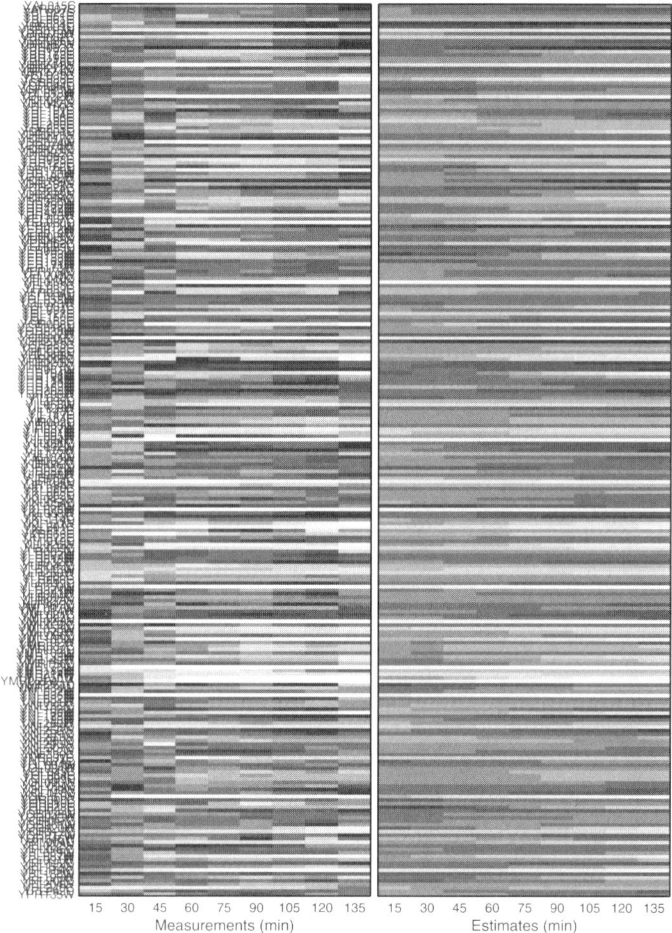

Fig. 11.2. The measured and simulated activation values of a group of stress-related genes when dtt is added to the cultivation medium. High activation values are represented by the light shade, whereas black refers to low activation

start to recover towards the original state. However, the expression of the genes represented by the first state variable stay on the new level, probably because the new pH value requires partially different gene activation also in the course of time.

It is also interesting to analyse the state sequences of the two hydrogen peroxide addition experiments included in both data sets. While in the Gasch data the final H_2O_2 concentration is 0.3 mM, Causton et al. use the value 0.4 mM. Otherwise the two experiments are identical. Figure 11.4 shows that also the state sequences estimated by SSI resemble each other. Naturally, the response changes for the stronger stimulation result to state signals with stronger deviations from the original state. It

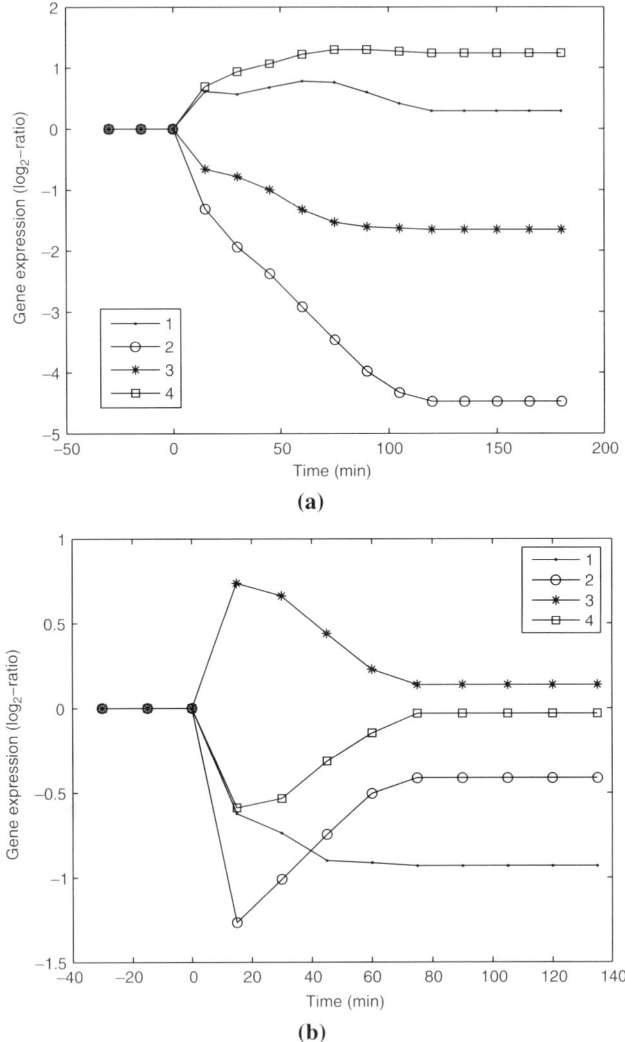

Fig. 11.3. The state sequences for (**a**) dtt and (**b**) alkali addition experiments

is also evident that the final states are reached faster when the required changes are smaller.

Cell Cycle

The second example deals with yeast gene expression during the cell cycle. In a cell cycle experiment the yeast cultivation is grown in steady conditions but the cells are synchronized to start the cell cycle simultaneously.Thus the collected microarray data contains the cyclic activation and deactivation of the genes participating

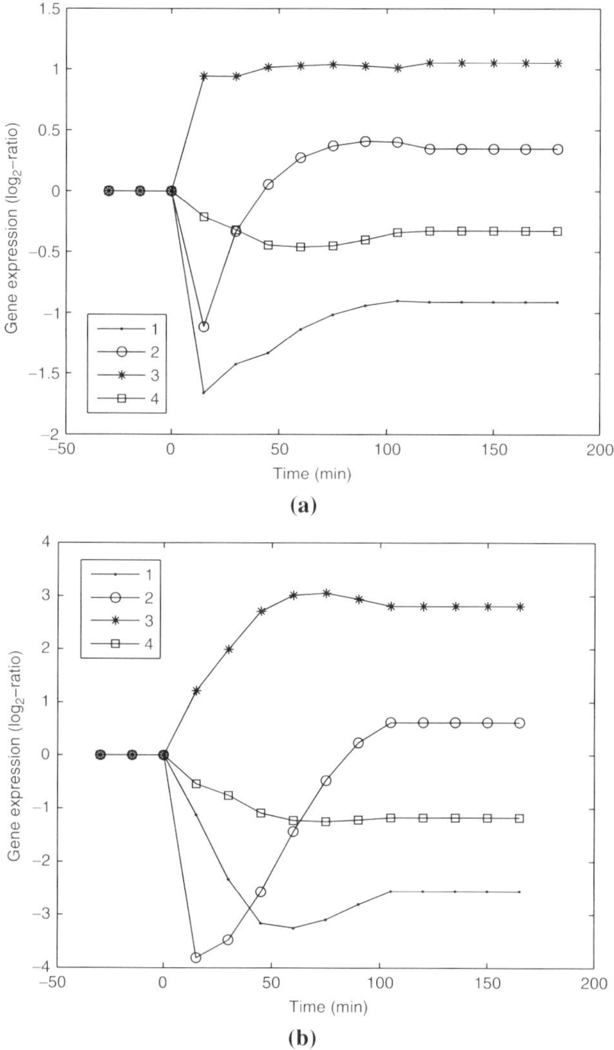

Fig. 11.4. The state sequences for the two hydrogen peroxide treatments: (**a**) Final H_2O_2 concentration 0.3 mM (Gasch et al.); (**b**) Final H_2O_2 concentration 0.4 mM (Causton et al.)

in the different phases of the cell cycle. As explained in Sect. 11.4.1, different synchronization methods were used in the included experiments, but the resulting cell cycle response should be comparable at least after the performed time scaling. Since the data were collected from four different experiments describing the same phenomenon, there were also data available for proper validation of the estimated model.

Because the cell cycle activity in the yeast genome is not (directly) caused by external stimulation, no deterministic input signal was available for the modeling. Consequently, the purely stochastic state space model (i.e. (11.6) with $\mathbf{u}(k) \equiv 0$) was a better suited model structure for this case. As a result, the modeling is completely concentrated on the connections and dependencies between the genes participating in the cell cycle.

The stochastic state space model parameters were estimated using the α-factor, cdc28 and elutriation data sets as estimation data. The two main singular values of the projection $\mathcal{O}_i$ were observed to dominate, so the model dimension $n = 2$ was selected accordingly. Since the model could not be simulated due to the lack of the deterministic input, validation was performed by predicting the response of the genes a certain number of samples ahead in time. The prediction was obtained by using Kalman filter (see e.g. [6]), which is the common prediction approach for state space models in dynamical system analysis.

Figure 11.5 shows the measured and predicted responses in the cdc15 cell cycle time series, which was left as independent validation data. Clearly already two latent variables can quite nicely reproduce the main behaviour of the genome. When analysing the state sequences related to one of the estimation experiments (Fig. 11.6), it can be noted that the cyclic activation scheme of the cell cycle genes is captured by the model. Since the oscillations of the two characteristic modes are in different phase, the basic oscillation of genes in all phases of the cell cycle can be described as a linear combination of the two state variables. The slight attenuation of the state sequences can be explained by the decrease of synchronization of the yeast cells in the cultivation; the length of cell cycle is not constant in the cells.

Because of the simplicity of the model, also the model parameters can be used in the analysis. Since the model now describes the gene activations as a linear combination of the two states, the coefficients of the state variables correlate with the activation phase of each gene. To demostrate this, the genes in Fig. 11.5 were sorted according to the coefficients in the first column of matrix $\mathbf{C}$. As expected, the order seems to correlate with the order in which the genes activate during the cell cycle, thus enabling e.g. the clustering of similar genes or linking of individual genes to different phases of the cycle.

11.5 Conclusions

Traditional gene expression studies often concentrate on the dependencies of a small group of case-specific genes, and ignore the wider analysis. This is of course relevant if one is interested only in these genes. However, to be able to properly describe the dynamics of a complete genome some day, the bottom-up strategy where these models should be combined to cover the whole system is facing serious difficulties; the complexity of the model would easily grow too high and the main behaviour of the genome would be buried under the small details.

As an alternative, the top-down approach for modeling gene expression dynamics that was presented in this article is strongly focused on revealing the general

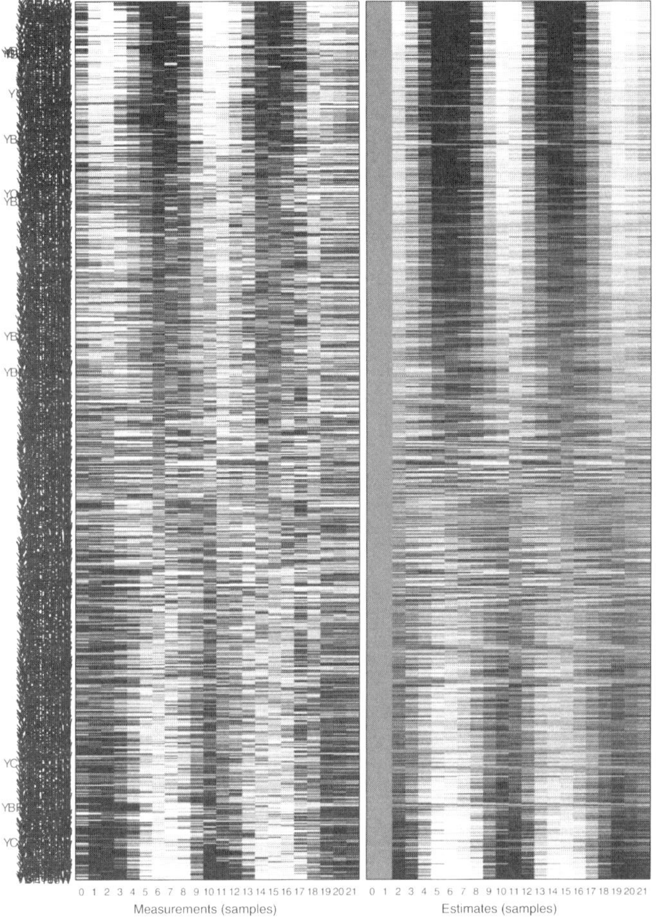

Fig. 11.5. The measured and two-step ahead predicted activation values of cell cycle genes in the cdc15 experiment

underlying properties of a large number of genes. It was shown that the SSI method could handle all of the over 4000 properly measured genes of the yeast genome and capture the common latent structure based on the stress experiment data. Also the cell cycle case suggested that the main behaviour of the genome is highly redundant, since already two latent variables could reproduce the basic activation pattern of the genes.

The presented two examples demonstrate that sticking to the statistical multi-regression tools and the usage of simple linear model structures is an advantage because of the good interpretability of the estimated models. The analysis of the state sequences clearly provides more information of the individual experiments and nicely summarizes the possible gene expression profiles of different genes.

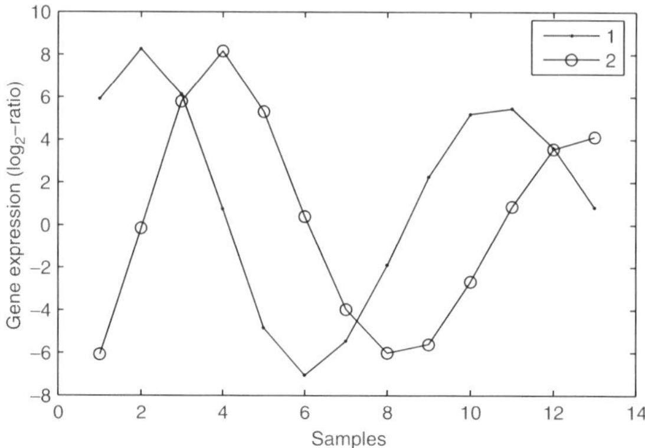

Fig. 11.6. State signals for the cdc28 cell cycle experiment

Additionally, the model parameters can be utilized in the analysis as shown by the ordering of the cell cycle genes according to their phase in the cycle.

11.6 Acknowledgments

This work was partly done in the SyMbolic project, funded by the National Technology Agency of Finland (TEKES) during 2004–2006. OH has also been supported by the Technology Promotion Foundation (TES). The authors would especially like to thank PhD Christophe Roos for his valuable comments and cooperation.

References

1. Basilevsky A (1994) Statistical factor analysis and related methods - theory and applications. John Wiley & Sons
2. Brown MPS, Grundy WN, Lin D, Christianini N, Sugnet CW, Furey TS, Ares M Jr., Haussler D (2000) Proc Natl Acad Sci USA 97(1):262–267
3. Causton HC, Ren B, Koh SS, Harbison CT, Kanin E, Jennings EG, Lee TI, True HL, Lander ES, Young RA (2001) Mol Biol Cell 12:323–337
4. Cho RJ, Campbell MJ, Winzeler EA, Stenmetz L, Conway A, Wodicka L, Wolfsberg TG, Gabrielian AE, Landsman D, Lockhart DJ, Davis RW (1998) Molecular Cell 2:65–73
5. Gasch AP, Spellman PT, Kao CM, Carmel-Harel O, Eisen MB, Storz G, Botstein D, Brown PO (2000) Mol Biol Cell 11:4241–4257
6. Grewal MS, Andrews AP (1993) Kalman filtering theory and practice. Prentice Hall
7. Haavisto O, Hyötyniemi H (2006) Neocybernetic modeling of a biological cell. In: Honkela T, Raiko T, Kortela J, Valpola H (eds) Proceedings of the Ninth Scandinavian Conference on Artificial Intelligence (SCAI 2006) 209–216
8. Haavisto O, Hyötyniemi H, Roos C (2007) Journal of Bioinformatics and Computational Biology 5(1):31–46
9. Holter NS, Maritan A, Cieplak M, Fedoroff NV, Banava JR (2001) Proc Natl Acad Sci USA 98(4):1693–1698

10. Holter NS, Mitra M, Maritan A, Cieplak M, Banavar JR, Fedoroff NV (2000) Proc Natl Acad Sci USA 97(15):8409–8414
11. Hyötyniemi H (2001) Multivariate regression – techniques and tools. Tech. Rep. 125, Helsinki University of Technology, Control Engineering Laboratory
12. Hyötyniemi H (2006) Neocybernetics in biological systems. Tech. Rep. 151, Helsinki University of Technology, Control Engineering Laboratory
13. Liang Y, Kelemen A (2005) International Journal of Bioinformatics Research and Applications 1(4):399–413
14. Ljung L (1999) System identification, theory for the user, second ed. Prentice Hall PTR
15. Maraziotis I, Dragomir A, Bezerianos A (2005) Gene Networks inference from expression data using a recurrent neuro-fuzzy approach. In: Proceedings of the IEEE 27th Annual Conference of the Engineering in Medicine and Biology Society 4834–4837
16. Ong IM, Glasner JD, Page D (2002) Bioinformatics 18(Suppl. 1):S241–S248
17. Rangel C, Angus J, Ghahramani Z, Lioumi M, Sotheran E, Gaiba A, Wild DL, Falciani F (2004) Bioinformatics 20(9):1361–1372
18. Raychaudhuri S, Stuart JM, Altman RB (2000) Principal components analysis to summarize microarray experiments: Application to sporulation time series. In: Proc of the fifth Pac Symp Biocomput 5:452–463
19. Spellman PT, Sherlock G, Zhang MQ, Iyer VR, Anders K, Eisen MB, Brown PO, Botstein D, Futcher B (1998) Mol Biol Cell 9:3273–3297
20. Troyanskaya O, Cantor M, Sherlock G, Brown P, Hastie T, Tibshirani R, Botstein D, Altman RB (2001) Bioinformatics 17(6):520–525
21. Van Overschee P, De Moor B (1996) Subspace Identification for Linear Systems. Kluwer Academic Publisher, Boston, Massachusetts
22. Wu F-X, Zhang WJ, Kusalik AJ (2004) Journal of Biological Systems 12(4):483–500

12

The Amine System Project: Systems Biology in Practice

Ismael Navas-Delgado[1], Raúl Montañez[2,3], Miguel Ángel Medina[2,3], José Luis Urdiales[2,3], José F. Aldana[1], and Francisca Sánchez-Jiménez[2,3]

[1] Computer Languages and Computing Science Department, University of Málaga, Spain (ismael,jfam)@lcc.uma.es
[2] Molecular Biology and Biochemistry Department, University of Málaga, Spain (raulemm,medina,jlurdial,kika)@uma.es
[3] Centre for Biomedical Research on Rare Diseases (CIBERER), Málaga, Spain

Summary. In this chapter we present an architecture for the development of Semantic Web applications, and the way it is applied to build an application for Systems Biology. Our working plan is designed to built an ontology-based system with connected biomodules that could be globally analysed, as far as possible. Supported by the advantages of the Semantic Web, we can keep the objective to work on the way to obtain an automated form to integrate both information and tools in our system.

Key words: Semantic Web, Systems Biology, Semantic Mediation, Amine

12.1 Introduction: Problem Statement and Development of a Systems Biology Pilot Project

A living organism is an open system that keeps a continuous interchange of chemical compounds, energy and information with its environment. This interchange involves a high number of elements (molecules) related among them in a dynamic hierarchical and modular way. Modules can be identified from the analysis of the interaction patterns. At molecular level, interacting networks include protein-protein interactions, metabolic pathways, and the different biosignalling pathways controlling intercellular cross-talk and regulation of gene expression [1]. These different local networks are also related among them. From the previous analysis, it is easily deduced that the integration of both structural and functional data concerning all of the involved elements, their spatial locations and their interrelationship patterns is an essential (but still a dawning) task for an efficient advance in biological knowledge. From the beginning of this century, new systemic approaches to the study of living organisms were proposed [2]. They are essential to let come into view the general rules that, as it also occurs in Physics, must govern the biological behaviour. It is our understanding that Systems Biology should include both the relationships among the elements

I. Navas-Delgado et al.: *The Amine System Project: Systems Biology in Practice*, Studies in Computational Intelligence (SCI) **94**, 277–292 (2008)
www.springerlink.com

of a biological system in a given steady- state and the responses of the system against any endogenous or exogenous perturbation, with the final aim to know not only the system itself but also its dynamics.

The available information required to get more holistic views of Molecular Biology problems increases daily, due to a large amount of data provided by Genome Projects and high-throughput technologies. Obviously, this is a great advantage for (and makes a great deal of) this scientific field. However, the required information is dispersed among different information repositories, which offer redundant, and sometimes ambiguous and/or controversial data. These facts can induce that processes of information retrieval lose both efficiency and fidelity. Another worth-mentioning disadvantage of the present trends and tools is the continuous overlapping of new information strata that frequently lead to cover up the previous information.

To sum up, it is clear that the development and support of intelligent integration methods for data searching, screening and analysis are absolutely required. These new tools should accomplish the following properties: i) to be in favour of reaching a consensus among the scientific community; ii) to be able to discriminate among redundant and/or wrong information; iii) to gain the possibility to access to information partially hidden for the web (a remarkable example of it, is for instance the access to data contained in printed papers); iv) to be able to grow towards augmented capabilities to be interconnected to other tools developed by the scientific community. Working in this sense, the discovery of new emergent properties of the systems will be allowed.

Under the name of "Amine System Project" (ASP), we have started working to construct a pilot system for the integration of biological information related to Biochemistry, Molecular Biology and Physiopathology of a group of compounds known as biogenic amines (http://asp.uma.es). Two general objectives can be distinguished in this project: i) development of new and more efficient tools for the integration of information stored in databases, with the aim to detect new emergent properties of this system; and ii) generation of in silico predictive models at different levels of complexity. It is being carried out by a multi-disciplinary group joining biochemists, molecular biologists and informaticians. In the following paragraphs, we present the biological context defined as our "system" and the reasons for this choice. Nevertheless, once the outcoming tools become validated, many of them could be easily adapted to study many other biological systems and to be compatible with many other bioinformatics tools and repositories.

Biogenic amines are low-molecular-weight compounds derived from amino acids. Members of the group have been working for the last 18 years on the different aspects of the amine metabolism and the molecular bases of the physiological effects caused by these compounds in different eukaryotic models, mainly mammalian cells [3–10]. Thus, the experience on the biological topic, as well as tools and facilities available for experimental validation of the in silico-derived hypotheses were important to define our system. We have been mainly devoted to those amines synthetised from cationic amino acids in mammalian organisms. These are: histamine (derived form histidine), and the polyamines putrescine, spermidine

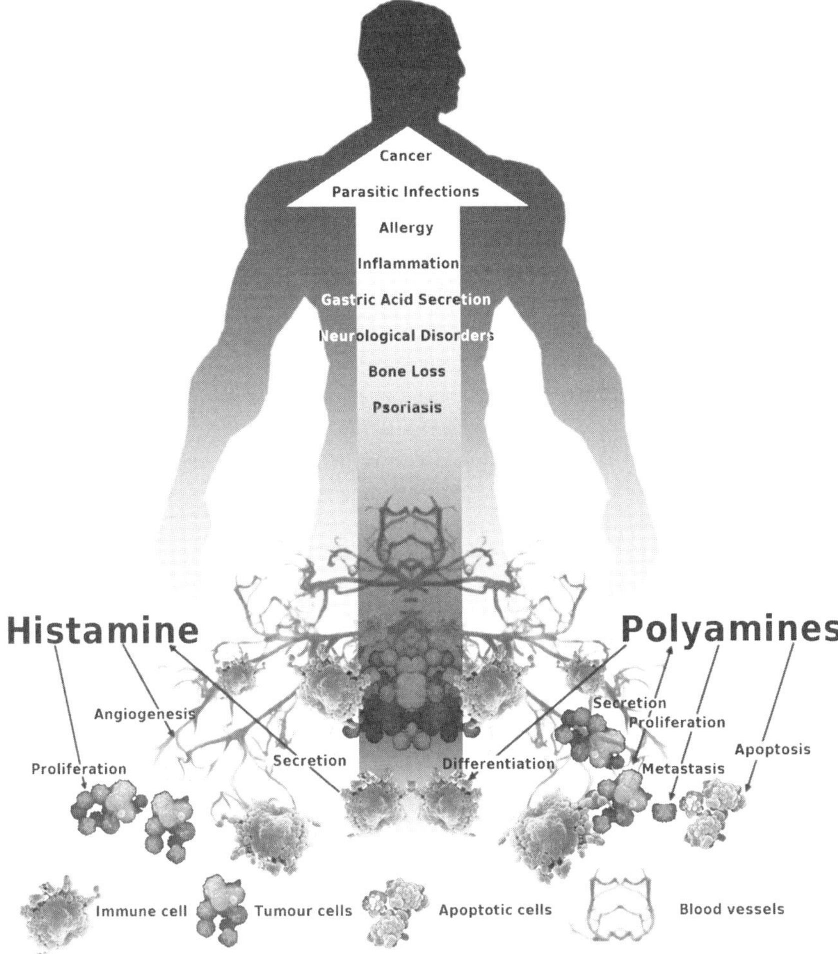

Fig. 12.1. Scheme of the most relevant biological processes modulated by histamine and polyamines at cellular level and their physiopathological consequences on the whole organism

and spermine (derived from arginine/ornithine and methionine). The biological processes modulated by histamine and polyamines in different human cell types and their physiopathological consequences on a human organism were recently reviewed by our group [11] and summarized in Figure 12.1.

Polyamines are essential polycations for every living cells known so far and play their major roles as modulators of the central mechanisms of cell life, growth and death [12]. Histamine is considered as an intercellular mediator of pleiotropic (and sometimes antagonic) effects elicited through different signalling pathways in different target cells: anaphylactic reactions and other inflammatory responses, gastric secretion, neurotransmission, bone-loss, fertility, angiogenesis and tumour growth.

This short summary of the molecular pathways and physiopathological processes related to histamine is enough to show how many biomolecular interactions are involved in its biological missions and how dispersed the amine-related information can be among many different repositories of specialised bibliography and databanks of many different biomedical areas. In some human cell types (for instance, mast cells or macrophages), both polyamines and histamine are essential elements for their specific physiological roles and it is proven that their metabolic pathways keep a molecular cross-talk at different levels [11]. In any case, pathological conditions associated to both polyamine and histamine affect an important percentage of the humanity at any stage of our lives. These circumstances guarantee the fruitfulness of any effort towards a more integrated analysis of the huge quantity of dispersed biochemical and phenomenological information that is required to generate new strategies for a better control of the polyamine/histamine-related diseases.

During the second part of the 20th century, an impressive quantity of high quality work was released from reductionistic approaches. It has provided information about almost every element involved in polyamine metabolism in different cell types. However, most of the attempts to use this information to drive intervention strategies has failed, since evolution has selected robust and sophisticated mechanisms to compensate alterations in the most essential pathways. Consequently, the scientific community will not fully profit from all these efforts until a more systemic view of the regulatory mechanisms associated to amine biochemistry is reached. The application of Systems Biology technologies could allow us to obtain a more extended, dynamics and fruitful level of knowledge on the causes and consequences of alterations in the amine-related pathways, that is, in the highly relevant physiopathological processes related to amine metabolism. Some examples of this assessment are the following examples of the application of our system, which are considered among our present aims: a) to obtain predictions on the structural and functional alterations of a given molecule produced by its interactions with others (protein, nucleic acid, metabolite or drug); b) to locate dynamic bifurcation points and putative hysteretical behaviour of the involved metabolic pathways being altered under pathological conditions or treatments; c) to determine the molecular structural and functional relationships among the amine-related biomolecules and the other cellular components. This emergent knowledge could suggest new strategies for their control and intervention, as explained in [13].

12.2 The Semantic Web: Practical Usage in Systems Biology

As mentioned before, retrieval of the impressive quantity of information disperse in different growing databases is essential for efficient use of the research investments and for advance of knowledge, not only in the amine field, but also in any biological/biomedical problem. The rapid increase in both volume and diversity of "omic" data further stress the necessity for development and adoption of data standards. A recent trend in data standard development has been to use eXtensible Markup Language (XML, http://www.w3.org/XML/) as the preferred mechanism to define data representations. However, XML cannot provide the necessary elements to achieve the level of interoperability required by the highly dynamic and integrated bioinformatics applications.

To solve this problem, an explicit description of the semantics in biological databases is required. This can be achieved by ontologies describing the biological systems. Ontologies provide a formal representation of the real world, shared by a sufficient amount of users, by defining concepts and relationships between them. In order to provide semantics to web resources, instances of such concepts and relationships are used to annotate them. These annotations over the resources, which are based on ontologies, are the foundation of the Semantic Web [14]. Given the size of the web, we have to deal with large amounts of knowledge. All this information must be represented and managed efficiently to guarantee the feasibility of the Semantic Web.

Knowledge representation and reasoning about this knowledge is a well-known problem for artificial intelligence researchers. Explicit semantics is defined by means of formal languages. Description Logics [15] is a family of logical formalisms for representing and reasoning about complex classes of individuals (called concepts) and their relationships (expressed by binary relations called roles). Description Logics are intended for formal knowledge representation and are based on a structured, decidable fragment of FOL (first Order Logic). The combination of formal knowledge representation altogether with the definition of formal but efficient reasoning mechanism is crucial for reasoning in Description Logics. Description Logics formalism allows the description of concepts, relationships and individuals (i.e. the knowledge base), and all of them together with complex concept formation and concept retrieval and realization provide a query/reasoning language for the knowledge base. Research in Description Logics deals with new ways to query a knowledge base efficiently.

The ongoing standards of current web-based ontology definition languages (such as OWL, http://www.w3.org/TR/owl-features/) are based on Description Logics. These languages provide mechanisms to define classes and properties and their instances. Web Ontology Language (OWL) is a markup language for publishing and sharing data using ontologies on the Internet. OWL is a vocabulary extension of the Resource Description Framework (RDF, http://www.w3.org/RDF/) and is derived from the DAML + OIL Web Ontology Language (http://www.w3.org/Submission/2001/12/). Together with RDF and other components, these tools make up the Semantic Web project. OWL was developed mainly because it has more facilities for expressing meaning and semantics than XML, RDF, and RDF-S, and thus OWL goes beyond these languages in its ability to represent machine interpretable contents on the web and perform reasoning over this knowledge.

OWL is seen as a major technology for the future implementation of a Semantic Web. OWL was designed specifically to provide a common way to process the content of web information. The language is intended to be read by computer applications instead of humans. Since OWL is written in XML, OWL information can be easily exchanged between different types of computers using different operating systems, and application languages. OWL's main purpose will be to provide standards that provide a framework for asset management, enterprise integration and the share and reuse of data on the Web (taking advantage of the reasoning capabilities that it provides being based on Description Logics).

Semantic Web technologies provide a natural and flexible solution for integrating and combining two levels of abstraction, the data level and the knowledge level, which are related by means of metadata. The information is annotated with semantic contents/metadata that commit with a domain ontology. The semantic interoperability among applications depends on their matching capability between information and knowledge schemas. Generally, this task is carried out at the implementation level, building a syntactic model shared among applications. Ontologies make possible to attain this objective by adding a semantic layer on the syntactic model with knowledge of what the information represents. In this way, some research areas are making a big effort to represent the knowledge they have by means of big ontologies that tend to become standards for representing information. However, these ontologies describe usually generic terms that explain the basis of the domain, and cannot be directly applied to annotate the data of common databases.

Thus, Bioinformatics researchers are developing several domain ontologies, representing big subjects in biology: protein ontology [16], sequence ontology [17] and gene ontology [18]. The main reasons to use an ontology are: to share common understanding of the structure of information among people or software agents, to enable reuse of domain knowledge, to make domain assumptions explicit, to separate domain knowledge from the operational knowledge and to analyze domain knowledge. If the available ontologies does not fulfill these requirements, then it is necessary to start the development of a new one.

12.3 Architecture and Future Prospects

As stated in the introduction of the "Amine System Project" we have started the building of a pilot system for the integration of biological information related to Biochemistry, Molecular Biology and Physiopathology (focusing our main interest on a group of compounds known as biogenic amines). This section introduces the architecture used to build this pilot system, and the way it will help researchers in the project context.

12.3.1 The Pilot: AMine Metabolism Ontology and the Semantic Directory Core

This section presents the Pilot developed for integrating dispersed resources, such as online databases, web services and data transformation tools.

This Pilot is based on a generic infrastructure (Semantic Directory Core, SD-Core), which is mainly used for registering and managing ontologies and their relationships with distributed resources (online databases, web services and data transformation tools).

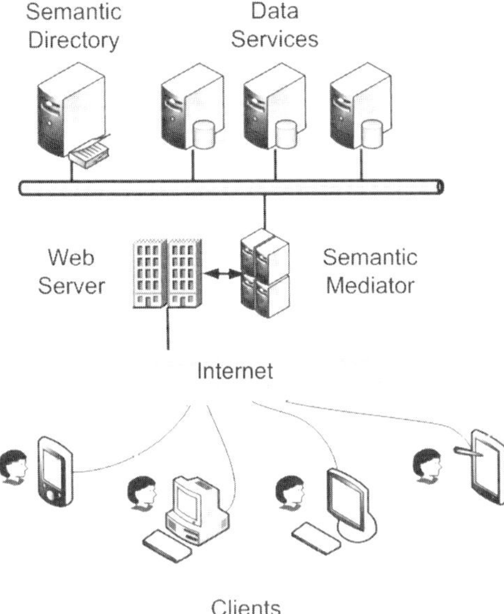

Fig. 12.2. Conceptual architecture of the Amine System Project. The kernel of the system is composed of a Semantic Directory and several Data Services. The Mediator and its Web interface provide an integrated access to the information

Previous works [19, 20] have allowed us to identify the minimum elements that can be useful for building Semantic Web applications, and they are the core of the proposed infrastructure (Figure 12.2). The internal elements of the SD-Core (Figure 12.3) are composed of a set of inter-related ontologies, which describe its semantics, and tools for taking advantage of this semantics. These ontologies include an ontology to manage metadata about ontologies registered in the SD-Core (Ontology Metadata Vocabulary, OMV), and an ontology to manage the metadata of registered resources and their relationships with registered ontologies (SD-Core Metadata Ontology, SDMO).

Tools to manage metadata represented as ontologies include from a simple OWL parser to a complex ontology reasoner. We make use of Jena (http://jena.sourceforge.net/) to access this knowledge in a first version that does not require the installation of any additional elements as a reasoner. However, the reasoning capabilities of Jena are limited, and it is not possible to infer new knowledge from the information registered in the system. For this reason, we have developed a version including the use of a reasoner, Racer [21], to improve the query results by taking advantage of the reasoning mechanisms it uses (concepts classification, concepts subsumption, complex concepts, etc.). However, Racer requires a license for being used, so the addition of this reasoner to the system has been carried out through the DIG API (http://dl.kr.org/dig/) (that provides Racer). In this way, the SD-Core can be changed for using another DIG compliant reasoner by installing and replacing Racer for other

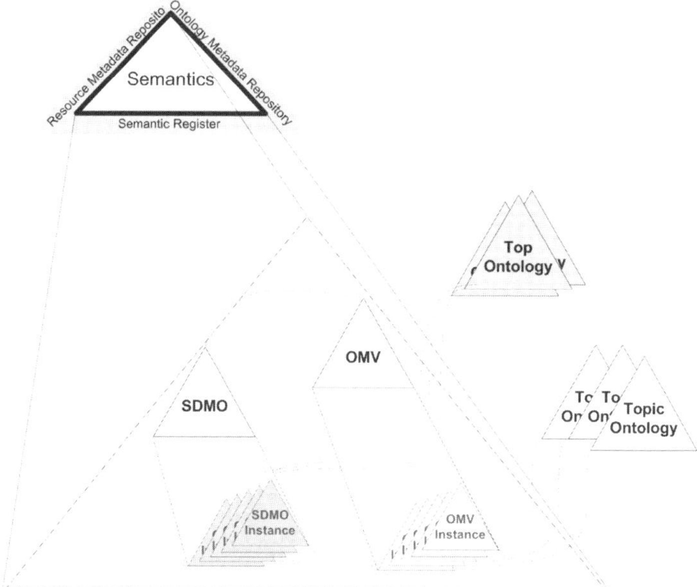

Fig. 12.3. Internal Elements of the SD-Core. The metadata used by SD-Core is represented as ontologies and managed by an ontology parser. The metadata about resources and ontologies is modified and accessed through three components. (Abbreviations are defined in the text)

Reasoner. Thus, the users interested in our proposal and who are not able to acquire a Racer license can make use of it.

When using any reasoner, for enhanced -DL based - reasoning mechanisms, it usually implies that the Web Server will have an overhead because of the reasoner activities. Thus, the way in which it has been included is by means of its installation in a remote machine thereby avoiding to overhead the server.

12.3.2 Ontology-Based Mediation

SD-Core provides necessary elements to deal with ontologies (and reasoning with them if a reasoner is included), but cannot fulfill all the requirements of the ASP project itself. The integration of information is a key requirement in Systems Biology and also in our project. For this reason we have adopted a mediator-based approach.

The main goal of mediation systems is to allow users to perform queries over heterogeneous databases, as if they were only one, using an integration schema. Mediators offer interfaces in order to query the system in terms of the integration schema. Thus, software developers can build applications that make use of distributed and heterogeneous databases as if they were a centralized database.

Internally, mediators transform user queries into a set of sub-queries that other software components (the wrappers), which encapsulate data sources' capabilities, will solve. Recently, the research has been devoted to the problem of Semantic Mediation, because it introduces the possibility of taking advantage of explicit semantics and reasoning mechanisms to provide better solutions.

Semantic mediation adds a few additional considerations to the logical information integration problems. In this scenario, sources not only export their logical schema, but also their conceptual model to the mediator, thus exposing their concepts, roles, classification hierarchies, and other high-level semantic constructs to the mediator. Semantic Mediation allows information sources to export their schema at an appropriate level of abstraction to the mediator. Mediators are applications that offer a transparent access to the data in distributed resources, being considered for the users as a single database. In this way, a Semantic Mediation system is a system that offers transparent access to the knowledge in distributed resources, being considered as a single knowledge-base. In this context Semantic Mediation systems are those in which the integrated resources are knowledge-bases (or resources enveloped to enable their access as a knowledge-base).

In our pilot, we focus on a intermediate kind of systems, Ontology-Based Mediation, in which data resources are kept unmodified and are registered to make their semantics explicit. The mediator takes advantage of this knowledge about the resources in order to better integrate the information (taking advantage of the semantics and reasoning mechanisms), but the resources do not change their interface allowing existing applications to keep using them.

The mediator can be developed from scratch, building all the required components to obtain the semantics and then use it to solve the integration problem. Nevertheless, the pilot uses the SD-Core for building the Ontology-Based Mediation System, thus avoiding the development of specific tools to deal with semantics.

In the pilot the sources' query capabilities are published as Web Services (called in our proposal Data Services). The goal of Data Services is to allow applications to access data repositories and data providing service functionalities by means of Web Services. The presented infrastructure (SD-Core) is used to register these services, defining their semantics with respect to previously registered ontologies.

The architecture of the proposed Ontology-Bases Mediator is composed of four main components, that can be provided as distributed components. The main advantage of this proposal is that the extension or modification of any part of the mediator will involve the modification of a single component, keeping the other components unchanged. The components are described as follows (see Figure 12.4):

- Controller: this component has as main task the interaction with the user interface, providing solutions as ontology instances for user queries (described in terms of one of the ontologies registered in the semantic directory). The queries are received as conjunctive queries in terms of one of the ontologies registered in the SD-Core.
- Query Planner: the task of this component is to find a query plan (QP), using the SD-Core, for a query described in terms of one of the ontologies registered in the Semantic Directory (O). The use of the SD-Core in combination with a reasoner will provide the mediator with the possibility of improving the results.

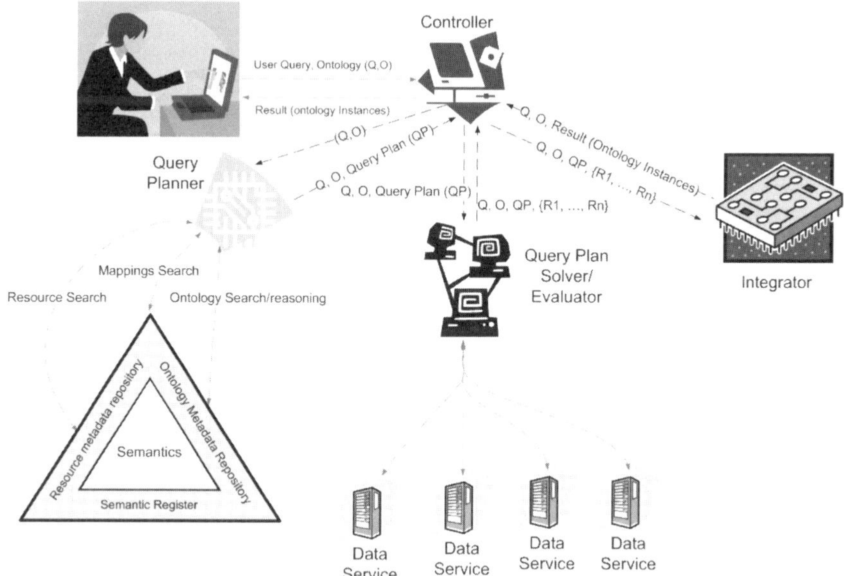

Fig. 12.4. Web Interface to make use of the mediator. The start point of this interface allows the user to search the protein for which he/she wants to know the 3D structure. Below the query for this first step is shown

Thus, when a query includes a concept, but it is not present in the resources, any of its sub-concepts could be used to solve de query.

- Query Solver: this component analyzes the query plan (QP), and performs the corresponding call to the data services implied in solving the sub-queries (SQ1, ..., SQn) of the query plan (R1, ..., Rn).
- Integrator: The results sent by data services (R1, ..., Rn) are composed by this component, obtaining a set of instances representing the result to the user query.

As far as any of the existing ontologies do not cope exactly with the requirements of our project we have started the development of a new ontology, the AMine Metabolism Ontology (AMMO). AMMO represents the minimum number of connections that can allow us to collect and to connect information from the different local biomodules considered in an eukaryotic cell. In practical terms, it is an ambitious information flow network, that would require many biocomputational efforts to be functional on the different Databases and Services associated to each of these concepts, as well as many validation efforts on experimental amine-related biological data (as expressed among our aims in the Introduction). At present, we are generating more specific ontologies, that can be considered as "local ontologies" (to keep a nomenclature in parallel to the situation *in vivo*). Of course, these local ontologies will be progressively integrated in AMMO.

Our present on-going efforts are focused on local ontologies recruiting information from Databases and Services on protein structures and interactions with other proteins or ligands. As it can be deduced from these lines, once the tools have been validated on the amine system, these resources could be applied to any other molecular biology topic. New information concerning predictive models, the AMMO evolution and validated tools and services generated from it will be available in the web page (http://asp.uma.es).

The mediator provides methods in order to send queries and retrieve the information. However, end-users will not have to implement an application to make use of the mediator and discover the advantages of our proposal. For this reason we have provided a first implementation of a front-end to test the mediator from a Web browser. This Web interface provides several use cases of biological interest in the ASP project. The queries (described in terms of the AMMO ontology) deal by the mediator in each use case are shown for expert users with knowledge in ontologies and conjunctive queries. Other users can run the application in order to get the result without getting intermediate results (see Figure 12.5).

12.3.3 Conclusion

As stated in the first section, any biological system presents several local biomolecular interaction networks that can be studied *in silico* by using different technologies, so that emergent information and a more dynamic picture of it can be obtained. Macromolecular interactions involved in gene regulation and signal transduction, intercellular communication and metabolic complexes can be studied by applying Graph Theory [22–25] on the results obtained from data mining from interaction databases, such as DIP (http://dip.doe-mbi.ucla.edu/), PPI (http://fantom21.gsc.riken.go.jp/PPI/), Transfac (http://www.gene-regulation.com/pub/databases.html) and others. The information provided by this technology is essential to detect elements having the major regulatory weight on the system (module hubs or connectors among different modules). The development of better tools able to join, screen and score the pre-existing information in databases is required in order to increase fidelity and efficiency of the emergent information from these approaches.

On the other hand, the behaviour of the different metabolic pathways, responsible for the interchange of compounds and energy with the environment, and their responses to different alterations (external stimuli, genetic changes and/or drug treatments) can be modelled *in silico* by following the rules of Enzymology for mathemathical formalization of enzyme kinetic and turn-over and Flux Control Theory [26, 27]. These technologies make possible a dynamic view of the evolution of the systems from an initial steady-state to the next, and provide information on the reactions with a major incidence on the flux of the pathway, which can change under different circumstances.

Even a single element (an enzyme, nucleic acid or others) can be considered as an interactive system, having a three-dimensional (3D) structure responsible for the information concerning its biological function. The 3D structure of a single macromolecule can be obtained by biophysical methods applied on purified versions of the molecule. However, these experimental approaches (for instance,

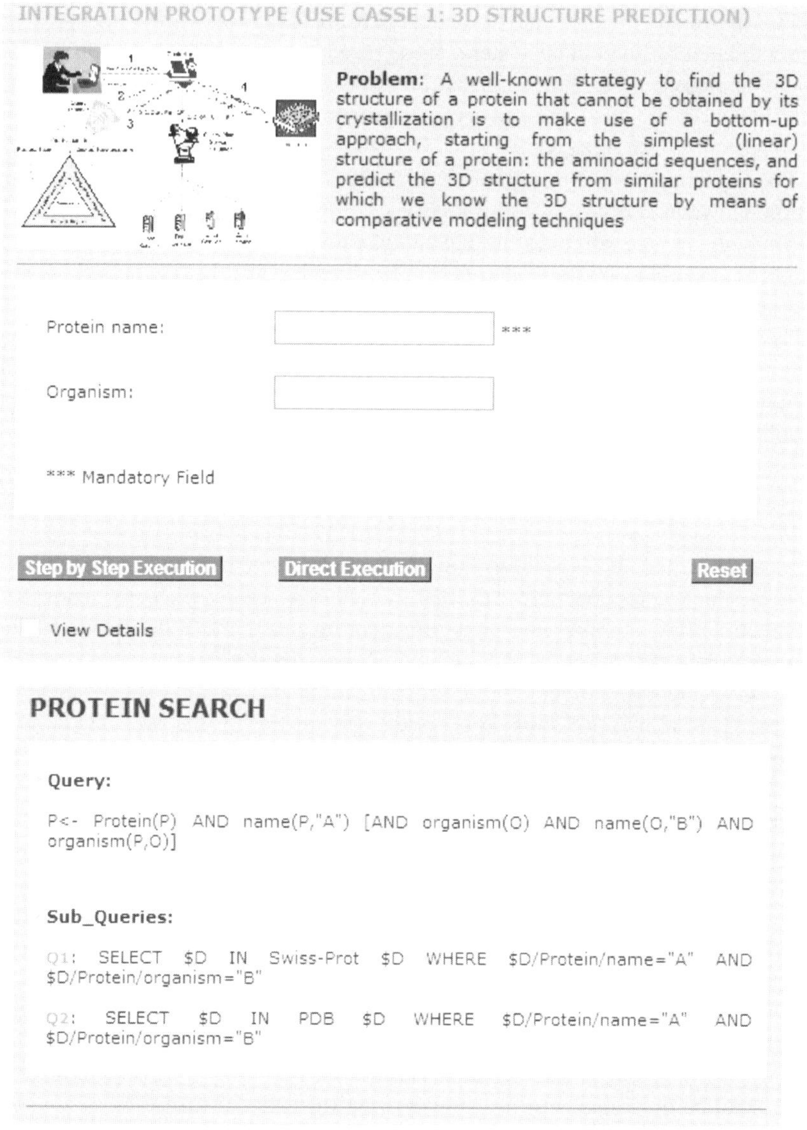

INTEGRATION PROTOTYPE (USE CASSE 1: 3D STRUCTURE PREDICTION)

Problem: A well-known strategy to find the 3D structure of a protein that cannot be obtained by its crystallization is to make use of a bottom-up approach, starting from the simplest (linear) structure of a protein: the aminoacid sequences, and predict the 3D structure from similar proteins for which we know the 3D structure by means of comparative modeling techniques

Protein name: ***

Organism:

*** Mandatory Field

Step by Step Execution Direct Execution Reset

View Details

PROTEIN SEARCH

Query:

```
P<- Protein(P) AND name(P,"A") [AND organism(O) AND name(O,"B") AND organism(P,O)]
```

Sub_Queries:

```
Q1: SELECT $D IN Swiss-Prot $D WHERE $D/Protein/name="A" AND $D/Protein/organism="B"
```

```
Q2: SELECT $D IN PDB $D WHERE $D/Protein/name="A" AND $D/Protein/organism="B"
```

Fig. 12.5. Web Interface to make use of the mediator. The start point of this interface allows the user to search the protein for which he/she wants to know the 3D structure. Below the query for this first step is shown

X-ray crystallography, NMR, and other spectroscopical techniques) frequently provide us just a static view of the molecule, losing information about conformational changes that are behind any biological function. Protein Data Bank (http://www.rcsb.org/pdb/home/home.do) stores information on the macromolecular 3D structures characterized so far. Molecular dynamics calculations applied to macromolecules can overcome this restriction, and it is considered nowadays a very promising technology for the characterization of biomolecular interactions and drug design. Even more, when the 3D structure of a given macromolecule cannot be obtained experimentally, biocomputational tools can allow us to predict its structure, under certain restrictions, with high-accuracy (for instance, ModWeb Database and Services, http://alto.compbio.ucsf.edu/modweb-cgi/main.cgi). Then, Molecular dynamics calculations can also be applied on these predicted structures to obtain information about dynamics during and after interactions/reactions with different ligands [28].

In our group, we have developed models at the 3 different levels mentioned above. For instance, a human transcription factor network model has been obtained, which clearly shows connectors between inflammation and cancer (two of the processes related to amine-metabolism) [29]. We have also developed predictive models on metabolic pathways related to polyamine metabolism in mammalian (including human) tissues, that can explain some of the phenomenological results obtained with transgenic animals and with different drug-treated models *in vivo* and *in vitro* [30, 31]. Finally, by applying protein modelling techniques, the first 3D model for the enzyme responsible of histamine synthesis in animals and humans was obtained, which has opened the possibility to design new and more specific anti-histaminic compounds [32].

All of these technologies involve capture and organization of information coming from different databases and services. Following our own interpretation of Systems Biology (see first section), we notice that all local biomodules of a system are connected and should be globally analysed, as far as possible. Reaching this goal is a long-term project that overpasses the activity of a single working group. Nevertheless, supported by the advantages of the Semantic Web (Section 12.2), we can keep the objective to work on the way to obtain an automated form to integrate both information and tools in our system. For this purpose, we have designed the presented architecture.

12.4 Acknowledgements

Supported by Projects CVI-267, CVI-657 and TIC-136 (Junta de Andalucía), and Projects SAF 2005-01812 and TIN2005-09098-C05-01 (Spanish Ministry of Education and Science).

Thanks are due to F. Villatoro, MG Claros, C. Rodríguez-Caso and AA. Moya-García, R. Fernandez-Santa Cruz, M.M Rojano and MM. Roldán for their helpful inputs and suggestions during evolution of ASP.

We thank the reviewers for their comments and help towards improving of the chapter.

References

1. Rodríguez-Caso, C. and Solé, R. (2006) Networks in cell biology, Fundamentals of Data Mining in Genomics and Proteomics, W. Dubitzky, M. Granzow, and D. Berrar, Eds. Kluwer Academic Publishers, vol. in press.
2. Kitano, H. (2002) Systems biology: a brief overview, Science, vol. 295, no. 5560, pp. 1662–1664, 1095–9203 (Electronic) Journal Article Review.
3. Medina, M.A, Urdiales, J.L., Matés, J.M., Núñez de Castro, I. and Sánchez- Jiménez, F. (1991) Diamines interfere with the transport of l-ornithine in ehrlich-cell plasma-membrane vesicles, Biochem J, vol. 280 (Pt 3), pp. 825–827, 0264- 6021 (Print) Journal Article.
4. Engel, N., Olmo, M.T., Coleman, C.S., Medina, M.A., Pegg, A.E. and Sánchez-Jiménez, F. (1996) Experimental evidence for structure-activity features in common between mammalian histidine decarboxylase and ornithine decarboxy- lase, Biochem J, vol. 320 (Pt 2), pp. 365–368, 0264-6021 (Print) Journal Article.
5. Fajardo, I., Urdiales, J.L., Medina, M.A. and Sánchez-Jiménez, F. (2001) Effects of phorbol ester and dexamethasone treatment on histidine decarboxylase and ornithine decarboxylase in basophilic cells, Biochem Pharmacol, vol. 61, no. 9, pp. 1101–1106, 0006-2952 (Print) Journal Article.
6. Fajardo, I., Urdiales, J.L., Paz, J.C., Chavarría, T., Sánchez-Jiménez, F. and Medina, M.A. (2001) Histamine prevents polyamine accumulation in mouse c57.1 mast cell cultures, Eur J Biochem, vol. 268, no. 3, pp. 768–773, 0014-2956 (Print) Journal Article.
7. Rodríguez-Caso, C., Rodríguez-Agudo, D., Sánchez-Jiménez, F. and Medina, M.A. (2003) Green tea epigallocatechin-3-gallate is an inhibitor of mammalian histidine decarboxylase, Cell Mol Life Sci, vol. 60, no. 8, pp. 1760–1763, 1420-682X (Print) Journal Article.
8. Rodríguez-Caso, C., Rodríguez-Agudo, D., Moya-García, A.A., Fajardo, I., Medina, M.A., Subramaniam, V. and Sánchez-Jiménez, F. (2993) Local changes in the catalytic site of mammalian histidine decarboxylase can affect its global conformation and stability, Eur J Biochem, vol. 270, no. 21, pp. 4376–4387, 0014-2956 (Print) Journal Article.
9. Fleming, J.V., Fajardo, I., Langlois, M.R., Sánchez-Jiménez, F. and Wang, T.C. (2004) The c-terminus of rat l-histidine decarboxylase specifically inhibits enzymic activity and disrupts pyridoxal phosphate-dependent interactions with l-histidine substrate analogues, Biochem J, vol. 381, no. Pt 3, pp. 769–778, 1470-8728 (Electronic) Journal Article.
10. Fleming, J.V., Sánchez-Jiménez, F., Moya-García, A.A., Langlois, M.R. and Wang, T.C. (2004) Mapping of catalytically important residues in the rat l-histidine decarboxylase enzyme using bioinformatic and site-directed mutagenesis approaches, Biochem J, vol. 379, no. Pt 2, pp. 253–261, 1470-8728 (Electronic) Journal Article.
11. Medina, M.A., Urdiales, J.L., Rodríguez-Caso, C., Ramírez, F.J. and Sánchez- Jiménez, F. (2003) Biogenic amines and polyamines: similar biochemistry for different physiological missions and biomedical applications, Crit Rev Biochem Mol Biol, vol. 38, no. 1, pp. 23–59, 1040-9238 (Print) Journal Article Review.
12. Cohen, S.S. (1998) A Guide to the Polyamines. New York: Oxford University Press.

13. Medina, M.A., Correa-Fiz, F., Rodríguez-Caso, C. and Sánchez-Jiménez, F. (2005) A comprehensive view of polyamine and histamine metabolism to the light of new technologies, J Cell Mol Med, vol. 9, no. 4, pp. 854–864, 1582-1838 (Print) Journal Article Review.
14. Berners-Lee, T., Hendler, J. and Lassila, O. (2001) The Semantic Web, Scientific American.
15. Baader, F., Calvanese, D., McGuinness, D.L., Nardi, D. and Patel-Schneider, P.F. (2003) The Description Logic Handbook: Theory, Implementation, and Applications. Cambridge University Press.
16. Hussain, F.K., Sidhu, A.S., Dillon, T.S. and Chang, E. (2006) Engineering trustworthy ontologies: Case study of protein ontology, CBMS'06: Proceedings of the 19th IEEE Symposium on Computer-Based Medical Systems. Washington, DC, USA: IEEE Computer Society, pp. 617–622.
17. Eilbeck, K., Lewis, S.E., Mungall, C.J., Yandell, M., Stein, L., Durbin, R. and Ashburner, M. (2006) The sequence ontology: a tool for the unification of genome annotations, Genome Biology, vol. 6, p. R44.
18. Ashburner, M., Ball, C.A., Blake, J.A., Botstein, D., Butler, H., Cherry, J.M., Davis, A.P., Dolinski, K., Dwight, S.S., Eppig, J.T., Harris, M.A., Hill, D.P., Issel-Tarver, L., Kasarskis, A., Lewis, S., Matese, J.C., Richardson, J.E., Ring-wald, M., Rubin, G.M. and Sherlock, G. (2000) Gene ontology: tool for the unification of biology. The Gene Ontology Consortium. Nat Genet, vol. 25, no. 1, pp. 25–29.
19. Navas-Delgado, I. and Aldana-Montes, J.F. (2004) A distributed semantic mediation architecture, Journal of Information and Organizational Sciences, vol. 28, no. 1–2, pp. 135–150.
20. Aldana-Montes, J.F., Navas-Delgado, I. and Roldan-Garcia, M.M. (2004) Solving Queries over Semantically Integrated Biological Data Sources, Int. Conf. on Web-Age Information Management (WAIM 2004). LNCS 3129.
21. Haarslev, V. and Möller, R. (2001) The Description Logic ALCNHR+ Extended with Concrete Domains: A Practically Motivated Approach, R. Goré, A. Leitsch, and T. Nipkow, editors, International Joint Conference on Automated Reasoning, IJCAR2001, June 18–23, Siena, Italy, pp. 29–44, Springer-Verlag.
22. Barabasi, A.L. (2002) Linked: The New Science of Networks. Cambridge: Perseus Books Group.
23. Lehner, B., Crombie, C., Tischler, J., Fortunato, A. and Fraser, A.G. (2006) Systematic mapping of genetic interactions in caenorhabditis elegans identifies common modifiers of diverse signaling pathways, Nat Genet, vol. 38, no. 8, pp. 896–903, 1061-4036 (Print) Journal Article.
24. Basso, K., Margolin, A.A., Stolovitzky, G., Klein, U., Dalla-Favera, R. and Califano, A. (2005) Reverse engineering of regulatory networks in human b cells, Nat Genet, vol. 37, no. 4, pp. 382–90, 1061-4036 (Print) Journal Article.
25. Vázquez, A., Dobrin, R., Sergi, D., Eckmann, J.P., Oltvai, Z.N. and Barabasi, A.L. (2004) The topological relationship between the large-scale attributes and local interaction patterns of complex networks, Proc Natl Acad Sci U S A, vol. 101, no. 52, pp. 17 940–945, 0027-8424 (Print) Journal Article.
26. Fell, D. (1996) Understanding the Control of Metabolism, ser. Frontiers in Metabolism. Ashgate Publishing.
27. JM, L., EP, G. and JA, P. (1991) Flux balance analysis in the era of metabolomics, Brief Bioinform, vol. 7, no. 2, pp. 140–150.

28. Garcia-Viloca, M., Gao, J., Karplus, M. and Truhlar, D.G. (2004) How enzymes work: analysis by modern rate theory and computer simulations. Science, vol. 303, no. 5655, pp. 186–195.
29. Rodríguez-Caso, C., Medina, M.A. and Solé, R.V. (2005) Topology, tinkering and evolution of the human transcription factor network, Febs J, vol. 272, no. 24, pp. 6423–6434, 1742-464X (Print) Journal Article.
30. Rodríguez-Caso, C., Montañez, R., Cascante, M., Sánchez-Jiménez, F. and Medina, M.A. (2007) Mathematical modeling of polyamine metabolism in mammals, J Biol Chem, vol. 281, no. 31, pp. 21 799–812, 0021-9258 (Print) Journal Article.
31. Montañez, R., Rodríguez-Caso, C., Sánchez-Jiménez, F. and Medina, M.A. (2007) In silico analysis if arginine catabolism as a source of nitric oxide or polyamines in endothelian cells, Amino Acids, in press.
32. Moya-García, A.A., Pino-Ángeles, A. and Sánchez-Jiménez, F. (2006) New structural insights to help in the search for selective inhibitors of mammalian pyridoxal 5'-phosphate-dependent histidine decarboxylase, Inflammation Res., vol. 55, Supplement 1, pp. S55–S56.

13

DNA Encoding Methods in the Field of DNA Computing

Aili Han[1,2] and Daming Zhu[2]

[1] Department of Computer Science and Technology, Shandong University at Weihai, Weihai 264209, China

[2] School of Computer Science and Technology, Shandong University, Jinan 250061, China

hanal@sdu.edu.cn

Summary. Bioinformatics studies the acquisition, process, store, distribution, analysis, etc of biological information so as to understand the meanings of biological data by means of mathematics, computer science and biological techniques. Some researches on Bioinformatics, such as the properties of DNA and the Watson-Crick's law, provide a probability of computing with DNA molecules. DNA computing is a new computational paradigm that executes parallel computation with DNA molecules based on the Watson-Crick's law. The procedure of DNA computing can be divided into three stages: encoding information, computation (molecular operations) and extraction of solution. The stage of encoding information is the first and most important step, which directly affects the formation of optimal solution. The methods of encoding information can be divided into two classes: the methods of encoding information in graphs without weights and the methods of encoding information in graphs with weights. The previous researches, which belong to the first class, such as Adleman's encoding method [1] for the directed Hamiltonian path problem, Lipton's encoding method [2] for the SAT problem, and Ouyang's encoding method [3] for the maximal clique problem, do not require the consideration of weight representation in DNA strands. However, there are many practical applications related to weights. Therefore, weight representation in DNA strand is an important issue toward expanding the capability of DNA computing to solve optimization problems. Narayanan et al [6] presented a method of encoding weights by the lengths of DNA strands. Shin et al [6] proposed a method of encoding weights by the number of hydrogen bonds in fixed-length DNA strand. Yamamoto et al [7] proposed a method of encoding weights by the concentrations of DNA strands. Lee et al [9] proposed a method of encoding weights by the melting temperatures of fixed-length DNA strands. Han et al [10, 11] proposed a method of encoding weights by means of the general line graph. They also gave a method of encoding weights [12] by means of the relative length graph and several improved DNA encoding methods [13–16] for the maximal weight clique problem, the traveling salesman problem, the minimum spanning tree problem and the 0/1 knapsack problem. In this chapter, I collect and classify the present methods of encoding information in DNA strands, which will benefit the further research on DNA computing.

Supported by the Natural Science Foundation of Shandong Province of China.

A. Han and D. Zhu: *DNA Encoding Methods in the Field of DNA Computing*, Studies in Computational Intelligence (SCI) **94**, 293–322 (2008)

www.springerlink.com

13.1 Introduction

Bioinformatics studies the acquisition, process, store, distribution, analysis, etc of biological information so as to understand the meanings of biological data by means of mathematics, computer science and biological techniques. Some researches on Bioinformatics, such as the properties of DNA, the Watson-Crick's law, provide a probability of computing with DNA molecules, so DNA computing is an applied branch of Bioinformatics. The results of researches on Bioinformatics will improve the capabilities of DNA computing.

DNA computing is a computational paradigm that uses synthetic or natural DNA molecules as information storage media, in which the techniques of molecular biology, such as polymerase chain reaction, gel electrophoresis, and enzymatic reaction, are used as computational operators for copying, sorting, and splitting/concatenating information, respectively. Based on the massive parallelism of DNA computing, many researchers tried to solve a large number of difficult problems. In 1994, Adleman [1] solved a 7-vertex instance of the directed Hamiltonian path problem by means of the techniques of molecular biology. This creative research opened up a new way to computation with DNA molecules. A major goal of subsequent research in the field of DNA computing is to understand how to solve NP-complete problems. To address this goal, Lipton [2] abstracted a parallel molecular model on the basis of Adleman's experiment and applied it to solve the SAT problem; Ouyang *et al* [3] solved the maximal clique problem by means of DNA molecules; Head *et al* [4] solved the maximal independent set problem using operations on DNA plasmids; Sakamoto *et al* [5] presented a molecular algorithm of Boolean calculation by means of DNA hairpin formation. These previous researches on DNA computing do not require the consideration of weight representation in DNA strands.

However, there are many practical applications related to weights, such as the shortest path problem, the traveling salesman problem, the maximal weight clique problem, the Chinese postman problem, and the minimum spanning tree problem. Therefore, weight representation in DNA strand is an important issue toward expanding the capability of DNA computing to solve optimization problems. There exist previous works to represent weights in DNA molecules. Narayanan et al [6] presented a method of encoding weights by the lengths of DNA strands. Shin et al [6] proposed a method of encoding weights by the number of hydrogen bonds in fixed-length DNA strand. Yamamoto et al [7] proposed a method of encoding weights by the concentrations of DNA strands. Lee et al [9] proposed a method of encoding weights by the melting temperatures of fixed-length DNA strands. Han et al [10, 11] proposed a method of encoding weights by means of the general line graph. They also gave a method of encoding weights [12] by means of the relative length graph and several improved DNA encoding methods [13–16] for the maximal weight clique problem, the traveling salesman problem, the minimum spanning tree problem and the 0/1 knapsack problem.

In this chapter, we collect and classify the present DNA encoding methods in the field of DNA computing, which will benefit the further research on DNA computing.

13.2 Preliminaries to DNA Computing

In order to easily understand DNA encoding methods and the corresponding DNA algorithms, we first present some basic knowledge related to DNA computing.

13.2.1 Orientation of DNA Molecule

When DNA molecules combine with each other to form a DNA strand, 5′-phosphate group of one nucleotide always combine with 3′-hydroxyl group of another nucleotide by phosphodiester bonds, shown as P in Fig. 13.1. This is called as 5′-3′ orientation or 3′-5′ orientation [17, 18]. The nucleotide with 5′ free-end being located at the most left end and 3′ free-end being located at the most right end is marked as 5′-$X_1X_2\ldots X_n$-3′, and the nucleotide with 3′ free-end being located at the most left end and 5′ free-end being located at the most right end is marked as 3′-$X_1X_2\ldots X_n$-5′, where X_i denotes one letter in the alphabet $\{A, G, C, T\}$. Take Fig. 13.1 as an example. The DNA strand shown in Fig. 13.1(a) is marked as 5′-*AGC*-3′, and the DNA strand shown in Fig. 13.1(b) is marked as 3′-*CGA*-5′. Note that 5′-*AGC*-3′ and 3′-*CGA*-5′ are the same DNA molecules. In this chapter, we use the following representation: The DNA molecule 5′-$X_1X_2\ldots X_n$-3′ is written as $X_1X_2\ldots X_n$, and 3′-$X_1X_2\ldots X_n$-5′ is written as $-X_1X_2\ldots X_n$ [10, 11, 13, 14]. Note that $-X_1X_2\ldots X_n = X_nX_{n-1}\ldots X_1$.

Definition 13.2.1.1 For any DNA strand s, let h represent a mapping function from each base to its complement, or $h(A)=T$, $h(G)=C$, $h(C)=G$, $h(T)=A$. The obtained DNA strand $h(s)$ is called the complement of s, and its reversal $-h(s)$ is called the reverse complement of s. The mapping function h is called the complementary mapping from s to s' [10, 11, 14].

Take the DNA strand $s = AGC$ as an example. The complement of s is $h(s) = TCG$, and the reverse complement is $-h(s) = -TCG = GCT$. Obviously, the DNA strand AGC can combine with $-TCG$ to form a double-stranded DNA (dsDNA) through hydrogen bonds, as shown in Fig. 13.2. It can be concluded that any DNA strand s can combine with its reverse complement $-h(s)$ to form a dsDNA through

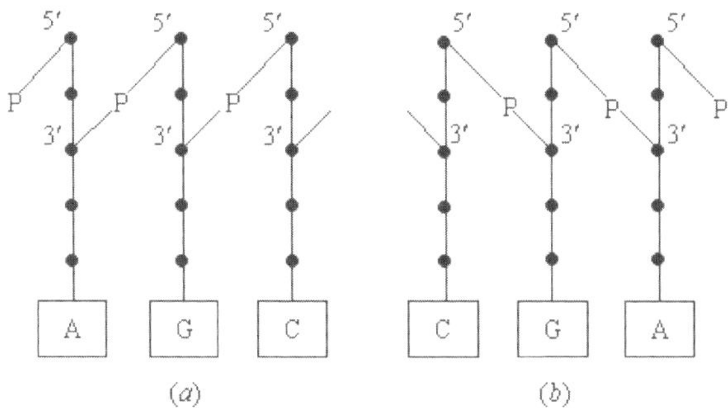

Fig. 13.1. Orientation of DNA molecule. (a) 5′-3′ orientation and (b) 3′-5′ orientation

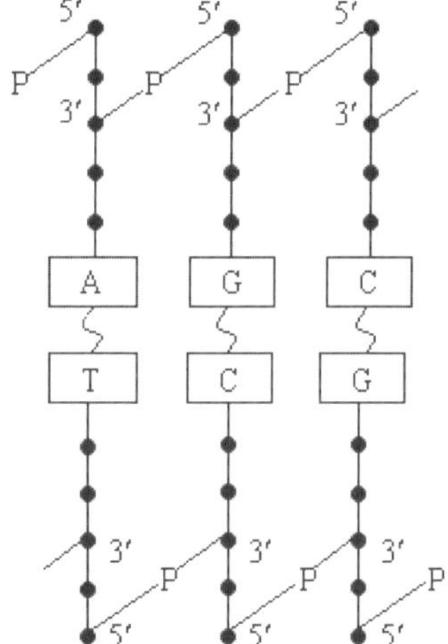

Fig. 13.2. A double-stranded DNA

hydrogen bonds. Note that DNA double strand consists of two DNA sequences $\alpha_1\alpha_2\ldots\alpha_k$ and $-\beta_1\beta_2\ldots\beta_k$ that satisfy the Watson-Crick's law, that is, for each $i = 1, 2, \ldots, k$, $k \in Z$, α_i and β_i must be complements, and the two complementary sequences anneal in an antiparallel fashion.

13.2.2 Basic Operations in DNA Computing

A tube [17] is defined as a multiset of words on the alphabet $\{A, C, G, T\}$. A multiset means a set in which the repeated words are regarded as different elements. For example, the multiset $\{AGC, AGC, GCTA\}$ has three elements, and the set $\{AGC, AGC, GCTA\}$ has two elements since the repeated words in a set are regarded as one element. Actually, a tube is a multiset of DNA strands. The basic operations [17] in DNA computing are as follows.

(1) *Merge*: For two tubes N_1 and N_2, it forms the unite $N_1 \cup N_2$ (multiset).

(2) *Amplify*: For a given tube N, it copies N into two shares (only for multiset).

(3) *Separate*: For a given tube N and a word w, $w \in \{A,C,G,T\}^*$, it generates two tubes: $+(N,w)$ and $-(N,w)$, where $+(N,w)$ consists of all the strands including w, and $-(N,w)$ consists of all the strands excluding w.

(4) $(N, \leq n)$: For a given tube N and an integer n, it generates a tube of all strands in N whose lengths are less than or equal to n.

(5) $B(N,w)$ and $E(N,w)$: For a given tube N and a word w, $B(N,w)$ generates a tube of all strands in N that begin with w, and $E(N,w)$ generates a tube of all strands in N that end with w.

(6) *Detect*: For a given tube N, it returns true if there exists at least one DNA strand in N, otherwise it returns false. This can be done through gel electrophoresis.

All the above operations can be implemented by means of the present biological techniques.

13.3 DNA Encoding Methods for the Problems Related to Graphs without Weights

Based on the massive parallelism of DNA computing, the previous researchers focused on NP-complete problems. These previous researches do not require the consideration of weight representation in DNA strands. Some of them are given in the following.

13.3.1 DNA Encoding Method for the Hamiltonian Path Problem

In 1994, Adleman [1] solved an instance of the directed Hamiltonian path problem by means of the molecular biology techniques. A 7-vertex graph was encoded in DNA strands and the operations were performed with standard DNA protocols and enzymes. This experiment demonstrates the feasibility of carrying out computations at the molecular level.

The Hamiltonian Path Problem

For a directed graph $G = (V, E)$, a path is called a Hamiltonian path if and only if it contains each vertex in G exactly once. A directed graph G with designated vertices v_{in} and v_{out} is said to have a Hamiltonian path if and only if there exists a path $e_1 e_2 \dots e_n$ that begins with v_{in}, ends with v_{out} and enters every other vertex exactly once.

Fig. 13.3 shows a graph which for $v_{in} = 0$ and $v_{out} = 6$ has a Hamiltonian path, given by the edges $0 \rightarrow 1$, $1 \rightarrow 2$, $2 \rightarrow 3$, $3 \rightarrow 4$, $4 \rightarrow 5$, $5 \rightarrow 6$. If the edge $2 \rightarrow 3$ were removed from the graph, the resulting graph with the same designated vertices would not have a Hamiltonian path. Similarly, if the designated vertices were changed to $v_{in} = 3$ and $v_{out} = 5$, there would be no Hamiltonian path.

For a directed graph $G = (V, E)$, the Hamiltonian path problem (HPP) is to determine whether there exists a Hamiltonian path in it, that is, to find a directed path that starts with a given vertex, ends with another one, and visits every other vertex exactly once. HPP has been proved to be NP-complete. There are well known algorithms for deciding whether a directed graph with designated vertices has a Hamiltonian path, but all known deterministic algorithms for HPP have exponential worst-case complexity. In 1994, Adleman [1] designed a non-deterministic DNA algorithm for HPP which runs in a polynomial time.

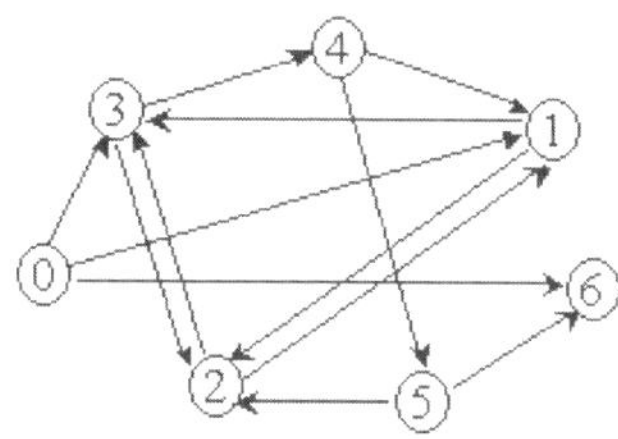

Fig. 13.3. The directed graph solved by Adleman

DNA Encoding Method

Given a directed graph $G = (V, E)$ with designated vertices $v_{in} = v_0$ and $v_{out} = v_6$, as shown in Fig. 13.3, Adleman [1] proposed the following DNA encoding method for solving the Hamiltonian path problem.

(1) Each vertex v_i in G was associated with a random 20-mer DNA strand denoted by s_i.

(2) For each edge $e_{ij} = (v_i, v_j)$ in G, a DNA strand s_{ij} was created which was $3'$ 10-mer of s_i (unless $i = 0$ in which case it was all of s_i) followed by $5'$ 10-mer of s_j (unless $j = 6$ in which case it was all of s_j).

The choice of random 20-mer DNA strand for encoding vertices in G is based on the following rationale [1]. First, choosing randomly several DNA strands from 4^{20} 20-mer DNA strands would unlikely share long common subsequences, which might result in unintended binding during the ligation step. Second, some deleterious features such as hairpin loops would unlikely arise in the several 20-mer DNA strands selected from 4^{20} 20-mer DNA strands. Finally, choosing 20-mers assured that binding between splint strands and edge strands would involve ten base pairs and would consequently be stable at room temperature.

DNA Algorithm

Given a directed graph $G = (V, E)$ with designated vertices $v_{in} = v_0$ and $v_{out} = v_t$. Based on the DNA encoding method, Adleman [1] designed the following DNA algorithm to solve the Hamiltonian path problem.

(1) *Merge*: All the DNA strands $h(s_i)$ and s_{ij} are mixed together in a single ligation reaction. Based on the Watson-Crick's law, generate random paths through the graph G.

(2) *Amplify*: The product of step 1 was amplified by polymerase chain reaction (PCR) using primers s_0 and $h(s_t)$. Thus, only those DNA molecules encoding paths that begin with v_0 and end with v_t were amplified. Keep only those paths that begin with v_{in} and end with v_{out} through the operations of $B(N, h(s_0))$ and $E(N, h(s_t))$.

(3) $(N, \leq 20n)$: Keep only those paths that enter exactly n vertices.

(4) $+(N,\ h(s_i))$: Keep only those paths that enter all the vertices at least once through the operation $+(N,\ h(s_i))$, $1 \leq i \leq n$, where n is the number of vertices in G

(5) *Detect*: If any paths remain, return *true*, otherwise return *false*. See the basic operations in section 13.2.2.

For more details, please see the reference [1].

13.3.2 DNA Encoding Method for the SAT Problem

Based on Adleman's experiment, Lipton [2] showed how to solve another famous NP-complete problem, the SAT problem. The advantage of the results is the huge parallelism inherent in DNA computing, which has the potential to yield vast speedups over conventional silicon-based computers.

The SAT Problem

Consider the Boolean formula $F = (x \vee y) \wedge (\neg x \vee \neg y)$, where the variables x and y are allowed to range only over the two values 0 and 1, $\vee$ is the logical OR operation, $\wedge$ is the logical AND operation, and $\neg x$ denotes the negation of x. Usually, one thinks of 0 as *false* and 1 as *true*. The SAT problem is to find the Boolean values for x and y that make the formula F true.

In general, a Boolean formula is of the form $C_1 \wedge C_2 \wedge \cdots \wedge C_m$, where $C_i (1 \leq i \leq m)$ is a clause; a clause is of the form $x_1 \vee x_2 \vee \cdots \vee x_k$, where x_i is a Boolean variable or its negation. The SAT problem is to find the values for the variables that make the formula have the value 1, that is, to find the values for the variables that make each clause have the value 1. The SAT problem has been proved to be NP-complete.

DNA Encoding Method

Given a Boolean formula F containing n variables $x_1, x_2, \ldots, x_n$, Lipton [2] designed the following DNA encoding method for the SAT problem.

(1) Construct a graph G with vertices a_1, x_1, x'_1, a_2, x_2, x'_2, $\ldots$, a_{n+1}, and with edges from a_i to both x_i and x'_i and from both x_i and x'_i to a_{i+1}, as shown in Fig. 13.4.

For the constructed graph G, each stage of a path has exactly two choices: If it takes the vertex with an unprimed label, or x_i, it encodes a 1; if it takes the vertex with a primed label, or x'_i, it encodes a 0. For example, the path $a_1x'_1a_2x_2a_3$ encodes the binary number 01. Obviously, each path starting with a_1 and ending with a_{n+1} in G encodes an n-bit binary number.

(2) Each vertex v_i $(1 \leq i \leq 3n+1)$ in G is assigned a random DNA strand s_i of length 20.

The DNA strand s_i corresponding to vertex v_i has two parts. The first half is denoted by s'_i and the second half is denoted by s''_i. That is, $s'_is''_i$ is the code corresponding to vertex v_i, $v_i \in \{a_1, x_1, x'_1, a_2, x_2, x'_2, \ldots, a_{n+1}\}$.

(3) For each edge $e_{ij} = (v_i, v_j)$, $-h(s''_is'_j)$ is used to encode it, where $-h(x)$ denotes the reverse complement of x.

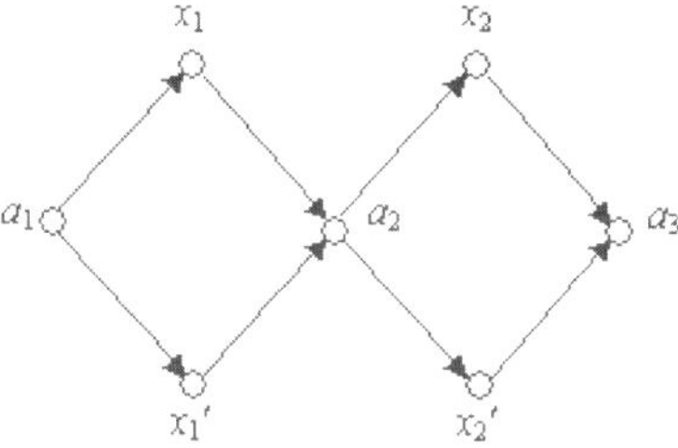

Fig. 13.4. The graph corresponding to a Boolean formula that contains 2 variables x_1, x_2

DNA Algorithm

For the DNA algorithm proposed by Lipton [2], let $S(t,i,a)$ denote all the sequences in tube t for which the ith bit is equal to a, $a \in \{0,1\}$. This is done by performing one extraction operation that checks for the sequence corresponding to x_i {if $a = 1$} or x_i' {if $a = 0$}. Consider the Boolean formula $F = (x \vee y) \wedge (\neg x \vee \neg y)$. The following steps can solve it.

(1) Construct a graph G with vertices a_1, x_1, x_1', a_2, x_2, x_2', a_3, and with edges from a_i to both x_i and x_i' and from both x_i and x_i' to a_{i+1}, $1 \le i \le 2$. Each vertex in G is assigned a random DNA strand of length 20. For each edge $e_{ij} = (v_i, v_j)$, $-h(s_i'' s_j')$ is used to encode it.

(2) Let t_0 be the tube including the sequences corresponding to $a_1x_1a_2x_2a_3$, $a_1x_1'a_2x_2a_3$, $a_1x_1a_2x_2'a_3$, $a_1x_1'a_2x_2'a_3$.

(3) Let tt_1 be the tube corresponding to $S(t_0,1,1)$. Let the reminder be tt_1', and tt_2 be $S(tt_1',2,1)$. Pour tt_1 and tt_2 together to form t_1.

(4) Let tt_3 be the tube corresponding to $S(t_1,1,0)$. Let the reminder be tt_3' and tt_4 be $S(tt_3',2,0)$. Again pour tt_3 and tt_4 together to form t_2.

(5) Detect DNA in the last tube t_2. If there is any DNA in t_2, the formula is satisfiable.

Now consider the SAT problem on n variables and m clauses. Suppose that, as is usual, each clause consists of a fixed number of variables or their negations. Let C_1, C_2, $\ldots$, C_m be the m clauses. A series of tubes t_0, t_1, $\ldots$, t_m are constructed so as to let t_k be the set of n-bit numbers in which each element x satisfies that $C_1(x) = C_2(x) = \cdots = C_k(x) = 1$, where $C_i(x)$ is the value of C_i on x.

(1) Construct a graph G with vertices a_1, x_1, x_1', a_2, x_2, x_2', $\ldots$, a_{n+1}, and with edges from a_i to both x_i and x_i' and from both x_i and x_i' to a_{i+1}, $1 \le i \le n$. Each vertex in G is assigned a random DNA strand of length 20. For each edge $e_{ij} = (v_i, v_j)$, the DNA strand $-h(s_i'' s_j')$ is used to encode it.

(2) Let t_0 be the tube including all n-bit sequences.

(3) Construct t_{k+1}, $k = 0, 1, \ldots, m-1$, step by step. Let C_{k+1} be the clause $x_1 \vee x_2 \vee \ldots x_l$, where x_i is a literal or its negation. For each literal x_i, if x_i is equal to x_j, then form $S(t_k, j, 1)$; if it is equal to $\neg x_j$, then form $S(t_k, j, 0)$. The reminder of each extraction is used for the next step. Pour all the reminders together to form t_{k+1}.

(4) Detect DNA in the tube t_m. If there is any DNA in t_m, the formula is satisfiable.

For more details, please see the reference [2].

13.3.3 DNA Encoding Method for the Maximal Clique Problem

In 1997, Ouyang [3] solved the maximal clique problem using the techniques of molecular biology. A pool of DNA molecules corresponding to the ensemble of six-vertex cliques was built, followed by a series of selection processes. The algorithm is highly parallel and has satisfactory fidelity.

The maximal Clique Problem

Mathematically, a clique is defined as a subset of vertices in a graph, in which each vertex is connected to all other vertices in the subset. The clique including the most vertices is called the maximal clique. The maximal clique problem asks: Given an undirected graph with n vertices and m edges, how many vertices are in the maximal clique? The corresponding decision problem has been proved to be NP-complete. Take the graph shown in Fig. 13.5(a) as an example. The vertices (5, 4, 3, 2) form the maximal clique, that is, the size of the maximal clique is four.

DNA Encoding Method

For an undirected graph $G = (V, E)$ with n vertices, Ouyang [3] designed the following DNA encoding method to solve the maximal clique problem.

(1) Each clique in G is represented as an n-bit binary number. A bit set to 1 represents the vertex being in the clique, and a bit set to 0 represents the vertex being out of the clique. Thus, the set of all the cliques in G is transformed into the ensemble of n-bit binary numbers, which is called the complete data pool [3].

(2) Each bit in a binary number corresponds to two DNA sections: one for the bit's value (V_i) and another for its position (P_i). The length of V_i is set to 10 base pairs if the value of V_i is equal to 0, and 0 base pair if the value of V_i is equal to 1. Thus, the ensemble of DNA strands representing the complete data pool is constructed.

(3) The restriction sequence is embedded within V_i if the value of V_i is equal to 1.

Thus, for a DNA strand representing an n-bit binary number, there are n value sections (V_0 to V_{n-1}) sandwiched sequentially between $n+1$ position sections (P_0 to P_n). The last position section P_n is needed for PCR amplification.

DNA Algorithm

In order to easily understand the DNA algorithm for the maximal clique problem, the definition of complementary graph [3] is given as follows.

Definition 13.3.3.1 For an undirected graph $G = (V, E)$, the graph $G' = (V, E')$ only containing all the connections absent in G is called the complementary graph of G.

Take the graph shown in Fig. 13.5(a) as an example, Fig. 13.5(b) is its complementary graph. According to definition 13.3.3.1, any two vertices connected in the

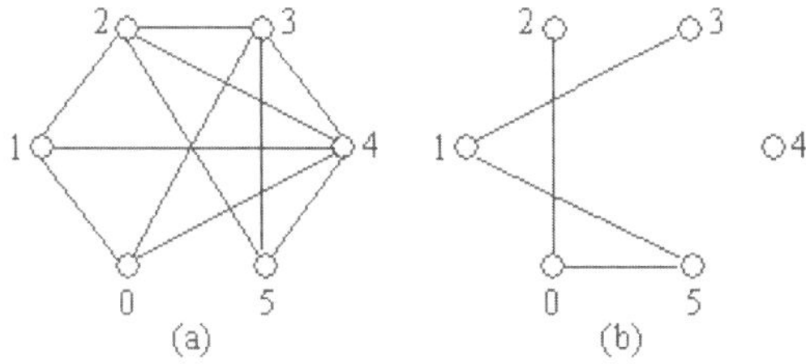

Fig. 13.5. A graph and its complementary graph. (a) An undirected graph G and (b) The complementary graph G' of G

complementary graph are disconnected in the original graph and therefore cannot be members of the same clique; this means that the corresponding bits cannot both be set to 1. Based on this, Ouyang [3] devised the following DNA algorithm for the maximal clique problem.

(1) Generate the random sequence of each P_i and V_i, and then construct the complete data pool by means of the technique of parallel overlap assembly (POA) [3].

The construction starts with $2n$ DNA strands: $P_0V_0^0P_1$, $P_0V_0^1P_1$, $-h(P_1V_1^0P_2)$, $-h(P_1V_1^1P_2)$, $P_2V_2^0P_3$, $P_2V_2^1P_3$, $-h(P_3V_3^0P_4)$, $-h(P_3V_3^1P_4)$, etc. Each DNA strand consists of two position motifs and one value motif, $P_iV_iP_{i+1}$ for even i and $-h(P_iV_iP_{i+1})$ for odd i, where the value of V_i can be 0 or 1. The $2n$ DNA strands were mixed together for thermal cycling [3]. During each thermal cycle, the position string in one DNA strand is annealed to the reverse complement of the next DNA strand. After a few thermal cycles, a data pool with all combinations of $V_0V_1V_2 \dots V_{n-1}$ was built. The POA procedure was followed by PCR, and the molecules with P_0 and P_n at their ends were exponentially amplified.

(2) Eliminate all the numbers containing connections in the complementary graph from the complete data pool. The remainder corresponds to all the cliques in the original graph.

Guided by the complementary graph, the data pool was digested with restriction enzymes. These enzymes break DNA at specific restriction sites, which were embedded within the sequences for $V_i = 1$. Consider 0-2 connection in the complementary graph, that is, there were xx...x0x0, xx...x0x1, xx...x1x0, and xx...x1x1 in the data pool. The data pool was divided into two tubes, t_0 and t_1. In t_0, the DNA strands containing $V_0 = 1$ were cut with *Afl* II. Thus, t_0 contained only xx...x0x0 and xx...x1x0. In t_1, the DNA strands containing $V_2 = 1$ were cut with *Spe* I. Thus, t_1 contained only xx...x0x0 and xx...x0x1. And then, t_0 and t_1 were put into tube t, which contained xx...x0x0, xx...x1x0 and xx...x0x1. That is, tube t did not contain xxx1x1.

(3) Sort the remaining data pool to find the data containing the largest number of 1's. The clique with the largest number of 1's tells us the size of the maximal clique.

For more details, please see the reference [3].

13.4 DNA Encoding Methods for the Problems Related to Graphs with Weights

The previous works deal with the problems related to graphs without weights. The significance of these researches is that they demonstrate how DNA can be used for representing information and solving the problems in the complexity class NP. However, there are many practical applications related to weights. Representation of weight information is one of the most important but also challenging problems in DNA computing. Some of the methods representing weights in DNA strands are given in the following.

13.4.1 Encoding Weights by the Lengths of DNA Strands

In 1998, Narayanan *et al* [6] proposed a DNA encoding method of representing weights for the shortest path problem. For a connected, weighted graph $G = (V, E)$, the shortest path problem is to find a path with minimum cost (weight) that begin with a specified vertex and end with another one.

DNA Encoding Method

For a connected, weighted graph $G = (V, E)$ with n vertices and m edges. Narayanan *et al* [6] designed the following DNA encoding method for the shortest path problem.

(1) For each vertex v_i ($i = 1, 2, \ldots, n$), assign a unique DNA sequence s_i with fixed length to encode it.

(2) Sort all the edges in G by distance (weight), and put their distances into a vector $\mathbf{D}$. For each distance d in $\mathbf{D}$, a DNA sequence s_d is randomly selected whose length l is associated with the location of distance d and a constant factor k. Consider $k = 3$ and $D = \{2, 5, 9, 10\}$, 2 is represented by a strand of length 3, 5 is represented by a strand of length 6, and so on.

(3) For each edge $e_{ij} = (v_i, v_j)$, the DNA strand s_{idj} is created in the following way: if $i = 1$, create the strand s_{idj} as ALL s_i + ALL s_d + HL s_j; if $i > 1$, create the strand s_{idj} as HR s_i + ALL s_d + HL s_j, where ALL represents the whole DNA strand, HL the left half, HR the right half, and + the join operation.

DNA Algorithm

For a connected, weighted graph $G = (V, E)$, the following steps adapted from Adleman's DNA algorithm [1] can extract the shortest path between the initial vertex v_1 and the destination vertex v_t.

(1) *Merge*: Put all the DNA strands $h(s_i)$ and s_{idj} into a tube, and perform a DNA ligase reaction in which random paths through G are formed.

(2) *Amplify*: The strands beginning with v_1 are amplified through a polymerase chain reaction using primers s_1 and $h(s_t)$.

(3) $B(N, s_1)$: Keep only the strands beginning with v_1.

(4) $E(N, s_t)$: Keep only the strands ending with v_t.

(5) All the obtained strands are sorted through gel electrophoresis. The shortest strand corresponds to the desired solution.

For more details, please see the reference [6].

13.4.2 Encoding Weights by the Number of Hydrogen Bonds

Shin *et al* [7] presented an encoding method that uses fixed-length codes for representing integer and real values. In this method, the relative values of *G/C* contents against *A/T* contents are taken into account to represent weights in the graph, which is based on the fact that hybridization between *G/C* pairs occurs more frequent than those between *A/T* pairs because there are 3 hydrogen bonds between G and C, whereas 2 hydrogen bonds between A and T. Generally, the ligation between DNA sequences is influenced by DNA length and the *G/C* contents [7]: The longer the sequences, the more often they get hybridized; the more G/C pairs the sequences have, the more probability they get hybridized.

Shin's method was applied to the traveling salesman problem. For a connected, weighted graph, the traveling salesman problem is to find a minimum cost (weight) path that begins with a specified vertex and ends there after passing through all other vertices exactly once.

DNA Encoding Method

For a connected, weighted graph $G = (V, E)$, the DNA encoding method proposed by Shin [7] is as follows.

(1) For each vertex v_j, which is the common vertex of edges e_{ij} and e_{jk}, the DNA sequence s_j consists of 4 components: 10 bp weight sequence $h(W''_{ij})$, 10 bp position sequence P'_j, 10 bp position sequence P''_j, and 10 bp weight sequence $h(W'_{jk})$, where x' denotes the first half of x, and x'' denotes the last half of x.

(2) For each edge $e_{ij} = (v_i, v_j)$, the DNA sequence s_{ij} also consists of 4 components: 10 bp link sequence $h(P''_i)$, 10 bp weight sequence W'_{ij}, 10 bp weight sequence W''_{ij}, and 10 bp link sequence $h(P'_j)$. The orientation of edge code is opposite to that of vertex code.

For more details, please see the reference [7].

DNA Algorithm

The DNA algorithm proposed by Shin [7] consists of two parts: a genetic algorithm for optimizing the DNA codes, and a molecular algorithm for simulating the DNA computing process.

Algorithm for genetic code optimization

The codes are optimized using the following genetic algorithm.

(1) For each vertex v_i, generate randomly the vertex position sequences P_i, $1 \leq i \leq n$.

(2) For each edge $e_{ij} = (v_i, v_j)$, generate the edge link sequences, that is, the reverse complements of the last half of the vertex position sequence P_i and the first half of the vertex position sequence P_j.

(3) For each weight w_{ij} on edge e_{ij}, generate randomly the edge weight sequences W_{ij}.

(4) Generate the vertex weight sequences according to the edge weight sequences.

(5) While ($generation\ g \leq g_{max}$) do {Evaluate the fitness of each code; Apply genetic operators to produce a new population}.

(6) Let the best code be the fittest one.

In the genetic algorithm, the amount of *G*/*C* contents in edge sequences is optimized in step 5 so as to let the edges with smaller weights have more *G*/*C* contents and thus have higher probability being contained in the final solution. The fitness function is to promote the paths with lower costs (path lengths) so that the minimum cost path could be found. Let N_{eij} denote the number of hydrogen bonds in edge e_{ij}, S_h denote the total number of hydrogen bonds in all edges, W_{eij} denote the weight on edge e_{ij}, and S_w denote the *sum* of the weight values. The fitness function is defined as follows: if $|N_{eij}/S_h - W_{eij}/S_w| \leq \theta$, then $F_i = |N_{eij}/S_h - W_{ei}/S_w|$; otherwise, $F_i = 0$, where the threshold value θ is determined by experiments.

Molecular algorithm

The molecular algorithm adopted the same as the iterative version of molecular programming [19]. The iterative molecular algorithm (IMA) iteratively evolves fitter sequences rather than simply filtering out infeasible solutions. This procedure is summarized as follows.

(1) Encoding: Determine the code sequence using the algorithm of genetic code optimization.

(2) While ($cycle\ c \leq c_{max}$) do {*Synthesis*: Produce candidate solutions by molecular operators; *Separation*: Filter out infeasible solutions by laboratory steps}.

(3) Keep only those paths that begin with V_{in} and end with V_{in}.

(4) Keep only those paths that enter exactly $n+1$ vertices, where n is the number of vertices in the graph.

(5) Keep only those paths that enter all the vertices at least once.

(6) Select the path that contains the largest amount of G/C pairs, which corresponds to the minimum cost path.

For more details, please see the reference [7].

13.4.3 Encoding Weights by the Concentrations of DNA Strands

Yamamoto *et al* [8] presented a method of encoding weights by the concentrations of DNA strands, and used it to the shortest path problem.

DNA Encoding Method

For a connected, weighted graph $G = (V, E)$, the DNA encoding method proposed by Yamamoto [8] is as follows.

(1) Each vertex v_i in G is associated with a 20-mer DNA sequence denoted by s_i.

(2) For each edge $e_{ij} = (v_i, v_j)$ in G, a DNA strand s_{ij} that is $3'$ 10-mer of s_i followed by $5'$ 10-mer of s_j is created. The relative concentration D_{ij} of s_{ij} is calculated by the following formula: $D_{ij} = (min/w_{ij})^{\alpha}$, where min represents the minimum weight in G, w_{ij} represents the weight on edge e_{ij}, and α is a parameter value.

DNA Algorithm

Based on the DNA encoding method, the DNA algorithm [8] for the shortest path problems is as follows.

(1) For each vertex v_i in G, set the concentration of $h(s_i)$ to a certain value. Note that the concentrations of all the DNA strands $h(s_i)$ are set to the same value.

(2) For each edge e_{ij} in G, calculate the relative concentration D_{ij} of s_{ij} according to the formula $D_{ij} = (min/w_{ij})^{\alpha}$.

(3) Put all the DNA strands $h(s_i)$ with the same concentration and all the DNA strands s_{ij} with different concentrations D_{ij} to construct random paths through G.

(4) Amplify the DNA paths that begin with the start vertex and end with the destination vertex.

(5) Determinate the DNA strand of encoding the shortest path.

For more details, please see the reference [8].

13.4.4 Encoding Weights by the Melting Temperatures of DNA Strands

Lee *et al* [9] introduced a DNA encoding method to represent weights based on the thermodynamic properties of DNA molecules, and applied it to the traveling salesman problem. This method uses DNA strands of fixed-length to encode different weights by varying the melting temperatures, in which the DNA strands for higher-cost values have higher melting temperatures than those for lower-cost values.

DNA Encoding Method

For an instance of the traveling salesman problem, Lee *et al* [9] gave the following method of encoding weights.

(1) Each city sequence is designed to have a similar melting temperature. That is, city sequences contribute equally to the thermal stability of paths.

(2) Cost sequences are designed to have various melting temperatures according to the costs. A smaller cost is represented by a DNA sequence with a lower melting temperature.

(3) Road sequences that connect two cities are generated using the sequences of departure cities, costs, and arrival cities. The first part of the road sequence is the complement of the last half of the departure city, the middle part represents the cost information, and the last part is the complement of the first half of the arrival city.

There are several empirical methods to calculate the melting temperatures. One of them is the *GC* content method that uses the content of *G* and *C* in DNA strand as a main factor determining melting temperature.

For more details, please see the reference [9].

DNA Algorithm

The DNA algorithm modifies the PCR protocol to employ temperature gradient in the denaturation step. The denaturation temperature is low at the beginning of PCR and then it increases gradually. With help of the denaturation temperature gradient PCR (DTG-PCR), the more economical paths of lower T_m can be amplified more intensively. The DNA algorithm presented by Lee *et al* [9] for the traveling salesman problem is as follows.

(1) Generate the answer pool through the operations of hybridization and ligation.

(2) Select the paths satisfying the conditions of the traveling salesman problem through the operations of PCR with primers and affinity-separation.

(3) Amplify the more economical paths through the operation of DTG-PCR.

(4) Separate the most economical path among the candidate paths.

(5) Read the final path through sequencing.

For more details, please see the reference [9].

13.4.5 Encoding Weights by Means of the General Line Graph

Han *et al* [10,11] proposed a DNA encoding method to represent weights and applied it to the Chinese postman problem, an instance of optimization problems on weighted graphs. For a weighted, undirected graph $G=(V,E)$, Han *et al* first convert it into its general line graph $G'=(V',E')$, and then design the DNA encoding method based on G'.

The Chinese Postman Problem

The Chinese postman problem is to find a minimum cost tour that a postman sets out from the post office, walks along each street to deliver letters, and returns to the post office. If the layout of streets is an Euler graph, the Euler tour is just what we want; otherwise, he needs to walk along some streets more than once. The Chinese postman problem can be abstracted as follows: For a connected, weighted, undirected graph $G=(V,E)$, $v_i \in V$, $1 \le i \le n$, $e_j \in E$, $1 \le j \le m$, where the weight on edge e_j is w_j, $w_j \ge 0$, $w_j \in Z$, it is to find a minimum cost (weight) tour that begins with a specified vertex and ends there after passing through all the given edges. That is, the Chinese postman problem is to find a shortest tour that goes through all the edges in G.

Construction of General Line Graph

Definition 13.4.5.1 For an undirected graph $G=(V,E)$, $v_i \in V$, $1 \le i \le n$, $e_j \in E$, $1 \le j \le m$, a mapping function f is constructed to satisfy: (1) For each edge $e_j \in E$, there exists only one vertex v'_j to satisfy $f(e_j)=v'_j$; (2) If any two edges e_i and e_j are adjacent, draw an undirected edge between v'_i and v'_j; (3) If v_i is with odd degree, maximum one self-loop is added to each of the vertices which are mapped from the edges linked to v_i. The function f is called the mapping function from edges to vertices, and the obtained graph is called the general line graph of G.

Take the weighted, undirected graph G shown in Fig. 13.6(a) as an example. The procedure of mapping from edges to vertices is as follows. (1) The edges e_1, e_2, $\ldots$,

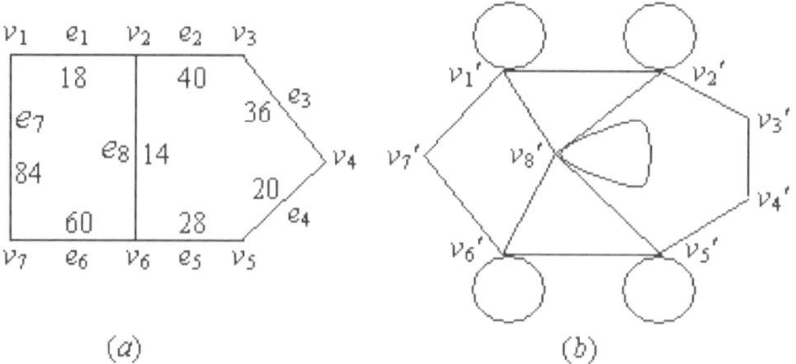

Fig. 13.6. A weighted graph G and its general line graph G'. (a) A weighted graph G and (b) The general line graph G' of G

e_8 are respectively mapped to the vertices v_1', v_2', ..., v_8'. (2) Adding the undirected edges. The vertex v_1' is linked to v_2', v_8', v_7' since the edge e_1 is adjacent to e_2, e_8, e_7. The vertex v_2' is linked to v_3', v_8', v_1' since the edge e_2 is adjacent to e_3, e_8, e_1. Similar operation is done for other vertices v_3', v_4', ..., v_8'. (3) Adding the self-loops. A self-loop is respectively added to v_1', v_2' and v_8' since v_2 is with odd degree and v_2 is linked to the edges e_1, e_2 and e_8. A self-loop is respectively added to v_5' and v_6' since v_6 is with odd degree and v_6 is linked to the edges e_5, e_6 and e_8. The obtained general line graph G' is shown in Fig. 13.6(b).

By means of the mapping function from edges to vertices, Han *et al* [10, 11] convert the problem of searching for the shortest tour that pass through each edge at least once into that of searching for the shortest tour that pass through each vertex at least once. Note that the shortest tour may be not only one. For example, the shortest tours in Fig. 13.6(a) are $v_1e_1v_2e_2v_3e_3v_4e_4v_5e_5v_6e_8v_2e_8'v_6e_6v_7e_7v_1$ and $v_1e_1v_2e_8'v_6e_8v_2e_2v_3e_3v_4e_4v_5e_5v_6e_6v_7e_7v_1$, where e_8' denotes the reversal of e_8. In order to easily observe, we use edge sequence to denote the shortest tour in G, such as $e_1e_2e_3e_4e_5e_8e_8'e_6e_7$, and use vertex sequence to denote the shortest tour in G', such as $v_1'v_2'v_3'v_4'v_5'v_8'v_8'v_6'v_7'$.

DNA Encoding Method

Given a connected, weighted, undirected graph $G = (V, E)$, $v_i \in V$, $1 \leq i \leq n$, $e_j \in E$, $1 \leq j \leq m$, where the weight on edge e_j is w_j, $w_j \geq 0$, $w_j \in Z$. If w_i is a real number, all the weights are multiplied by a certain integer (i.e. 10) and then they are rounded into integers. The main idea of the DNA encoding method proposed by Han *et al* [10, 11] is as follows: The given graph G is firstly converted into its general line graph $G' = (V', E')$, $v_i' \in V'$, $1 \leq i \leq m$, where v_i' is mapped from e_i. For each vertex v_i', use DNA strand s_i of length w_i to encode it. For each edge $e_{ij}' = (v_i', v_j')$, use the DNA strand s_{ij}, which is the reverse complement of the last half of s_i and the first half of s_j, to encode it. Note that the DNA strands to encode vertices are of different lengths. The detailed encoding method [10, 11] for the Chinese postman problem is as follows:

(1) All the edges e_j ($1 \leq j \leq m$) are mapped to vertices v'_j. If e_i and e_j are adjacent, an undirected edge is drawn between v'_i and v'_j. If v_i is with odd degree, maximum one self-loop is added to each of the vertices which are mapped from the edges linked to v_i.

(2) For each vertex v'_i, use DNA strand s_i of length w_i to encode it.

(3) For each edge $e'_{ij} = (v'_i, v'_j)$, use the reverse complement of the last half of s_i and the first half of s_j to encode it. Specifically, s_i and s_j are firstly divided into two substrands with equal length, or $s_i = s'_i s''_i$, $s_j = s'_j s''_j$. And then use the DNA strand $s_{ij} = -h(s''_i s'_j)$ to encode edge $e'_{ij} = (v'_i, v'_j)$, where s_{ij} is with the length of $|s_i|/2$+$|s_j|/2$. Here, suppose that weights in the weighted graph are all even. If there exists one or more weights are odd in a practical problem, all the weights are multiplied by 2. Thus, a half of the optimal solution is the desired results.

Note that, for any undirected edge $e'_{ij} = (v'_i, v'_j)$, if walk from v'_i to v'_j, the code s_{ij} is the reverse complement of s''_i and s'_j, or $-h(s''_i s'_j)$; if walk from v'_j to v'_i, the code s_{ji} is the reverse complement of $(-s_j)''$ and $(-s_i)'$, or $s_{ji} = -h((-s_j)''(-s_i)') = h(s''_i s'_j) = -s_{ij}$. That is, only need one code $s_{ij} = -h(s''_i s'_j)$ to encode edge $e'_{ij} = (v'_i, v'_j)$.

Take the weighted graph G shown in Fig. 13.6(a) as an example. We specifically analyze the proposed DNA encoding method. First of all, the general line graph G' is converted from the given graph G, as shown in Fig. 13.6(b). For the vertices v'_1, $v'_2, \ldots, v'_8$ in the general line graph G', the following DNA strands $s_1, s_2, \ldots, s_8$ with the lengths of 18, 40, 36, 20, 28, 60, 84, 14 are respectively selected to encode them.

$s_1 = CAGTTGACATGCAGGATC$

$s_2 = CAACCCAAAACCTGGTAGAGATATCGCGGGTTCAACGTGC$

$s_3 = TAGTACTGATCGTAGCAACCTGGTACCAAGCTTGAC$

$s_4 = CGCATGCAGGATTCGAGCTA$

...

$s_8 = TGGTTTGGACTGGT$

For each edge $e'_{ij} = (v'_i, v'_j)$, the DNA strand $s_{ij} = -h(s''_i s'_j)$ is used to encode it. For example, the code of $e'_{12} = (v'_1, v'_2)$ is as follows.

$s_{12} = -h(s'_1{}' s'_2) = h(TGCAGGATCCAACCCAAAACCTGGTAGAG)$

$=ACGTCCTAGGTTGGGTTTTGGACCATCTC$

Obviously, s_{12} is with the length of $18/2+40/2 = 29$. The joint among the DNA strands of encoding edge $e'_{12} = (v'_1, v'_2)$ and vertices v'_1, v'_2 is shown in Fig. 13.7.

On the basis of Fig. 13.7, the next is to extend rightward to s_3 or s_8, and to extend leftward to s_7 or s_8. The extension of the DNA strand $s_1 s_2$ rightward is shown in Fig. 13.8, and its extension leftward can similarly be drawn up.

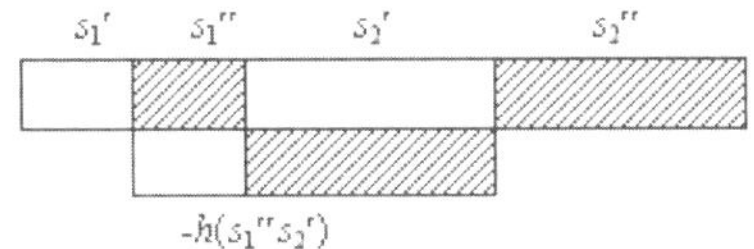

Fig. 13.7. Joint of DNA strands s_1, s_2 and s_{12}

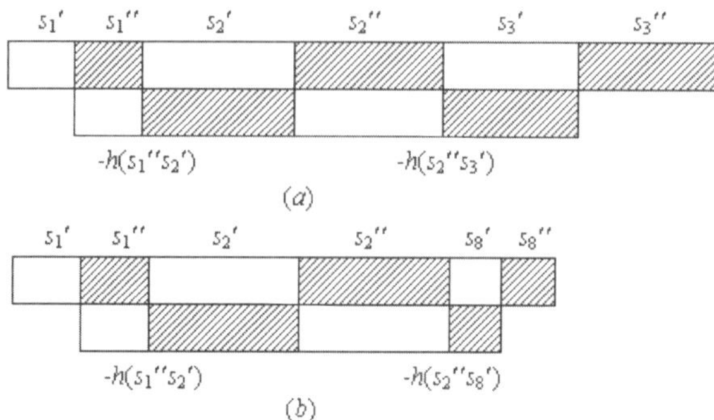

Fig. 13.8. Extension rightward of DNA strand s_1s_2. (a) Joint of DNA strands s_1, s_2, s_3, s_{12}, s_{23} and (b) Joint of DNA strands s_1, s_2, s_8, s_{12}, s_{28}

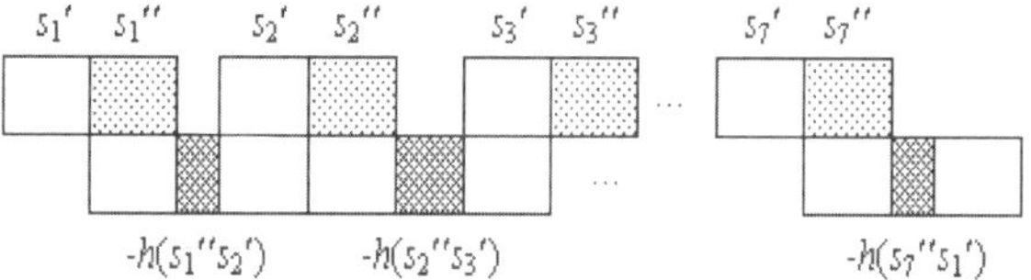

Fig. 13.9. Alternant DNA strand and double strand

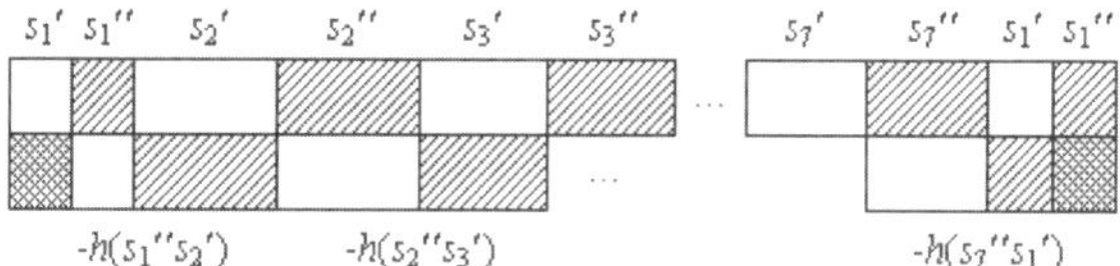

Fig. 13.10. Double-stranded DNA corresponding to the optimal solution

On the basis of Fig. 13.8(a), the next is to extend rightward to s_4, and to extend leftward to s_7 or s_8, and so on. Thus, with the help of the property of reverse complementation between vertex codes and edge codes, the DNA strands may extend continually to form various random paths including the optimal solution.

In the DNA encoding method [10, 11], the paths generated in a single ligation reaction are double-stranded DNA instead of alternant DNA strand and double strand. An alternant DNA strand and double strand is shown in Fig. 13.9, and a double-stranded DNA is shown in Fig. 13.10. It is well known that the stable structure of DNA molecules is DNA double strand. In an alternant DNA strand and double strand, the part of DNA strand can combine with other molecules through hydrogen bonds based on the Watson-Crick's law. Since DNA double strand are more stable than alternant DNA strand and double strand, the proposed DNA encoding method can more easily generate the optimal solution. In addition, the proposed DNA encoding method uses DNA strands of different lengths to encode different vertex. It also has characteristics of easy encoding and low error rate. But when the values of the weights are very large, the lengths of the DNA strands are very long which result in higher space complexity. For more details, please see the reference [10, 11].

DNA Algorithm

For the general line graph G' converted from the given graph G, suppose that v'_1 is the original vertex just as well because, for any shortest tour C going through all the edges in G', the length of the route beginning with v'_1 and ending there along C is equal to that of the route beginning with v'_i $(i \neq 1)$ and ending there along C.

In order to easily generate the optimal solution, the DNA algorithm proposed by Han *et al* [10, 11] searches for the shortest path instead of the shortest tour. The reason is that the length of a shortest tour that begins with v'_1 and end there after passing through all the edges in G' is equal to that of a shortest path that begins with v'_1 and end there after passing through all the edges in G'. Moreover, the polymerase chain reaction (PCR) in the biological techniques is generally carried out on a linear template; there is no circular template so far. The detailed DNA algorithm [10, 11] for the Chinese postman problem is as follows.

(1) *Merge*: The DNA strands s_i and s_{ij} $(1 \leq i, j \leq m)$ are mixed together in a single ligation reaction. Based on the Watson-Crick's law, generate various DNA molecules corresponding to the random paths.

(2) *Amplify*: The product of step 1 is amplified by polymerase chain reaction (PCR) using primers $-h(s''_1)$ and $-h(s'_1)$. Thus, only those DNA molecules encoding paths that begin with v'_1 and end with v'_1 were amplified.

(3) $B(N, s_1)$: Separate all the paths with the departure vertex v'_1, or separate all the DNA molecules with $5'$ end being s_1.

(4) $E(N, s_1)$: Separate all the paths with the arrival vertex v'_1, or separate all the DNA molecules with $3'$ end being s_1.

(5) $+(N, s_i)$: For each vertex v'_i $(2 \leq i \leq m)$, separate all the paths including v'_i.

(6) Separate the shortest path through gel electrophoresis.

(7) Determinate the nucleotides sequence of the shortest path, which corresponds to the optimal solution.

For more details, please see the reference [10, 11].

13.4.6 RLM: Relative Length Method of Encoding Weights

Han [12] presented a method of encoding weights in DNA strands for the problems related to graph with weights, which is referred to the relative length method (RLM), and applied it to the traveling salesman problem. The RLM method can directly deal with weights of either real numbers or integers, even very small and very big positive weights, and the lengths of DNA strands used in the RLM method are not proportional to the values of weights.

Definitions Involved in the RLM Method

Definition 13.4.6.1 For a weighted graph $G = (V, E)$, $v_i \in V$, $1 \leq i \leq n$, $e_j \in E$, $1 \leq j \leq m$, all the weights are sorted in a nondecreasing order, and the equal weights are at the same position. Thus, all the weights are divided into p groups $(p \leq m)$ according to their ranking. The p groups are numbered from 1 to p, respectively. The group number is called the order number of the weight.

Definition 13.4.6.2 For a weighted graph $G = (V, E)$, $v_i \in V$, $e_{ij} \in E$, $1 \leq i, j \leq n$, where the weight on edge e_{ij} is w_{ij}, all the weights w_{ij} are remarked as $w_{ij,k}$, where k is the order number of w_{ij}. For each remarked weight $w_{ij,k}$, we add $k-1$ nodes on edge e_{ij}. The obtained graph G' is called the relative length graph of G.

Obviously, if the weight w_{ij} is remarked as $w_{ij,k}$, the edge e_{ij} will be divided into k segments. The bigger the order number, the more the segments of the edge. That is, the segment number of an edge represents the relative length of the edge. Note that the segment number of an edge is not directly proportional to the weight on the edge. For example, the segment numbers of edges with weights 2, 1000 and 1002 are 1, 2 and 3, respectively.

RLM Method of Encoding Weights

With the help of the relative length graph, Han [12] devised a method of encoding weights in DNA strands for the traveling salesman problem. For a weighted graph $G = (V, E)$ with n vertices and m edges, $v_i \in V$, $e_{ij} \in E$, $1 \leq i, j \leq n$, where the weight on edge e_{ij} is w_{ij}, the RLM method [12] is as follows.

(1) All the weights are divided into p groups ($p \leq m$) according their order numbers, and each weight w_{ij} is remarked as $w_{ij,k}$ if it belongs to the kth group ($1 \leq k \leq p$).

(2) For each remarked weight $w_{ij,k}$, we add $k-1$ nodes on edge e_{ij}. The added nodes are marked as $v_{eij,1}$, $v_{eij,2}$, ..., $v_{eij,k-1}$, respectively. The obtained graph G' is the relative length graph of G.

(3) For each vertex and each added node, we use DNA strand s_i of length $2c$ ($c \in Z$, $c \geq 5$) to encode it. The DNA strand s_i is divided into two sub-strands with equal length, or $s_i = s_i' s_i''$. See the DNA encoding method in section 13.4.5.

(4) For each edge e_{ij} (including the edges that are connecting the nodes newly added in step 2), we use DNA strand $s_{ij} = -h(s_i'' s_j')$ to encode it, where $-h(s)$ denotes the reverse complement of s. Thus, when the vertex-node codes and the edge codes are mixed together, they can combine with each other to form dsDNAs since any DNA strand s can combine with its reverse complement $-h(s)$ to form dsDNA.

Take the graph G shown in Fig. 13.11(a) as an example. All the weights are sorted in a nondecreasing order, or 1.2, 2, 2.5, 3. Thus, they are divided into 4 groups, and the order numbers of weights 1.2, 2, 2.5, 3 are 1, 2, 3, 4, respectively. Therefore, the weights w_{12}, w_{13}, w_{14}, w_{15}, w_{23}, w_{34}, w_{35}, w_{45} are remarked as $w_{12,1}$, $w_{13,3}$, $w_{14,4}$, $w_{15,2}$, $w_{23,4}$, $w_{34,2}$, $w_{35,1}$, $w_{45,3}$, respectively. For each remarked weight $w_{ij,k}$, we add $k-1$ nodes on edge e_{ij}. For example, we add two nodes on e_{13} since w_{13} is remarked as $w_{13,3}$, we add three nodes on e_{14} since w_{14} is remarked as $w_{14,4}$, and so on. The obtained graph G' is the relative length graph of G, as shown in Fig. 13.11(b).

For each vertex v_i in Fig. 13.11(b), we use DNA strand s_i of length 10 ($c = 5$) to encode it. Here, the added nodes are also viewed as vertices. Consider the vertex v_1 and the node $v_{e13,1}$. The DNA strands $s_1 = TTAGCGCATG$, $s_{e13,1} = GTTACGTGAG$ are selected to encode them, respectively.

For each edge e_{ij}, we use DNA strands $s_{ij} = -h(s_i'' s_j')$ to encode it. The edge linking the vertex v_1 and the node $v_{e13,1}$ are encoded by the following DNA strand.

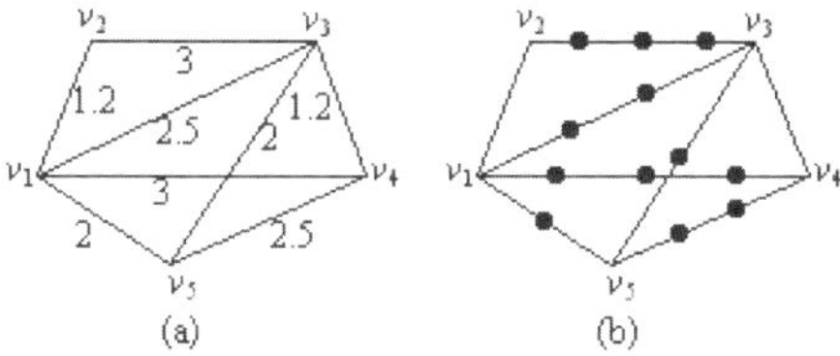

Fig. 13.11. A weighted graph its relative length graph. (a) A weighted graph G and (b) The relative length graph G' of G

$se_{1,e13,1} = -h(s_1''s_{e13,1}') = -h(GCATGGTTAC) = GTAACCATGC$

Thus, the DNA strands s_1, $s_{e13,1}$ and $se_{1,e13,1}$ can combine with each other. Based on the property of reverse complementation between the vertex-node codes and the edge codes, the dsDNAs may extend continually to form various random paths including the optimal solution. For more details, please see the reference [12].

DNA Algorithm

Given a weighted graph $G = (V, E)$ with n vertices and m edges, $v_i \in V$, $e_{ij} \in E$, $1 \leq i, j \leq n$, where the weight on edge e_{ij} is w_{ij}, $w_{ij} \geq 0$. Suppose that v_1 is the start vertex just as well. The DNA algorithm [12] for the traveling salesman problem using the RLM encoding method is as follows.

(1) Construct the relative length graph G' of the given graph $G = (V, E)$. For each vertex or node in G', use DNA strand s_i of length $2c$ to encode it. For each edge e_{ij} in G', use the DNA strand $s_{ij} = -h(s_i''s_j')$ to encode it.

(2) *Merge*: All the DNA strands s_i and s_{ij} are mixed together in a single ligation reaction. Based on the Watson-Crick's law, randomly form various dsDNAs corresponding to the random paths.

(3) $B(N, s_1)$: Separate all the paths beginning with the start vertex v_1, or separate all the DNA molecules with $5'$ end being s_1.

(4) $E(N, s_1)$: Separate all the paths ending with the destination vertex v_1, or separate all the DNA molecules with $3'$ end being s_1.

(5) $+(N, s_i)$: For each vertex v_i $(2 \leq i \leq n)$, separate all the paths including v_i.

(6) Separate the shortest path by means of gel electrophoresis.

(7) Determinate the nucleotide sequence of the shortest path. Suppose that the nucleotide sequence corresponds to v_1, v_{ei}, v_{ei+1}, ..., v_2, v_{ej}, v_{ej+1}, ..., v_1. Delete the nodes v_{ex} from the vertex sequence, the obtained vertex sequence v_1, v_2, ..., v_1 corresponds to the optimal solution.

The RLM method [12] is an improvement on the previous work [6]. The main improvements are as follows. (1) The lengths of DNA strands used in the RLM method are not proportional to the values of weights, which makes the RLM method can easily encode weights of positive real numbers or integers, even very small or very large number. That is, the weights that can be encoded by the RLM method may be in a very broad range since weights are encoded in DNA strands only according to their order numbers. For example, if the weights are 30, 1.8, 400, their order numbers are 2, 1, 3, respectively. Thus, the edges are respectively divided into 2, 1, 3 segments, and the DNA strands to encode them are with the lengths of 20, 10, 30, respectively. (2) The RLM method can distinguish the paths with almost same weights, such as 1000 and 1001, because with the help of the relative length graph, the difference between the lengths of DNA strands used to encode paths is always above or equal to $2c$ ($c \geq 5$). For more details, please see the reference [12].

13.4.7 Method of Encoding Nonlinear Solutions

Han *et al* [16] presented a DNA encoding method for the minimum spanning tree problem, an instance of optimization problems on weighted graphs. The minimum spanning tree problem cannot be directly solved based on the molecular biology techniques because the degrees of some vertices in a minimum spanning tree may be above to 2, which cannot be directly represented by linear DNA strands.

The Minimum Spanning Tree Problem

For a connected, weighted, undirected graph $G = (V, E)$, a spanning tree is a tree that contains all vertices of G, the weight of the spanning tree is the sum of the weights on edges in it, and the minimum spanning tree (MST) is a spanning tree with minimum weight. The MST problem is to find a MST for a connected, weighted, undirected graph.

The MST problem is very important because there are many situations in which MST must be found. Whenever one wants to find the cheapest way to connect a set of terminals, such as cities, electrical terminals, computers, or factories, by using roads, wires, or telephone lines, the solution is a MST for the graph with an edge for each possible connection weighted by the cost of that connection. The MST problem has been studied since the fifties, there are many exact algorithms for it. Han *et al* [16] presented a DNA solution to the MST problem.

DNA Encoding Method

In order to clearly describe the DNA encoding method [16] for the MST problem, we first give the definition of recognition code.

Definition 13.4.7.1 For a connected, weighted graph $G = (V, E)$ with n vertices, a DNA strand used to distinguish a vertex from others is called the recognition code of the vertex.

The length l of recognition code should satisfy: $4^{l-1} < n \leq 4^l$, or $l = \lceil log_4 n \rceil$, where 4 stands for the number of letters in the alphabet $\{A, T, G, C\}$.

For a connected, weighted graph $G = (V, E)$, $v_i \in V$, $e_{ij} \in E$, where each weight w_{ij} on edge e_{ij} is an integer, the DNA encoding method proposed by Han *et al* [16] for the MST problem is as follows.

(1) Let $l = max\{\lceil log_4 n \rceil, 6\}$. For each vertex v_i, use DNA strand r_i of length l to encode it. Here, l denotes the length of recognition code, and 6 is an empirical value which indicates the minimum length of recognition code.

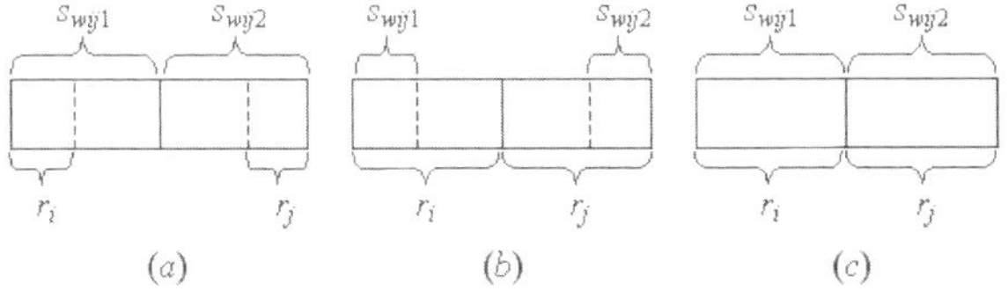

Fig. 13.12. The DNA strand s_{ij} of encoding edge e_{ij}. (a) In the case of w_{ij} being larger than l, (b) In the case of w_{ij} being less than l and (c) In the case of w_{ij} being equal to l

If $n > 4^6$, the recognition codes of length $\lceil log_4 n \rceil$ are needed to distinguish each vertex from others; otherwise, each vertex can be distinguished from others using the recognition codes of length 6. Here, select $l = 6$ instead of $l < 6$ because too short recognition codes would result in high error rate. For example, if the number of vertices in G is 4 and $l = 1$, each vertex can be distinguished from others but in the DNA algorithm, when the recognition codes combine with one part of the DNA strands corresponding to edges, they may combine with another part of the DNA strands because they are too short and easy to be successfully matched to several parts of the DNA strands based on the Watson-Crick's law.

(2) For each edge e_{ij}, the DNA strand s_{ij} of length $2p = 2 \times max\{w_{ij}, l\}$ are used to encode it. Here, the first l letters of s_{ij} are the same as r_i, and the last l letters of s_{ij} are the same as r_j. In addition, the first w_{ij} letters of s_{ij} is marked as s_{wij1}, and the last w_{ij} letters of s_{ij} is marked as s_{wij2}. Note that, when w_{ij} is larger than l, the DNA strand s_{ij} is with a length of $2p = 2 \times w_{ij}$, as shown in Fig. 13.12(a). Here, r_i or r_j should not be the substring of the center part of s_{ij}. When w_{ij} is less than l, the DNA strand s_{ij} is with a length of $2p = 2 \times l$, as shown in Fig. 13.12(b). And when w_{ij} is equal to l, the DNA strand s_{ij} used to encode edge e_{ij} is shown in Fig. 13.12(c).

(3) For any two adjacent edges e_{ij}, e_{jk}, add one DNA strand s_{aijk} as an additional code, which is the reverse complement of s_{wij2} and s_{wjk1}, or $s_{aijk} = -h(s_{wij2}s_{wjk1})$. Obviously, the additional code s_{aijk} is with a length of $w_{ij} + w_{jk}$. Thus, the DNA strands s_{ij} and s_{jk} can combine with the additional code s_{aijk} to form a fragment of dsDNA. Note that, for the edges e_{ij}, e_{ji}, add one DNA strand $s_{aiji} = -h(s_{wij2}s_{wji1})$ as an additional code, which is with a length of $2w_{ij}$.

For more details, please see the reference [16].

DNA Algorithm

The MST problem cannot be directly solved based on molecular biology techniques because some degrees of vertices in a MST may be above to 2, which cannot be directly represented by linear DNA strands. Take the graph G shown in Fig. 13.13(a) as an example, a MST of G is given in Fig. 13.13(b). The degrees of vertices v_2 and v_5 in the MST are 3, which cannot be directly represented by linear DNA strands. In order to generate DNA strands of encoding a MST, each edge in the MST is copied to form an Euler graph G', as shown in Fig. 13.13(c). The Euler cycle in G' can be found out by means of the molecular biology techniques, and the MST can be obtained from the Euler cycle.

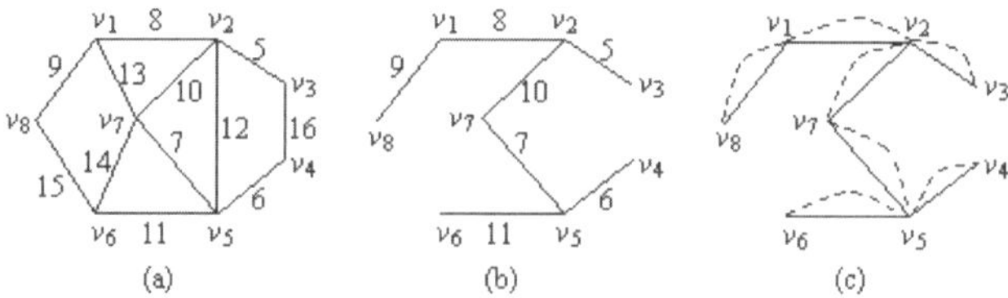

Fig. 13.13. A weighted graph and its minimum spanning tree and Euler graph. (a) A weighted graph G, (b) A minimum spanning tree T of G and (c) The Euler graph G' of T

For a connected, weighted graph $G = (V, E)$, $v_i \in V$, $1 \leq i \leq n$, $e_{ij} \in E$, where each weight w_{ij} on edge e_{ij} is an integer, the DNA algorithm proposed by Han *et al* [16] for the MST problem is as follows.

(1) Let $l = max\{\lceil log_4 n \rceil, 6\}$. For each vertex v_i, use DNA strand r_i of length l to encode it. For each edge e_{ij}, let $p = max\{w_{ij}, l\}$, use DNA strands s_{ij} of length $2p$ to encode it, in which the first l letters are the same as r_i, the last l letters are the same as r_j, and the center part does not include the substrings r_i or r_j. In addition, the first w_{ij} letters of s_{ij} is marked as s_{wij1}, and the last w_{ij} letters of s_{ij} is marked as s_{wij2}.

(2) For any two adjacent edges e_{ij}, e_{jk}, add one DNA strand s_{aijk} as an additional code, which is the reverse complement of s_{wij2} and s_{wjk1}, or $s_{aijk} = -h(s_{wij2}s_{wjk1})$. Note that, for any edge e_{ij}, add one DNA strand $s_{aiji} = -h(s_{wij2}s_{wji1})$ as an additional code.

(3) *Merge*: All the DNA strands s_{ij} and s_{aijk} $(1 \leq i, j \leq n)$ are mixed together. Based on the Watson-Crick's law, generate randomly various part dsDNAs.

(4) *Denature*: All the part dsDNAs are converted into the DNA strands by means of heating.

(6) $-(N, s_i)$: All the DNA strands with any additional code are discarded. Let m denote the number of edges in G, and s_i denote one additional code. All the DNA strands without any additional code can be obtained through the following DNA program: For i:=1 to m do $\{-(N, s_i)\}$. Note that in the DNA encoding method, all the upper DNA strands do not include any additional code.

(7) $+(N, r_i)$: Separate the DNA strands in which the number of recognition codes being at $5'$ end is equal to n, and the number of recognition codes being at $3'$ end is also equal to n. Let cr_i denote the reverse complement of the recognition code r_i, or $cr_i = -h(r_i)$, $1 \leq i \leq n$. All the DNA strands generated in step 6 are mixed with cr_i so as to make the DNA strands combine with cr_i to form part dsDNAs. Thus, all the DNA strands including r_i can be obtained by the operation $+(N, r_i)$. The number of the recognition codes in the DNA algorithm is n, so all the DNA strands with n recognition codes can be obtained through the following DNA program: For i:=1 to n do $\{+(N, r_i)\}$.

(8) Separate the DNA strands with the minimum weight through gel electrophoresis.

(9) Determinate the nucleotide sequence of the DNA strand with minimum weight, which corresponds to the Euler cycle. The minimum spanning tree can be obtained from the Euler cycle.

For more details, please see the reference [16].

13.4.8 DNA Encoding Method for the Maximal Weight Clique Problem

Based on Ouyang's DNA algorithm [3] for the the maximal clique problem, Han *et al* [13] proposed an DNA encoding method for the maximal weight clique problem. For an undirected, weighted graph $G = (V, E)$, $v_i \in V$, $e_{ij} \in E$, where the weight on vertex v_i is w_i, $w_i \geq 0$, the maximal weight clique problem (MWCP) is to find a subset of mutually adjacent vertices, i.e. a clique, which has the largest total weight. Suppose that each weight w_i on vertex v_i is an integer. If one of the weights is a real number, all the weights are multiplied by a certain integer (i.e. 10) and then they are rounded into integers.

DNA Encoding Method

For an undirected, weighted graph $G = (V, E)$ without parallel edges, that is, there is maximum one edge between any two vertices in G, the DNA encoding method proposed by Han *et al* [13] for the maximal weight clique problem is as follows.

(1) For each vertex $v_i \in V$, use two DNA strands s_{i1} and s_{i2} to encode it. The DNA strand s_{i1} consists of three parts: s'_{i1}, s_{wi} and s''_{i1}, where s_{wi} is with a length of w_i, and s'_{i1} or s''_{i1} is with a length of 10, that is, the DNA strand $s_{i1} = s'_i s_{wi}\, s''_i$ is with a length of $20 + w_i$. The strand s_{i2} is the reverse complement of s_{wi}, or $s_{i2} = -h(s_{wi})$. Obviously, s_{i2} can combine with the center part of s_{i1} to form a fragment of dsDNA. After encoding each vertex, the restriction sequences are embedded at both sides of s_{wi} and s_{i2}. The codes of vertex v_i are shown in Fig. 13.14(a).

(2) For each edge $e_{ij} \in E$, use the DNA strand $se_{ij} = -h(s''_{i1} s'_{j1})$ to encode it, which is the reverse complement of the last part of s_{i1} and the first part of s_{j1}. Obviously, the DNA strand se_{ij} is with a length of 20. Thus, the DNA strands corresponding to vertices v_i and v_j can combine with the DNA strand corresponding to edge e_{ij} to form a stable dsDNA, as shown in Fig. 13.14(b).

For more details, please see the reference [13].

DNA Algorithm

For an undirected, weighted graph $G = (V, E)$, $v_i \in V$, $e_{ij} \in E$, where the weight on vertex v_i is w_i, $w_i \geq 0$, let n_c denote the number of edges in the complementary graph G' of G. The DNA algorithm proposed by Han *et al* [13] for MWCP is as follows.

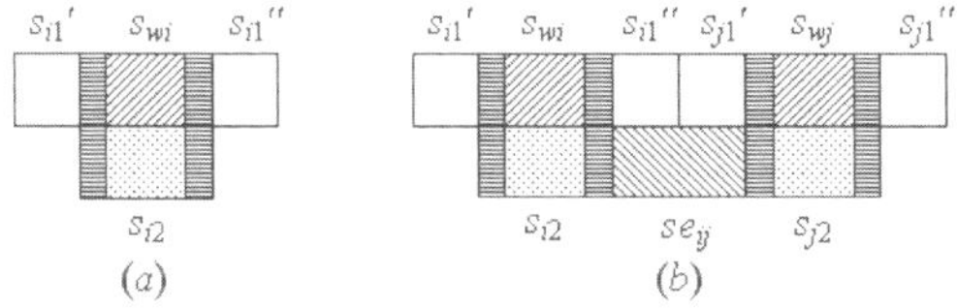

Fig. 13.14. Joint of the vertex codes and the edge code. (a) DNA strands s_{i1} and s_{i2} of encoding vertex v_i and (b) Joint of DNA strands s_{i1}, s_{i2}, s_{j1}, s_{j2} and se_{ij} of encoding vertices v_i, v_j and edge e_{ij}

(1) For each vertex $v_i \in V$, use two DNA strands s_{i1} and s_{i2} to encode it. The DNA strand $s_{i1} = s'_{i1}s_{wi}s''_{i1}$ is with a length of $20+w_i$, and the strand $s_{i2} = -h(s_{wi})$ is with a length of w_i. For each edge $e_{ij} \in E$, use the DNA strand $se_{ij} = -h(s''_{i1}s'_{j1})$ to encode it.

(2) *Merge*: All the DNA strands s_{i1}, s_{i2} and se_{ij} are mixed together in a single ligation reaction. Based on the Watson-Crick's law, generate randomly the various dsDNAs.

(3) *Amplify*: The dsDNAs starting with v_1 and ending with v_n are amplified through polymerase chain reaction (PCR). Let $sumw = \Sigma_{i=1 \to n} w_i$. Only those dsDNAs whose length is equal to or less than *sumw* are saved. The set of the saved dsDNAs is called the complete data pool.

(4) *Digest*: All the dsDNAs in the complete data pool are digested with restriction enzymes. The enzymes break DNA at specific restriction sites, which were embedded within the sequences for v_i ($1 \leq i \leq n$). See step 2 in the DNA algorithm for MCP in section 13.3.3.

(5) (N, $\leq sumw$): Separate all the dsDNA whose length is less than or equal to *sumw*. By n_c sequential restriction operations with different enzymes, all the DNA fragments connected by the edges in the complementary graph G' are digested. Each time of digesting the DNA fragments, let $sumw = sumw - min\{w_i, w_j\}$.

(7) Separate the longest dsDNAs in the remaining data pool through gel electrophoresis.

(8) Determinate nucleotides sequence of the longest dsDNAs, which corresponds to the optimal solution.

The DNA algorithm for MWCP [13] is an improvement on Ouyang's algorithm for MCP [3]. The main improvements are as follows. (1) On the basis of Ouyang's DNA computing model, Han *et al* [13] add weight representation in DNA strands. (2) In Ouyang's algorithm for MCP, the space complexity is $O(n^n)$, where n denotes the number of vertices in the given graph. In Han's algorithm for MWCP, the space complexity is $O(d^n_{max})$, where d_{max} denotes the maximum of vertex degrees in the given graph. (3) In Ouyang's algorithm, all the combinations of vertices are in the complete data pool, whereas in Han's algorithm, those vertices disconnected by edge in G are not in the complete data pool. For more details, please see the reference [13].

13.4.9 DNA Encoding Method for the 0/1 Knapsack Problem

Han *et al* [15] presented a DNA encoding method for the 0/1 knapsack problem. Given a set of n items and a knapsack of capacity c, where each item i has a profit p_i and a weight w_i, the 0/1 knapsack problem is to select a subset of the items which satisfies: the sum of weight does not exceed the knapsack capacity c and the sum of profit is maximal.

DNA Encoding Method

For an instance of the 0/1 knapsack problem, let $\mathbf{I} = \{1, 2, \ldots, n\}$, $\mathbf{P} = \{p_1, p_2, \ldots, p_n\}$, $\mathbf{W} = \{w_1, w_2, \ldots, w_n\}$, and the knapsack capacity $c = c_0$. Suppose that each profit p_i and each weight w_i are integers. If one of the profits is not an integer, all the profits are multiplied by a certain integer (i.e.10) and then they are rounded into integers. If one of the weights is not an integer, all the weights and the knapsack capacity are multiplied by a certain integer and then they are rounded into integers. Also suppose that $w_i \geq p_i$ for all i. If there is any $w_i < p_i$, all of the weights and the capacity are multiplied by a certain integer. The DNA encoding method proposed by Han [15] for 0/1 knapsack problem is as follows.

(1) For each item i $(1 \leq i \leq n)$, use two DNA strands s_{i1} and s_{i2} of different length to encode it. The DNA strand $s_{i1} = s'_{i1}s_{pi}s''_{i1}$ is with a length of w_i, where the center part s_{pi} is with a length of p_i, the first part s'_{i1} is with a length of $\lfloor (w_i - p_i)/2 \rfloor$, and the last part s''_{i1} is with a length of $w_i - p_i - \lfloor (w_i - p_i)/2 \rfloor$. The DNA strand s_{i2} is the reverse complement of s_{pi}, or $s_{i2} = -h(s_{pi})$. Thus, s_{i2} can combine with the center part of s_{i1} to form a fragment of dsDNA.

(2) For any two items i and j $(1 \leq i, j \leq n)$, add one DNA strand s_{aij} as an additional code, which is the reverse complement of the last part of s_{i1} and the first part of s_{j1}, or $s_{aij} = -h(s''_{i1}s'_{j1})$. Thus, the DNA strands of encoding items i and j can combine with the additional code s_{aij} to form dsDNA.

For more details, please see the reference [15].

DNA Algorithm

Based on the DNA encoding method, the DNA algorithm proposed by Han [15] for the 0/1 knapsack problem is as follows.

(1) For each item i $(1 \leq i \leq n)$, use the DNA strands $s_{i1} = s'_{i1}s_{pi}s''_{i1}$ and $s_{i2} = -h(s_{pi})$ to encode it. For any two items i and j $(1 \leq i, j \leq n)$, add one DNA strand $s_{aij} = -h(s''_{i1}s'_{j1})$ as an additional code.

(2) *Merge*: All the DNA strands s_{i1}, s_{i2} and s_{aij} $(1 \leq i, j \leq n)$ are mixed together. Based on the Watson-Crick's law, generate randomly various dsDNAs.

(3) $(N, \leq c)$: All the dsDNAs whose length is above to the knapsack capacity c are discarded.

(4) *Denature*: The remaining dsDNAs are converted into the DNA strands by means of heating.

(5) $+(N, h(s_{aij}))$: All the DNA strands without any additional code are discarded. Let w denote the reverse complement of an additional code, or $w = -h(s_{aij})$, $w \in \{A, C, G, T\}^*$, $1 \leq i, j \leq n$. All the DNA strands generated in step 4 are mixed with w so as to make the DNA strands combine with w to form part dsDNAs. Thus, the DNA strands without any additional code can be separated away. That is, by means of the operation $+(N, w)$, all the DNA strands without any additional code can be separated away. Note that in the encoding method, all the upper DNA strands do not include any additional code.

(6) Delete the additional codes from the remaining DNA strands. All the DNA strands with additional code are put in a tube. Let $s = uyv$ denote one of the DNA strands with additional codes, where u, y, v represent a fragment of one DNA strand, respectively. Put the DNA strands $-h(u)$ and $-h(v)$ into the tube. After annealing,

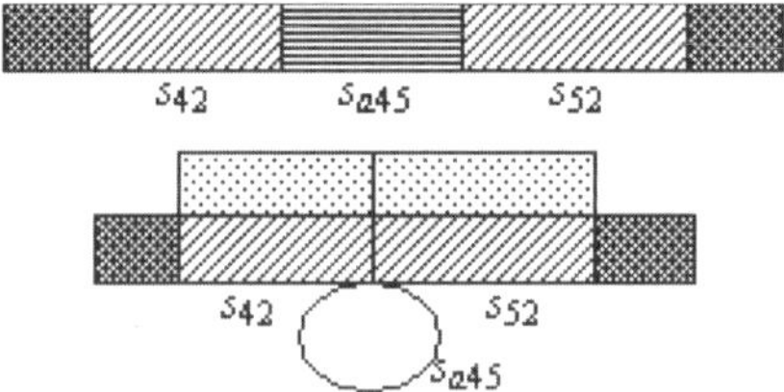

Fig. 13.15. Deletion of the Additional Codes

the strand u combine with $-h(u)$, the strand v combine with $-h(v)$, and fold y, as shown in Fig. 13.15. And then the restriction enzymes are put into the tube to delete y.

(7) Separate the DNA strands with the maximum profit by means of gel electrophoresis.

(8) Determinate the nucleotides sequence of the DNA strand with the maximum profit, which corresponds to the optimal solution.

The DNA algorithm [15] for the 0/1 knapsack problem has the following characteristics: (1) The length of DNA strand s_{i1} which is used to encode item i are equal to the weight w_i, and its center part s_{pi} is with a length of the profit p_i. Thus, the length of the dsDNAs generated in the DNA algorithm is equal to the sum of the weights. By means of the operation $(N, \leq c)$, all the dsDNAs whose length is above to the knapsack capacity c can be discarded. (2) It uses one additional code to link the DNA strands of encoding two items, and the DNA strands s_{i2} and s_{j2} are still linked to the additional code after the dsDNAs are denatured. Since the additional codes can be deleted from the strand s_{i2} $(1 \leq i \leq n)$ by means of the deletion operation and the length of the remaining fragment of s_{i2} is equal to the sum of the profits, so the fragment with the maximum profit can be separated by means of gel electrophoresis which corresponds to the optimal solution. For more details, please see the reference [15].

13.5 Conlusion

Bioinformatics studies the biological information by means of mathematics, computer science and biological techniques. The results of these researches provide a probability of computing with DNA molecules. As an applied branch of the thriving multidisciplinary research area of Bioinformatics, DNA computing has characteristics of higher parallelism and lower costs. Based on the massive parallelism of DNA computing, many researchers tried to solve a large number of difficult problems. These researches demonstrate how DNA can be used for representing information and solving the computational problems and enrich the theories related to DNA computing, in which the methods of representing weights in DNA strands are one of the most important but also challenging issues in DNA computing. Some methods of encoding weights in DNA strands are given in this chapter, which will benefit the further researches on DNA computing, and the rapid development of Bioinformatics will certainly improve the capabilities of DNA computing.

References

1. Adleman L M (1994) Molecular Computation of Solutions to Combinatorial problems. Science 266:1021–1024
2. Lipton R J (1995) DNA solution of hard computational problems. Science 268:542–545
3. Ouyang Q, Kaplan P D, Liu S, et al (1997) DNA solution of the maximal clique problem. Science 278:446–449
4. Head T, Rozenberg G, Bladergroen R S, et al (2000) Computing with DNA by operating on plasmids. Biosystems 57:87–93
5. Sakamoto K, Gouzu H, Komiya K, et al (2000) Molecular computation by DNA hairpin formation. Science 288:1223–1226
6. Narayanan A, Zorbalas S, et al (1998) DNA algorithms for computing shortest paths. In: Proceedings of the Genetic Programming, Morgan Kaufmann 718–723
7. Shin S Y, Zhang B T, Jun S S, et al (1999) Solving traveling salesman problems using molecular programming. In: Proceedings of the Congress on Evolutionary Computation. IEEE Press 994–1000
8. Yamamoto M, Matsuura N, Shiba T, et al (2002) Solutions of shortest path problems by concentration control. Lecture Notes in Computer Science 2340:203–212
9. Lee J Y, Shin S Y, Park T H, et al (2004) Solving traveling salesman problems with DNA molecules encoding numerical values. BioSystems 78:39–47
10. Han A, Zhu D (2006) DNA Encoding Method of Weight for Chinese Postman Problem. In: Proceedings of 2006 IEEE Congress on Evolutionary Computation. IEEE Press 2696–2701
11. Han A, Zhu D (2007) DNA Computing Model Based on a New Scheme of Encoding Weight for Chinese Postman Problem. Computer Research and Development 44:1053–1062
12. Han A (2006) RLM: A New Method of Encoding Weights in DNA Strands. In: Proceedings of the Sixth International Conference on Hybrid Intelligent Systems. IEEE Press 118–121
13. Han A, Zhu D (2006) A New DNA-Based Approach to Solve the Maximum Weight Clique Problem. Lecture Notes in Computer Science 4115:320–327
14. Han A, Zhu D (2006) A New DNA Encoding Method for Traveling Salesman Problem. Lecture Notes in Computer Science 4115:328–335
15. Han A, Zhu D (2006) DNA Computing Model for the Minimum Spanning Tree Problem. In: Proceedings of the 8th International Symposium of Symbolic and Numeric Algorithms for Scientific Computing. IEEE Press 372–377
16. Han A (2006) DNA Computing Model for the 0/1 Knapsack Problem. In: Proceedings of the Sixth International Conference on Hybrid Intelligent Systems. IEEE Press 122–125
17. Paun G, Rozenberg G, Salomaa A (1998) DNA Computing: New Computing Paradigms. Springer, Berlin. Translated by Xu Jin, Wang Shudong, Pan Linqiang (2004) Tsinghua University Press, Beijing
18. Setubal J, Meidanis J (1997) Introduction to Computational Molecular Biology. Cole Publishing Company, Thomson. translated by Zhu H, et al (2003) Science Press, Beijing
19. Zhang B T, Shin S Y (1998) Molecular algorithms for efficient and reliable DNA computing. In: Genetic Programming, Morgan Kaufmann 735–742
20. Xu J, Zhang L (2003) DNA Computer Principle, Advances and Difficulties (I): Biological Computing System and Its Applications to Graph Theory. Journal of Computer Science and Technology 26: 1–10
21. Yin Z (2004) DNA Computing in Graph and Combination Optimization. Science Press, Beijing

22. Wang L, Lin Y, Li Z (2005) DNA Computation for a Category of Special Integer Planning Problem. Computer Research and Development 42:1431–1437
23. Chen Z, Li X, Wang L, et al (2005) A Surface-Based DNA Algorithm for the Perfect Matching Problem. Computer Research and Development 42:1241–1246
24. Braich R S, Chelyapov N, Johnson C, et al (2002) Solution of a 20-variable 3-SAT problem on a DNA computer. Science 296:499–502
25. Lancia G (2004) Integer Programming Models for Computional Biology Problems. Journal of Computer Science and Technology 19:60–77
26. Ibrahim Z, Tsuboi Y, Muhammad M S, et al (2005) DNA implementation of k-shortest paths computation. In: Proceedings of IEEE Congress on Evolutionary Computation. IEEE press 707–713
27. Jonoska N, Kari S A, Saito M (1998) Graph structures in DNA computing. In: Computing with Bio-Molecules–Theory and Experiments. Penn State 93–110

Index